Intelligent Systems Reference Library

Volume 107

Series editors

Janusz Kacprzyk, Polish Academy of Sciences, Warsaw, Poland
e-mail: kacprzyk@ibspan.waw.pl

Lakhmi C. Jain, Bournemouth University, Fern Barrow, Poole, UK, and
University of Canberra, Canberra, Australia
e-mail: jainlc2002@yahoo.co.uk

About this Series

The aim of this series is to publish a Reference Library, including novel advances and developments in all aspects of Intelligent Systems in an easily accessible and well structured form. The series includes reference works, handbooks, compendia, textbooks, well-structured monographs, dictionaries, and encyclopedias. It contains well integrated knowledge and current information in the field of Intelligent Systems. The series covers the theory, applications, and design methods of Intelligent Systems. Virtually all disciplines such as engineering, computer science, avionics, business, e-commerce, environment, healthcare, physics and life science are included.

More information about this series at http://www.springer.com/series/8578

Kazumi Nakamatsu · Roumen Kountchev
Editors

New Approaches in Intelligent Control

Techniques, Methodologies and Applications

 Springer

Editors
Kazumi Nakamatsu
School of Human Science and Environment
University of Hyogo
Himeji
Japan

Roumen Kountchev
Department of Radio Communications and
 Video Technologies
Technical University of Sofia
Sofia
Bulgaria

ISSN 1868-4394   ISSN 1868-4408   (electronic)
Intelligent Systems Reference Library
ISBN 978-3-319-81212-0   ISBN 978-3-319-32168-4   (eBook)
DOI 10.1007/978-3-319-32168-4

This Springer imprint is published by Springer Nature
The registered company is Springer International Publishing AG Switzerland

Preface

Recently various intelligent techniques, for example neural network computing, fuzzy reasoning, genetic algorithms, etc., have been developed in theories and practice, also applied to many intelligent systems all over the world. This volume is titled "New Approaches in Intelligent Control" and introduces some new approaches in intelligent control area from both the viewpoints of theory and application, which consists of 11 contributed chapters by prominent authors from all over the world including this introductory chapter. This volume also divides one dual side with another volume titled "New Approaches in Image Analysis" (Eds. Roumen Kountchev and Kazumi Nakamatsu). Each chapter in this volume constitutes one self-contained monograph, and includes summary, conclusion and future works in terms of its main themes. Some of the chapters introduce specific case studies of various intelligent control systems and others focus on intelligent theory based control techniques with small applications. The most remarkable specificity of this volume is that the last three chapters deal with intelligent control based on paraconsistent logics.

The rest of this chapter introduces the summaries of all contributed chapters.

Design of Fuzzy Supervisor-Based Adaptive Process Control Systems

The modern industrial processes are difficult to model and control by classical means for their nonlinearity, inertia, model uncertainty, and varying parameters. The adaptive fuzzy logic controllers (AFLCs) improve the system performance but are computationally hard to design and embed in programmable logic controllers (PLCs) for wider industrial applications.

In this chapter, a design approach for simple AFLCs is suggested, based on main controllers linear, FLC or parallel distributed compensation (PDC), and fuzzy logic supervisors (FLSs) for online auto-tuning of their gains or scaling factors. The effect is a continuous adaptation of the control surface in response to plant changes. Approximation of the designed AFLC to a PDC equivalent on the basis of neuro-fuzzy and optimization techniques enables the stability analysis of the AFLC system using the indirect Lyapunov method and also its PLC implementation. The AFLC is applied for the real-time control of the processes in a chemical reactor, a dryer, and a two-tank and an air-conditioning systems, decreasing overshoot, settling time, control effort, and coupling compared to classical FLC and linear control systems.

Intelligent Carpooling System: A Case Study for Bacău Metropolitan Area

Mobility is one of the most basic features of our modern society. People and goods move around the entire Earth in a continuous and broad attempt to fulfill economic, safety, and environmental goals. The Mobility Management or Transportation Demand Management is a collection of strategies for encouraging more efficient traffic patterns towards achieving specific planning objectives. For example, people can choose to switch from peak hours to non-peak time, or to cycle instead of using car. Administrative regulations could introduce incentives or reimbursements when alternative commuting modes are used. Governmental policies could include fuel tax increases or pay-as-you-drive freeway taxes or car insurances.

The goal of this chapter is to present several alternative travel modes, their social impact and their utility. As an example, we present two applications for shared-use mobility in the metropolitan area of Bacau, Romania. The applications integrate diverse computing languages with platforms, standards, and technologies. The experimental results are encouraging, allowing us to consider that seamless integration of hybrid management systems for transportation could have tremendous economic and social impact at global scale.

Naval Intelligent Authentication and Support Through Randomization and Transformative Search

The problem addressed, in this chapter, pertains to how to represent and apply knowledge to best facilitate its extension and use in problem solving. Unlike deductive logics (e.g., the predicate calculus), an inherent degree of error is allowed so as to greatly enlarge the inferential space. This allowance, in turn, implies the application of heuristics (e.g., multiple analogies) to problem solving as well as their indirect use in inferring the heuristics themselves.

This chapter is motivated by the science of inductive inference. Examples of state-space search, linguistic applications, and a focus methodology for generating novel knowledge (components) for wartime engagement for countering (cyber) threats (WAMS) are provided.

Big Data Approach in an ICT Agriculture Application

The advent of big data analytics is changing some of the current knowledge paradigms in science as well in industry. Even though, the term and some of the core methodologies are not new and have been around for many years, the continuous price reduction of hardware and related services (e.g., cloud computing) are making the application of such methodologies more affordable to almost any research area in academic institutions or company research centers. It is the aim of this chapter to address these concerns because big data methodologies will be extensively used in the new ICT agriculture project, in order to know how to handle them, and how they could impact normal operations among the project members, or the information flow between the system parts. The new paradigm of big data and its multiple benefits have been used in the novel nutrition-based vegetable production and distribution system in order to generate a healthy food recommendation to the end user and to provide different analytics to improve the system efficiency. Also, different version of the user interface (PC and Smartphone) was designed keeping in mind features like easy navigation, usability, etc.

Intelligent Control Systems and Applications on Smart Grids

This chapter discusses advances in intelligent control systems and their applications in micro-energy grids. The first section introduces a PID fuzzy model reference learning controller (FMRLC) implemented in the control loop of static VAR compensator (SVC) to stabilize the voltage level in the island mode microgrid. FMRLC performance is compared to the conventional PI controller. The introduced results show that SVC with FMRLC has better capabilities to compensate for microgrid nonlinearity and continuous adaptation for dynamic change of the microgrid's connected loads and sources. The second section discusses performance optimization of micro-energy grid. A recent heuristic optimization technique, called backtracking search optimization algorithm (BSA), is proposed for performance optimization of micro-energy grids with AC/DC circuits, where it is used for the selection of the scheme parameters of PWM pulsing stage used with a novel distributed flexible AC transmission system (D-FACTS) type called green plug-energy economizer (GP-EE). The following section includes simulation models and results to illustrate the merits of the proposed intelligent control designs and their use to achieve high-performance micro-energy grids.

Control Through Genetic Algorithms

Many real-world applications require automatic control. This chapter addresses genetic algorithms to achieve the control, based on their numerous advantages for the difficult problems. First of all, an unitary approach of the control through the perspective of the systems theory is presented. There are described examples of control in biology, economy, and technical areas in order to highlight the general system behaviors: preventive control, reactive control, or combined control. In this chapter, fundamentals of genetic algorithms theory are featured: genetic representation, genetic operators, how it works, and why it works. Further, two process control systems based on genetic algorithms are described: a chemical process control involving mass transfer, where the genetic algorithms are used in the system identification for a NARMAX model, an important issue with respect to model-based control and a job shop scheduling process in manufacturing area where the genetic algorithm is the tool to model the optimization process control.

Knowledge-Based Intelligent Process Control

In the last decades, the number of process control applications that use intelligent features has increased. This is mainly due to the complex and critical character of the process to be controlled. The intelligent process control systems work better than conventional control schemes in the domains of fault diagnosis (detection, cause analysis, and repetitive problem recognition); complex control schemes; process and control performance monitoring and statistical process control; real-time quality management; control system validation, startup and normal or emergency shutdown. Conventional control technologies use quantitative processing while knowledge-based integrates both qualitative and quantitative processing (having as target the increase of efficiency). This chapter presents an overview of intelligent process control techniques, from rule-based systems, frame-based systems (object-oriented approach), hybrid systems (fuzzy logic and neural network). The focus is on expert systems and their extension, the knowledge-based systems. Finally, an industrial case study is presented with conclusions to knowledge-based systems limitations and challenges associated to real-time implementation of the system.

Ciphering of Cloud Computing Environment Based New Intelligent Quantum Service

Cloud computing environment is a new approach to the intelligent control of network communication and knowledge-based systems. It drastically guarantees scalability, on-demand, and pay-as-you-go services through virtualization environments. In a cloud environment, resources are provided as services to clients over the internet in the public cloud and over the intranet in the private cloud upon request. Resources' coordination in the cloud enables clients to reach their resources anywhere and anytime. Guaranteeing the security in cloud environment plays an important role, as clients often store important files on remote trust cloud data center. However, clients are wondering about the integrity and the availability of their data in the cloud environment. So, many security issues, which are pertinent to client data garbling and communication intrusion caused by attackers, are attitudinized in the host, network, and data levels. In order to address these issues, this chapter introduces a new intelligent quantum cloud environment (IQCE) that entails both intelligent quantum cryptography-as-a-service (IQCaaS) and quantum advanced encryption standard (QAES). This intelligent environment offers more secured data transmission by providing secret key among cloud's instances and machines. It is implemented using System Center Manager (SCM) 2012-R2, which in turn is installed and configured based on bare-metal Hyper-V hypervisor. In addition, IQCaaS solves the key generation, the key distribution, and the key management problems that emerge through the online negotiation between the communication parties in the cloud environment.

Paraconsistent Logic and Applications

In this work we summarize some of the applications of so-called Paraconsistent logics; mainly one class of them, the paraconsistent annotated logics. Roughly speaking, such systems allow inconsistencies in a nontrivial manner in its interior; so it is suitable to handle themes in which inconsistencies become a central issue, like pattern recognition, non-monotonic reasoning, defesable reasoning, deontic reasoning, multi-agent systems including distributed systems, collective computation, among a variety of themes.

Annotated Logics and Intelligent Control

Annotated logics are a kind of paraconsistent (and generally paracomplete) logic, whose origin is paraconsistent logic programming. Later, these logics have been extensively studied by many researchers and applied to many areas, in particular, artificial intelligence and computer science. Annotated logics are also suited as the foundations for intelligent control in that they can properly deal with both incomplete and inconsistent information. The chapter addresses the aspects of annotated logics as a control language for intelligent systems. After reviewing the motivation and formalization of annotated logics, we give an application to robotics to show how they can be used for intelligent control.

Paraconsistent Annotated Logic Program EVALPSN and Its Application to Intelligent Control

We have already proposed a paraconsistent annotated logic program called EVALPSN. In EVALPSN, an annotation called an extended vector annotation is attached to each literal. In order to deal with before-after relation between two time intervals, we also have introduced a new interpretation for extended vector annotations in EVALPSN, which is named before-after(bf)-EVALPSN.

In this chapter, we introduce paraconsistent annotated logic programs EVALPSN /bf-EVALPSN and their application to intelligent control, especially logical safety verification based control with simple examples. First, the background and overview of EVALPSN are given, and paraconsistent annotated logics $P\tau$ and the basic annotated logic program are recapitulated as the formal background of EVALPSN/bf-EVALPSN with some simple examples. Then EVALPSN is formally defined and its application to traffic signal control is introduced. EVALPSN application to pipeline valve control is also introduced with examples. Bf-EVALPSN is formally defined and its unique and useful reasoning rules are introduced with some examples. Lastly, this chapter is concluded with some remarks.

Himeji, Japan   Kazumi Nakamatsu
Sofia, Bulgaria   Roumen Kountchev
2015

Contents

Contributors

Abdelazeem A. Abdelsalam Faculty of Engineering, Department of Electrical Engineering, Suez Canal University, Ismailia, Egypt

Jair Minoro Abe Graduate Program in PE, ICET, Paulista University, São Paulo, Brazil; Institute of Advanced Studies, University of São Paulo, São Paulo, Brazil

Alireza Ahrary Faculty of Computer and Information Sciences, Sojo University, Kumamoto, Japan

Seiki Akama C-Republic Corporation, Kawasaki, Japan

Thouraya Bouabana-Tebibel LCSI Laboratory, École Nationale Supérieure d'Informatique, Algiers, Algeria

Mădălina Cărbureanu Department of Automatic Control Engineering, Petroleum-Gas University of Ploiesti, Ploiesti, Romania

Gloria-Cerasela Crişan Department of Mathematics, Informatics and Educational Sciences, "Vasile Alecsandri" University of Bacău, Bacău, Romania

Constantin-Sebastian Damian Department of Mathematics, Informatics and Educational Sciences, "Vasile Alecsandri" University of Bacău, Bacău, Romania

Ahmed S. Eldessouky Electrical Engineering Department, Canadian International College, New Cairo, Egypt

El-Sayed M. El-Horbaty Faculty of Computer and Information Sciences, Ain Shams University, Cairo, Egypt

Hossam A. Gabbar Energy Safety & Control Laboratory, University of Ontario Institute of Technology, Oshawa, ON, Canada

Omer K. Jasim Mohammad Al-Ma'arif University College, Anbar, Iraq

R. Dennis A. Ludena Department of Computer and Information Sciences, Sojo University, Kumamoto, Japan

Sanda Florentina Mihalache Department of Automatic Control Engineering, Petroleum-Gas University of Ploiesti, Ploiesti, Romania

Kazumi Nakamatsu School of Human Science and Environment, University of Hyogo, Himeji, Japan

Elena Nechita Department of Mathematics, Informatics and Educational Sciences, "Vasile Alecsandri" University of Bacău, Bacău, Romania

Elena Simona Nicoara Petroleum-Gas University of Ploiesti, Ploiesti, Romania

Sergiu-Mădălin Obreja Department of Mathematics, Informatics and Educational Sciences, "Vasile Alecsandri" University of Bacău, Bacău, Romania

Marius Olteanu Petroleum-Gas University of Ploiesti, Ploiesti, Romania

Mihaela Oprea Department of Automatic Control Engineering, Petroleum-Gas University of Ploiesti, Ploiesti, Romania

Ahmed M. Othman University of Ontario Institute of Technology, Oshawa, ON, Canada; Zagazig University, Zagazig, Egypt

Nicolae Paraschiv Petroleum-Gas University of Ploiesti, Ploiesti, Romania

Stuart H. Rubin SSC-PAC 71730, San Diego, CA, USA

Abdel-Badeeh M. Salem Faculty of Computer and Information Sciences, Ain Shams University, Cairo, Egypt

Adel Sharaf Sharaf Energy Systems Inc., Fredericton, NB, Canada

Snejana T. Yordanova Faculty of Automation, Technical University of Sofia, Sofia, Bulgaria

Chapter 1
Design of Fuzzy Supervisor-Based Adaptive Process Control Systems

Snejana T. Yordanova

Abstract The modern industrial processes are difficult to model and control by classical means for their nonlinearity, inertia, model uncertainty and varying parameters. The adaptive fuzzy logic controllers (AFLCs) improve the system performance but are computationally hard to design and embed in programmable logic controllers (PLCs) for wider industrial applications. In this chapter a design approach for simple AFLCs is suggested, based on main controllers—linear, FLC or parallel distributed compensation (PDC), and fuzzy logic supervisors (FLSs) for on-line auto-tuning of their gains or scaling factors. The effect is a continuous adaptation of the control surface in response to plant changes. Approximation of the designed AFLC to a PDC equivalent on the basis of neuro-fuzzy and optimization techniques enables the stability analysis of the AFLC system using the indirect Lyapunov method and also its PLC implementation. The AFLC is applied for the real time control of the processes in a chemical reactor, a dryer, a two-tank and an air-conditioning systems, decreasing overshoot, settling time, control effort and coupling compared to classical FLC and linear control systems.

Keywords Adaptive fuzzy logic control · Design · Genetic algorithms · Inertial processes · Nonlinear dynamic systems · Stability · Supervisor

List of Abbreviations

ADC	Analog-to-digital converter
AFLC	Adaptive fuzzy logic control/controllers
DAC	Digital-analog converter
DAQ	Data acquisition
FIM	Fuzzy inverse model
FL	Fuzzy logic
FU	Fuzzy unit
FLC	Fuzzy logic control/controllers

S.T. Yordanova (✉)
Faculty of Automation, Technical University of Sofia, Sofia 1000, Bulgaria
e-mail: sty@tu-sofia.bg

© Springer International Publishing Switzerland 2016   1
K. Nakamatsu and R. Kountchev (eds.), *New Approaches in Intelligent Control*,
Intelligent Systems Reference Library 107, DOI 10.1007/978-3-319-32168-4_1

FLS	Fuzzy logic supervisor
FSOC	Fuzzy self-organizing control
GA	Genetic algorithms
HVAC	Heating ventilation and air-conditioning
IMC	Internal model controller
ISE/IAE	Integral squared error/integral of absolute error
KBM	Knowledge-based modifier
LMI	Linear matrix inequality
MF	Membership function
MRFLC	Model reference FLC
NF	Neuro-fuzzy (model, structure, control etc.)
PDC	Parallel distributed compensation
PD/PI/PID	Proportional plus derivative/proportional plus integral/proportional plus integral plus derivative (for a linear control algorithm)
PLC	Programmable logic controllers
PWM	Pulse-width modulation
RH	Relative humidity
SFU	Supervisor fuzzy unit
SISO/TISO/TITO/MIMO	Single-input-single-output/two-input-single-output/ two-input-two-output/multi- input-multi output
2I	Two-input
ScF	Scaling factor (normalization/denormalization gains of a FLC)
TSK	Takagi-Sugeno-Kang fuzzy model
ZN	Ziegler-Nichols linear plant model (also a three parameter model)

1.1 Introduction

Modern processes grow in complexity and their modeling and control confront a large number of problems originating from the new demands posed by the increasing competition and standards of living for high quality, environmental and market considerations, energy and raw material savings [1–4]. The requirements imposed on process control affect different levels of process automation and result in fast and extensive changes in operating conditions and operation modes, and introduction of many inner loops and utility feedbacks to decrease waste of material and reduce energy consumption—all leading to highly nonlinear and time-varying systems of high order with many interactions between the control loops. There arises the need for dynamic and reactive responses at all levels of automation with development of relevant modeling and control techniques that describe the nonlinear time-varying plant behaviour of the new production methods and units, accounting also for plant model uncertainty and inertia, in a more unified way combining knowledge of system experts, measurements and operational experience.

Linear controllers satisfy the contemporary high demands for process control in a close area around a given operation point and still keep a leading position in industrial applications. Fuzzy logic controllers (FLCs) outperform them in system performance, robustness, smooth and energy efficient control, simple structure and computations that ease the industrial real time implementations [5]. Both linear classical and modern fuzzy logic (FL) controllers have standard design and evolve to improve their real time on-line self-tuning and adaptation in order to respond to plant changes and keep the high system performance standards.

Fuzzy logic has proven its superiority over the classic control techniques in building not only of direct and supervisory FLCs but also of on-line adaptation, estimation and observation mechanisms for both linear and FL controllers. Industrial companies develop FLC in programmable logic controllers (PLCs) [6] and also supply their linear controllers with extensions for fuzzy assisted auto-tuning [7–9]. The modern PID-controller concept has been enriched with nonlinear and state-dependent functions by sophisticated integration of continuous linear and discrete control (anti-windup, proportional kick up, retarded integral action, etc.). Its fusion with the fuzzy concept adds intelligent features and lays the foundation of the widely spread nowadays easy to design PID FLCs and fuzzy logic supervisors (FLS) for on-line auto-tuning of the main classical PID or PID-FLC controller from a higher level (using different information) in order to improve the performance of the closed loop system. The tuning considers parameters and scaling factors (ScFs), membership functions (MFs), rule bases, etc.

The idea of representing heuristic knowledge by linguistic labels and fuzzy sets implemented in linguistic rules as an easy way to deal with modeling uncertainties and complexity of mathematical relationships was first launched in 1965 by Lofti Zadeh. The fuzzy inference process involves MFs, fuzzy logic operators and knowledge rules which solving imitates the human way of decision making [1–4]. The MFs allow the representation of a degree of membership to a fuzzy set, associated to a linguistic label, for a given input numerical value. The rules 'If-Then' introduce the expert knowledge in a computable way by means of operators, usually "and" and "or". Since then the fuzzy set and fuzzy logic theory has been successfully applied to modeling, prediction and control of a great number of processes [1, 4, 5, 10]. With time it has been evolved to consider stability and robustness issues [11], optimization techniques to avoid trial-and-guess decisions and expert subjective concepts in synergism with artificial neural networks and genetic algorithms (GA) [2, 11]. Nowadays the adaptation idea and its equivalent in certain aspects FLS-based on-line auto-tuning, launched in [1, 4, 11–14], are subjects of intensive research work. The wide industrial applications set on the FLC tuning/adaptation also the demands of their real time implementation using PLC or embedded techniques—simplicity of the algorithms to guarantee reliability, easy design and programming, fast computations and low memory requirements.

The fuzzy system design for process control and modeling steps on the dominating model-free approach (Mamdani, Larsen and Sugeno fuzzy models) and the more recent and advanced model-based approach (Takagi-Sugeno-Kang (TSK) dynamic fuzzy models) [15]. The TSK plant model determines dynamic TSK-FLC, based on the principle of parallel distributed compensation (PDC) [15, 16]. A third group of evolving cloud-based fuzzy models [17], derived from streaming data, is emerging nowadays. They are not used in this chapter as their industrial applications, however, are still mainly in recognition problems and in control of mechanical systems (robot, helicopter, etc.) though some simulation results in process control also appear.

The aim of the investigations in this chapter is to present a novel general engineering approach for design of simple process adaptive FLCs (AFLCs) on the basis of FLS, which are suitable for easy embedding in industrial PLCs and wide engineering applications and to prove the closed loop stability and its advantages in real-time control of laboratory plants close to industrial environment. Various structures of FLSs are designed and applied for auto-tuning of the ScFs of model-free (Mamdani, Larsen) and model-based (TSK-PDC) main FLCs or of the parameters of linear PI and PID controllers in order to achieve high system performance and energy efficiency when controlling a nonlinear inertial plant with variable parameters. Genetic algorithms are employed to tune the main controllers and the FLSs [18–21]. Training and optimization techniques using Sugeno neuro-fuzzy (NF) structures [11, 13, 22] and GAs [11] are suggested to simplify the structure of the two-level controller for the purposes of system stability analysis and easy PLC programming and embedding for real time operation. The design approach is demonstrated and assessed by simulations in: the TSK-PDC FLS control of the product concentration in a chemical reactor; the PI FLC FLS control of two levels in a two-tank system; the NF control as a simplified equivalent to a FLC-FLS of the biogas production rate and the system stability analysis via phase trajectories approach in [11]; the FLC-PDC air temperature control in a laboratory dryer; the Mamdani FLC-FLS and the Sugeno NF biogas production rate control in anaerobic biological waste degradation process [11–13]. Real time investigations confirm the improvements for biogas production rate control and study the FLS-based system stability observing the phase trajectories in [14], for two-variable FLC of temperature and humidity control in a laboratory model of a room air-conditioning system in [23] and for air temperature control in a dryer in [11, 19, 21].

The applications are developed in MATLAB^{TM} environment. The approach effectiveness is estimated in comparison with systems only with main FLCs or with main linear PI/PID controllers. The performance indices assessed are the settling time, the overshoot, the robustness (preserved performance indices for different plant parameters in the different operation points), the control magnitude and smoothness and the introduced measure for energy efficiency.

The further organization of the chapter is as follows. Section 1.2 analyzes the state-of-the art in the AFLC design methods and formulates the problem to be solved. In Sect. 1.3 the adaptive properties of a FLS-based FLC are discussed and a procedure for the design of an adaptive two-level structure of a main controller and a FLS is developed. Section 1.4 presents a TSK plant modeling method, based on experimental data and GA optimization. Section 1.5 is devoted to the GA optimization of the parameters of the FLS-based AFLC. In Sect. 1.6 case studies for application of the FLS-based AFLC design approach for the control of various plant variables are presented. There different main controllers and structures of FLSs are combined and a real time control of laboratory plants is investigated and assessed. In Sect. 1.7 an approximation approach for building a functional equivalent of the designed FLS-based AFLC with simple structure of one fuzzy unit for the purpose of its PLC implementation and the system stability analysis is suggested. There the derived Lyapunov stability conditions of the fuzzy closed loop system with the equivalent PDC-based FLC are presented. Section 1.8 summarizes the results achieved and outlines the future research.

1.2 State-of-the-Art and Problem Formulation

The FLC design, based on heuristic information or numerical input-output data, though a better solution for modeling and control of complex processes may encounter difficulties in specifying and tuning the MFs and the rule base in order to ensure desired system performance. The on-line available data can be used so that the controller is automatically designed and continually learns in order the system performance objectives to be met at changes in the plant and the operation conditions. This idea inspires the development of adaptive/learning FLCs (AFLCs)—self-tuning, self-organizing (FSOC—with rule-base adaptation), gain scheduling and model reference (MRFLC), following the classical adaptive control theory principles [24].

1.2.1 Adaptive FLCs Based on Classical Adaptive Control Theory

The AFLC consists of a process monitor and an adaptation mechanism. The process monitor estimates the system performance in the direct (performance-based) adaptive control, or the plant parameters—in the identification-based adaptive control. The parameter estimators accept a given plant model structure commonly Ziegler Nichols (ZN) three parametric, time-series and fuzzy. They can be integrated with the adaptation mechanism directly to produce the correction of the control [25] or the fuzzy relation matrix. For system performance indicators, estimated from on-line measurements, are selected the overshoot, the rise time, the settling time, the decay ratio, etc., or the integral squared error (ISE), the integral of absolute error (IAE) or the time-weighted IAE, etc., the average or the maximum system absolute or squared error over several previous sample periods, the level of fulfillment of a function of goals and constraints or a combination of the above. In [26] the system performance is a normalized performance index, defined as function of the residual of current, predicted and mean square error. An AR model is suggested for on-line prediction of the error given the prediction horizon, the sliding window length and the model order. The reference model in MRFLC sets system performance specifications, which determines MRFLC as a specific type of performance-based AFLC.

The adaptation mechanism includes a fuzzy inverse model (FIM) and a knowledge-based modifier (KBM), and ensures system stability and desired adaptation convergence rate usually using gradient or Lyapunov methods. The FIM computes the necessary control action or correction on the basis of the process output [1] using the system Jacobian matrix or its estimate. For single-input-single-output (SISO) plants it is a unity scalar and the correction in control is the deviation of the output from the desired output. The control that has caused this deviation is to be corrected accounting for the plant order or time delay which often is difficult or computationally hard to estimate.

A fuzzy model-based or a model-free component is introduced to perform: (1) the processing of the error between reference model and system outputs in MRFLC; (2) the gain scheduler; (3) the process monitor and the FIM; (4) the main controller; (5) a combination of several of the above.

The adaptation, used for improvement of the system performance at plant changes, adjusts [1]:

- the scaling factors—this changes uniformly the FLC sensitivity, resolution, gains, expands/shrinks the MFs universes of discourse [11, 20, 27];
- the MFs (usually the peaks or the singletons)—this changes the FLC gain in a specific area of the universe of discourse;
- the rule bases—by identification of the rules which provoke the bad system performance, and their updating, usually modifying the consequents (singletons) or the rule-base relation matrix accounting for the plant time delay.

The FLC adaptation can adjust several of these groups. It acts as self-tuning if modifying the ScFs and the MFs, or self-organizing if modifying the rules.

1.2.1.1 Model Reference FLCs

A MRFLC for on-line adaptation in reference tracking control is developed in [28] on the basis of a TSK plant model with unknown parameters and a main PDC-FLC with local state feedbacks that ensure system asymptotic stability. In [29] a MRFLC strategy is suggested for adjusting the gains of the main linear PI controller for level. The MRFLC in [30] is devised as a "learning" system that improves its performance by interacting with its environment and memorizing past experience. The main controller is a PD FLC. The learning mechanism performs on-line automatic synthesis and adaptation of the centres of its output MFs. The FIM has inputs the deviation of the plant from the reference model output and its derivative, and output—the necessary changes in the control that will lead to zero deviation. The KBM changes locally the rule base—the centres of the output MFs of the main PD FLC only for the active rules at the current sample time in order to modify the control action by the correction considering the delay between plant input and output. Several reference models and FIMs for different operation points can be designed. A special auto-tuning mechanism can also be added for the input MFs and the scaling gains. Different modifications are suggested but they complicate the design and the feasibility. In [31, 32] self-evolving cloud-based autonomous learning MRFLC are developed and tested on a simulated and a pilot hydraulic plants and simulated industrial environment. Adaptive laws are proposed to tune the parameters in the consequent part autonomously using only the data density and selecting representative prototypes/focal points from the control hyper-surface acting as a data space. It is proven that it is possible to generate and self-tune/learn a non-linear controller structure and evolve it in on-line mode with no prior knowledge of the plant or off-line training.

1.2.1.2 Fuzzy Self-Organizing Controllers

FSOC is first suggested in [33]. A FSOC is developed in [34]. The plant is assumed to be nonlinear of infinite order and unknown dynamic with discrete-time repre-sention, determined by the bounds of the pseudo partial derivatives. The FLC output singletons β in the rules conclusions and the initial values are tuned to minimize both squared tracking error and control effort by gradient search algo-rithm. The weighted constants and the learning rate are determined from conver-gence and stability criteria using Lyapunov function. MFs and bounds for the pseudo partial derivatives in the description of the plant are set by trial and error. Two FSOCs for the control of water level and air pressure are designed in [35]. The main FLC rule base is on-line modified and memorized by a learning mechanism starting with no rules and no information about the plant. The knowledge is accumulated with the control. A FIM with inputs system error and its derivative produces output—the necessary rule-base correction passed to a KBM and then to a rule-base initializer that constructs the updated main FLC rule base. Many sym-metrical uniformly distributed in normalized universe of discourse MFs are used. The FSOC is experimentally tested and outperforms the linear PI/PID in accurate tracking of reference at changes of the plant dynamics. In [36] a FSOC is presented for nonlinear time-varying plant with known time delay and a gain that does not change its sign. No off-line pre-training is required and high control performance is achieved through a three-stage algorithm—a coarse fuzzy tuning of the output ScF, a correction of the fuzzy rule consequents (singletons), starting from zero position, as a function of the degree of activation of the rule and the system error at a previous time considering the plant time delay, a bounded correction of the position of the peaks of the triangular inputs MFs from requirement of homogeneous dis-tribution of the ISE throughout the defined by the MFs operating regions. A FSOC with two-level structure for tuning the rules of a FLC is presented in [37] and for adjusting the gains of a PID controller during system operation on the basis of a complex fuzzy decision mechanism is described in [38].

1.2.1.3 Self-Tuning FLCs

The self-tuning FLCs for auto-tuning of the main linear controller gains or the ScFs and MFs parameters of the main FLCs are very widely spread for their simplicity and good performance [39–42]. They resemble in structure the FLS but have the same inputs as the main controller—the system error and its derivative.

1.2.1.4 Gain-Scheduling FLC

FL gain scheduling control for coarse and fine FLC adaptation of the ScFs is suggested in [43] and for PID parameters tuning is developed in [44].

1.2.1.5 Problems of Process Adaptive FLC for Industrial Application

The main problems of the AFLC based on the classical adaptive control principles conclude in:

- Computationally hard algorithms (high order sparse relation matrices computation and storage, many MFs, high memory required to store previous experience, ill-posed problems in computation of FIM), difficult for embedding in PLCs for industrial real time control applications;
- Not enough testing of AFLC algorithms in real time on-line control in industrial environment, where: (1) the plant varies its parameters at reference changes, disturbances or in steady state (changing the feeding gas fuel type, or changing the feeding raw material—petrol, cement with different composition, etc. of a distillation/rectification column or a cement kiln, catalytic reactor with exothermic processes, etc.); (2) noisy data may decrease convergence rate; (3) several runs needed "to learn" and to compute ISE are impossible. Simulation tests of adaptation algorithms without validation are not reliable.
- Complicated design—stability and convergence are rarely considered;
- Trial-and-error and intuition-driven decisions about necessary input data (thresholds against persistent oscillations, initial values, plant model structure, order or/and inertia, time window for computing performance, bounds for tuning parameters—MFs centres, singletons, reference model design, etc.), rule-base design and ScFs tuning of FIM and other components, etc.;
- Specific application-bound design approaches.

In the industrial practice the approaches that gain popularity are those that ensure an easy and transparent design and a reliable work by computational and structural simplicity. The complexity can be reduced and the good results preserved by exploiting more intensively only one type of adaptation technique with less fuzzy rules and adjusting parameters.

1.2.2   FLS-Based Adaptive FLCs

The FLSs are widely spread for their good effect on the system performance using simple means—easy and unified design, scanty knowledge about the plant, non-complex structure and low computational effort which facilitates the feasibility for industrial applications in noisy environment and satisfies the real time restrictions. They supervise and auto-tune on-line in a nonlinear manner the main controllers— linear, FLCs and AFLCs [1, 21] from upper control level on the basis of available system data—measurements and expert knowledge, which are different from the used by the main controller, and from extra sources (operator or other systems). Usually the FLS has system performance measures as inputs. It is therefore clas-sified as a direct AFLC [20, 21]. The FLS can act also as a gain scheduler or a fuzzy switcher between fixed rule-bases (i.e., course and fine), or an adjusting mechanism of a FLC or an AFLC, etc. It can tune as well the plant model or the adaptation mechanism of an AFLC. This general role of the FLS makes attractive its appli-cation for adaptive control in order to improve system performance without knowledge about the plant and at significant, random and unknown plant changes. Using the FLS the necessary local modifications in the control surface can be ensured by on-line continuous adjustment only of the ScFs or gains, which effect is equivalent to tuning of rules, MFs and other more complicated techniques. In [11, 12, 14, 45–47] by means of FLSs ScFs are tuned to optimize different performance indices. FLSs tune other parameters—the weight of reference in the error in the proportional component of the PID controller in [48], a parameter in a derived Ziegler–Nichols (ZN) like relationship in [49]. MIMO FLCs are tuned to com-pensate interaction between loops and nonlinearity in an air-conditioning system in [23], MIMO PI controllers' set points for an evaporator are FLS adjusted in [50]. Linear controllers are tuned in various applications [51, 52], based on dead-beat control principle and a fuzzy predictor of the control action in order to reach a zero error at the next sampling instant [53], and on-line [21, 47] with FLS structure determined to minimize system error, derivative of error, control effort and overshoot.

Different FLS's parameters (MFs number, peaks and parameters, rules and their number, ScFs, weights, etc.) are usually GA optimized on-line or off-line in: (1) FLSs for PI/PID FLCs auto-tuning [18]; (2) FLSs for PID auto-tuning employing various fitness functions—systematically determined in a multi-objective optimization as a weighted sum of criteria [55], defined on system performance and energy efficiency measures [21], based on IAE (integral absolute error) [48], ISE (integral squared error) [50], integral squared relative error and control action and an estimate of the maximal overshoot with application to level control [18], etc. In off-line mode GAs use system simulation and a plant model, in on-line—there are restrictions related to ensuring system stability, reduction of the time for tuning by evaluation of the fitness function from one experiment or in steady state for a small population size and a small number of generations.

The main problems of the FLS-based AFLC are the need of analysis of system stability and simplification of the controller's structure of several fuzzy units (FUs) to a structure of one FU to enable an easy PLC embedding for real time operation in industrial environment. These drawbacks can be resolved by approximation of the two-level main controller-FLS structure to a simpler one-level.

1.2.3 Problem Formulation

The FLS-based adaptive system to be designed is shown in Fig. 1.1, where y_r is the reference for the controlled plant output variable $y(t)$, $e(t) = y_r - y(t)$ is the system error, $u(t)$ is the control action and $f(t)$ is the disturbance. The system consists of a nonlinear plant, a main controller and a FLS. The plant is a laboratory installation equipped with industrial measuring transducers and final control elements. The control algorithm together with graph-scopes, noise filters, measurement converters to physical variables, scaling, pulse-width modulation (PWM) and reference step generators is embedded in a MATLAB^{TM} Simulink model for real time operation via data acquisition (DAQ) interfacing board with analog-to-digital converter (ADC), digital-to-analog converter (DAC) and digital inputs/outputs [54–57].

The main controller is based on the PID concept and its derivatives—PI, PD, etc. as it is well studied and widely applied in the industrial practice. It can be a linear PID or a fuzzy—model-free Mamdani or Sugeno, a single-input-single-output (SISO) or a two-input-single-output (TISO) PID-FLC or a TSK model-based PID-PDC. The main FLC or PDC is designed on standard rule bases and a small number of standard MFs in normalized universes of discourse. Some examples of these main controllers are depicted in Fig. 1.2, where by arrows are denoted the gains or ScFs that are auto-tuned, d/dt—produces the derivative of error $\dot{e}(t)$, 1/s—is an integrator.

The investigations are not limited to the defined types of main controllers as it is also possible to use any linear or nonlinear controller—e.g. a linear or a nonlinear PDC-Smith predictor or a linear internal model controller (IMC) or a nonlinear PDC-IMC [58, 59], etc., which can be easily designed using expert knowledge about the plant and have gains and/or ScFs to be on-line nonlinearly tuned from a second level FLS.

Fig. 1.1 Fuzzy adaptive system on the basis of two-level controller with FLS

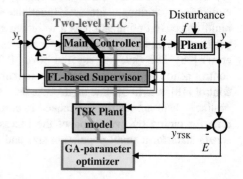

(a)

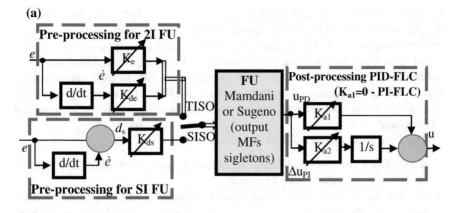

(b)

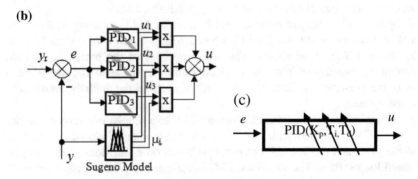

Fig. 1.2 Example of main controllers in the two-level controller with FLS: **a** a TISO/SISO (input the signed distance d_s) incremental PID-FLC; **b** a PID-PDC with three rules with local linear PID controllers and a Sugeno model for recognition of each PID area and fuzzy blending of their control actions u_i; **c** an ordinary linear PID

The FLS is built of several SISO or TISO FUs—one for a tuning parameter with a few standard MFs in absolute universes of discourse and rules derived from the relationship between performance estimates and tuned parameters. Inputs to the FUs are the performance estimates, computed from reliable on-line measurements, with known universes of discourse and desired Labels 'Norm', defined by the corresponding MFs. The FUs outputs adjust the gains and ScFs of the main controller—gains for a main linear PID and for the local linear incremental PIDs of a main PID-PDC, and ScFs for a main PID-FLC and a main PID-PDC.

All the restrictions set are related to the requirement for a simple in structure two-level controller with easy design, a small number of parameters to be tuned with great effect on the system performance and reduced trial-and-error decisions.

GA optimization is applied off-line: (1) to derive a TSK plant model from experimental data for the plant input-output if it is not initially available; (2) to optimize the FLS-based AFLC tuning parameters using closed loop system simulations based on the TSK plant model; (3) to derive a functional equivalent but simpler in structure AFLC which enables AFLC system stability analysis and easy programming of the AFLC on industrial PLCs for real time applications.

The problem is to develop a design procedure for a FLS-based AFLC using GA optimization and Lyapunov stability approaches, to test its easy application to various laboratory plants and to prove the simple AFLC superiority to an ordinary FLC and a linear controller via closed loop system experimentation in industrial environment.

The main tasks can be defined as follows:

- To outline a two-level FLS-based AFLC design procedure using small amount of expert information about the plant;
- To develop a TSK plant model from experimental plant input-output data, collected in the course of the real time operation of the closed loop system with a model-free designed main controller, by GA parameter optimization of an accepted plant model structure or training a NF structure;
- To optimize the tuning of the two-level FLS-based AFLC by using off-line GAs and simulations of the two-level FLS-based AFLC closed loop system built on the derived TSK plant model. The step responses of the tuned closed loop system are investigated in real time control of the plant in industrial environment to prove performance improvement with respect to the initially empirically tuned system;
- To approximate the two-level FLS-based AFLC to a simpler one-level controller of one FU using experimental data from the real time tuned closed loop control of the plant and off-line GA optimization based on simulations of the simpler closed loop system. The simplified FLC is validated in parallel open loop work to the two-level FLS-based AFLC in real time control of the plant;
- To analyze the stability of the closed loop system on the basis of the simplified AFLC and to estimate the facilities for embedding of the simplified AFLC in a PLC;
- To assess and compare the closed loop system performance and energy efficiency in case of AFLC control, standard FLC and linear control via simulation and real time investigations using laboratory plants.

1.3 Procedure for Design of Adaptive Two-Level FLS-Based FLC

The procedure suggested is based on [11, 16, 18–23].

The general block diagram of the Main Controller—FLS structure used is shown in Fig. 1.3, where a variety of combinations of main controllers—SISO FLC (SI FLC) or TISO FLC (two input—2I FLS) and PDC (for local linear PID the tuning parameters K_p, K_i and K_d correspond to the ScFs K_{de}, K_e and K_u respectively), FLS

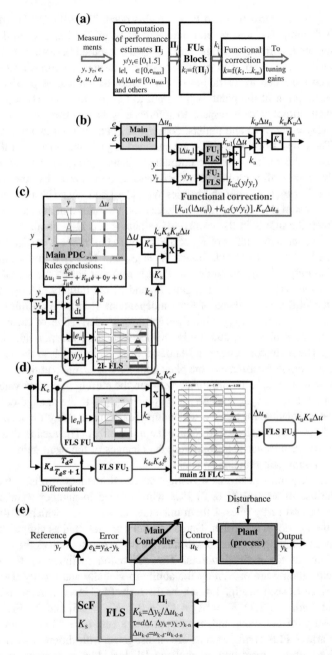

Fig. 1.3 Main controller—FLS structures: **a** a general FLS structure; **b** a FLS functional correction example; **c** a PDC-FLS example; **d** a 2I FLC FLS example; **e** a closed loop system with FLS-based AFLC using estimation of current plant gain K_k and of the plant delay τ

structures, types of correction and FUs are illustrated. The MIMO FU structure of the FLS is decomposed into several independent SI or 2I FUs, easy to design.

The functional correction on the basis of two computed performance indices is demonstrated in Fig. 1.3b. The magnitude of the normalized control rate (change $|\Delta u_n|$), which is related to restriction of the control action and respectively of the energy consumption of the plant, is the input to FU_1 of the FLS ($\Pi_1 = |\Delta u_n|$). The relative plant output y with respect to reference y_r is an estimate of the system overshoot ($\Pi_2 = y/y_r > 1$) and undershoot ($y/y_r < 1$) and is used as input to FU_2 of the FLS. The outputs of the two FUs are summed to yield the scale k_u of the output ScF K_u. Thus the main controller normalized incremental control action Δu_n is nonlinearly continually on-line modified and denormalized by the multiplier $[k_{u1}(|\Delta u_n|) + k_{u2}(y/y_r)] \cdot K_u$, which changes with $|\Delta u_n|$ and y/y_r. The FUs input MFs define the norms of these performance indices and the rules aim at ensuring outputs that will keep the inputs in the norms—Π_1—near 0 and Π_2—about 1.

In Fig. 1.3c the auto-tuning of K_u of a PDC main controller is carried out by the help of a FLS of one 2I FU with inputs $\Pi_1 = |e_n|$ and $\Pi_2 = y/y_r$ and a tuning gain K_s. The FLS output is the correction $K_s.k_u(|e_n|, y/y_r)$ that scales the signal $K_u\Delta u$ in the range of the system stability margins. K_u and K_s are initially GA optimized. In Fig. 1.3d the FLS is constructed of three independent SI FUs—FU_1 minimizes the normalized system error, FU_2 minimizes the system derivative of error and FU_3 minimizes the control rate by auto-tuning respectively the corresponding ScFs—K_e, K_{de} and K_a. The main controller is a Mamdani 2I FLC. In Fig. 1.3e the input to the FLS FU is a computed estimate of the plant gain $K_k = \Delta y_k/\Delta u_{k-d}$ at time moment t_k, where $\Delta y_k = y_k - y_{k-n}$ is the difference between the current measured value y_k and a past measured value y_{k-n}, and $\Delta u_{k-d} = u_{k-d} - u_{k-d-n}$ is the change of the plant input that is responsible for this Δy_k accounting for the plant delay $\tau = d\Delta t$. The past moment $n\Delta t$ ($n < d$) is selected to reduce the noise effect. The idea is that the FLS adapts the main controller gain following the changes of the plant gain, so that the open loop system gain remains constant.

A general requirement in this investigation for all main-controller—FLS structures is the use of several SI or 2I FUs with reduced in number standard easily designed MFs and fuzzy rules, known universes of discourse, which facilitates the design of the two-level controller. Gains and ScFs are selected as simple means for continuous auto-tuning in order to optimize the performance indices they are linked to keep them at the defined norms. The adaptation facilities of the dynamic gain/ScF auto-tuning are proven by the ability to globally and locally re-shape the control surface as seen in Fig. 1.4, where the main FLC is an incremental PI FLC with output du/dt and ScFs $K_s = [K_e = K_{de} = 0.1, K_{du} = 0.5]$ and the FLS has two Gaussian MFs for the input and the output and a rule base "Small input—Great output". Various FLS configurations are investigated with different FLS inputs—combinations of the measured normalized $|e|$, $|de|$, $|du|$ at current or previous moments and computed on their basis various performance indices, and outputs—scales k_s of ScFs K_s (k_e of K_e for e, k_{de} of K_{de} for de and k_{du} of K_{du} for du), or gains

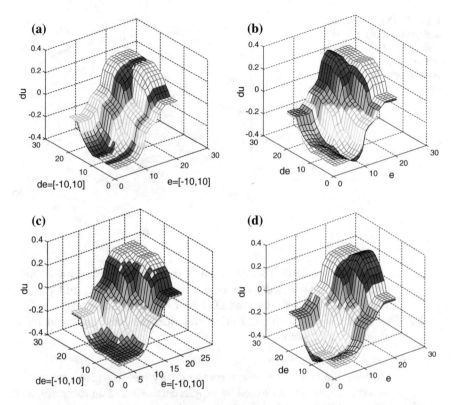

Fig. 1.4 The effect of FLS auto-tuning on the control surface (mesh—FLC, surf—FLC-FLS) for FLS: **a** input de and output k_{de}; **b** input e and output k_a; **c** input du and output k_a; **d** input de and output k_a

of linear controllers. The different configurations change different local areas in the control surface. The control surface of the main FLC without FLS (mesh) is laid over the control surface of the main FLC with FLS (surf-surface colored). For comparison in Fig. 1.5 are overlaid the control surfaces of a FLC without FLS for two different fixed ScFs and the effect observed is a uniform shrinking of the whole control surface for the reduced ScF value. This is a proof that continuous auto-tuning of a ScF results in an adaptation of the control action.

The FLS-based adaptive controller design follows the steps below.

Initial data: sample period, expert estimation of the range of plant inputs and outputs, system error, derivative of error, control action, number and ranges of linearization zones if available.

1. Design of the main controller

    1.1. Selection of the type of the main controller—Mamdani or Sugeno type SI or 2I PI/PID-FLC, PDC, other type FLC, linear controller

Fig. 1.5 The effect of the
ScF value on the FLC control
surface of a FLC without
supervisor (mesh—$K_a = 1$,
surf—$K_a = 0.5$)

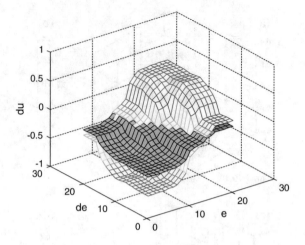

In case of an available TSK plant model or a nonlinear plant model from which a TSK plant model can be easily derived [15], the FLC can be selected of a PDC type, else the FLC is of Mamdani/Sugeno type

1.2 Design of the main controller using empirical methods and expert information about the plant

(a) For a linear controller
   The design is based on linear control theory methods given the type of control algorithm, estimated by experts parameters of an accepted/given linear plant model (usually ZN), and design criteria [5, 11].

(b) For a fuzzy logic controller

   • Determination of the FLC or PDC structure—input and output linguistic variables and ranges, pre- and post-processing and ScFs; for PDC—type of the local linear controllers to be designed and design criteria
   Design of the MFs and the fuzzy rules—3 to 7 MFs of triangle, trapezoidal or Gaussian shape with standard parameters—orthogonal, symmetrical, or singletons (usually only for the output), normalized universes of discourse, and standard rule bases

   • Initial tuning of the ScFs and for PDC—of the local linear controllers' gains, and linear controller gains from empirical, robust and stability and robust performance or other given criteria [1, 4, 5, 11, 15]

   • Estimation of ranges for tuning parameters which ensure system stability

   • Experimental study of the step responses of the closed loop system with the designed main controller in real time control of the plant. In case of a Mamdani/Sugeno FLC, collection of plant input-output data for further TSK or NF plant modeling—the plant input signals should be rich in magnitudes and frequencies so that the outputs reflect the plant nonlinearity in operation under various conditions.

1.3 TSK or NF plant modeling if a plant model is not available

The TSK plant model structure is determined from expert assessed linearization operation zones, then the parameters are obtained via GA optimization minimizing the error between the TSK plant model response and the recorded plant response in the real time control to the same plant input. The plant model is validated with independent experimental data from the used in modeling.

In case linearization zones are not identified as initial data a Sugeno NF model is trained using the collected data in the closed loop system real time control after processing—normalization, noise filtration, elimination of correlation.

A plant model—nonlinear, TSK or trained Sugeno NF, is required in: (1) simulations; (2) off-line simulations-based GA optimization of the tuning parameters of the FLS-based AFLC; (3) approximation of the two-level FLS-based AFLC to a one-level one-FU simple structure AFLC.

2. Design of the FLS structure

2.1 Identification of system performance estimates to be continuously optimized, based on available noise-free reliable measurements, with clearly defined universes of discourse and normal term and convenient for on-line computation and use

The performance estimates can include:

- a single performance measure—absolute such as settling time t_s, rising time t_r, maximal deviation y_m, error e, derivative of error $\dot{e}$, control effort u, control rate Δu, etc., or relative such as overshoot σ, relative error with respect to reference y_r, etc.
- an integral performance indicator—computed for some period of time ISE or relative ISE with respect to reference, IAE, integral of control action or relative to maximal control action, etc.
- a combination or a function of performance measures and restrictions to signals and parameters
- a performance-related current plant gain assessment

2.2 Investigation of the impact of the main controller's gains/ScFs on the various performances for high/low sensitivity, correlation, contradiction effects, etc.

2.3 Selection of the type of correction of the main controller's gains/ScFs in on-line adaptation—additive, multiplicative or functional

2.4 Selection of the one-FU-based tuning channels FLS—system performance measures to be improved or a plant gain estimate as FLS inputs—tuning parameters of the main controller as FLS outputs—ScFs, gains of local linear controllers in PDC and gains of linear controllers, etc.

2.5. Determination of the configuration of FLS FUs, each related with tuning of a separate ScF or gain—number of FUs, type of each FU—SISO or TISO, inputs-outputs, universes of discourse given as initial data for the design procedure

3. Design of each FU of the FLS and empirical tuning of the FU's output ScF

   3.1. Determination of a minimal possible number of standard inputs MFs with simple shape in the known range of the selected performance estimate with defined normal term

   3.2. Derivation of the rule base for each FU or selection of a proper standard rule base according to the desired relationship inputs-output, e.g. from empirical knowledge it is established that "**If** K_k is 'low' (operation region 1) **and** (overshoot)$_k$ is 'high' (not allowed for this region) **Then** control correction Δu_k is 'negative big'"

   3.3. Initial tuning of the FU output scaling gain K_s using empirical methods in order to ensure system stability during adaptation and effectiveness of correction, e.g. in additive correction the FU output has to be in the range of the main controller auto-tuned signal

4. Connection of the designed FLS to the main controller in the closed loop system for processing of measurements, computation of performance estimates—the FUs inputs, computation of the gains/ScFs corrections from the FUs outputs

5. GA optimization of the ScFs and gains of both the main controller and the FLS via simulations

   Bounds for the optimized in GA AFLC parameters are computed from the values from the empirical initial tuning. The fitness function can combine several relative integral performance measures, requirements for smooth and economic control and estimates for energy efficiency [11, 18–21]. Its evaluation is based on simulation investigations of the FLS-based AFLC closed loop system with the plant model. The reference step changes are random in magnitude and in duration to facilitate that the AFLC learns the plant nonlinearity and time-variance and ensures proper adaptation.

6. Approximation of the optimally tuned FLS-based AFLC to a functional equivalent with simple structure of one FU

   6.1. Determination of the simple AFLC structure

      (a) In case of available or obtained TSK plant model
         A PDC structure can be assumed as the simple one-FU functional equivalent of the FLS-based AFLC. It is determined by the TSK plant model.

      (b) In case of a Sugeno NF model
         A NF controller is accepted

6.2. Parameter optimization of the accepted structure

The parameters are computed to minimize an accepted functional of the error between the outputs of the optimally tuned FLS-based AFLC closed loop system and of PDC/NF-AFLC system. The PDC/NF-AFLC system output is simulated using the plant model. The FLS-based AFLC system output can either be simulated or recorded from the real-time control. The two systems have identical step changes in references. To better learn the adaptive features of the FLS-based AFLC the step reference changes have to be random and rich in magnitude and in duration which is easier, more safety for the plant and faster in simulation investigations of the two systems.

(a) For PDC structure

The PDC parameters are off-line GA optimized. The fitness function is the accepted functional for minimization.

(b) For NF structure

The NF structure is trained to minimize the accepted functional.

6.3. Validation of the simplified PDC/NF-AFLC via simulations in open loop parallel operation to the FLS-based AFLC working in the closed loop system for references not used in the approximation and smaller sample period.

The simplified AFLC has the same input and its output is compared with the output of the original FLS-based AFLC, which controls the plant input. If the two outputs are close, the simplified AFLC replaces the FLS-based AFLC and is connected to control the plant in a closed loop.

7. Experimental validation in real time control of the simple PDC/NF-AFLC

7.1. Completion of the FLS-based AFLC in the MATLAB^{TM} Simulink model and connection via the DAQ board to the real world plant—the laboratory installation.

7.2. Validation of the simplified AFLC structure in real time open loop parallel operation to the originally designed FLS-based AFLC.

8. Study of the designed PDC/NF-AFLC system stability by derivation of TSK-PDC-based Lyapunov stability conditions and their solution using Linear Matrix Inequality (LMI) numerical approach

9. Assessment of simplified AFLC system adaptation capacity, robustness, energy efficiency, smooth and economic control in real time investigations and comparison with the closed loop systems with different designed controllers—ordinary FLC or PDC without FLS, linear, etc.

10. PLC completion of the simplified PDC/NF-AFLC and preparation for engineering applications. A general PDC/NF-AFLC can be firmly programmed and only parameters related to the different applications can be additionally specified [19, 20, 40].

1.4 TSK Plant Modeling

The derivation of the TSK plant model is based on experimental plant input-output data. For MIMO plants these data are collected for each SISO main and cross channel and separate TSK models computed and united [11, 23]. The experimental data collected should consider input signals within the range of plant operation with a great variety of magnitudes and frequencies in order the TSK model to learn plant nonlinearity, inertia, time-variance, and other specific characteristics. They can be also processed to filter noise, reduce data by neglecting correlated data, normalized or standardized.

In this section an approach for derivation of a TSK plant model is presented on the example of a laboratory convection dryer with input analog control action u and output measured temperature y. The TSK plant model structure is determined on the basis of experimental observations of the plant step responses in different operation points. Three overlapping linearization zones are distinguished with the increase of the plant output y—each for similar adjacent step responses. The step responses suggest that an average Ziegler-Nichols (ZN) plant model can be assumed for each zone $P_i(s) = K_i \cdot e^{-\tau_i s} \cdot (T_i s + 1)^{-1}$, $i = 1 \div 3$, since it is simple with parameters K_i—gains, T_i—time constants and τ_i—time delays of clear physical nature and easy to assess from plant step responses. The different parameters for each zone ZN model correspond to the suggestion that the plant is nonlinear and a FLC is suitable for its control.

The TSK plant model structure is shown in Fig. 1.6. It is built of a dynamic part of three parallel channels and a Sugeno Model—fuzzy blender for the outputs of the channels y_i of the dynamic part to yield the final plant output y. The Sugeno Model has input y in a given range, here it is $[20, 80]°C$, which is partitioned by 3 MFs [S M B] for Small, Medium and Big to correspond to the number of linearization zones of operation of the plant. The MFs are allocated according to the defined by experts three overlapping zones. The plant output y is used to identify the degree of belonging μ_i of the current measured temperature to each of the linearization zones i. Therefore the Sugeno Model is built of 3 outputs $i = 1 - 3$, each with 3 singletons as MFs μ_{ij}, $j = 1 - 3$, allocated at 0 for $i \neq j$ and at 1 for $i = j$. The three parallel channels of the dynamic part have a common input—the plant input u. Each channel consists of two time-lags in series—the first with the physical meaning of the time lag in the ZN model, and the second—the linear term of the Taylor's series expansion of the time delay element in the ZN model—$e^{-\tau_i s} \approx (\tau_i s + 1)^{-1}$. The inertia of the plant is increased by the common time lag—$(K = 1, T = 1)$ at the plant model output, where the initial condition (initial or ambient air temperature $y(0)$ in the dryer) is added.

The TSK plant model parameters—gains K_{ITSK} and time constants T_{ITSK}, $l = 1 \div 6$, are computed by a GA optimization. The initial condition and the peaks of the input MFs can also be included in the parameters to be optimized. The accepted fitness function is minimization of the integral squared relative error

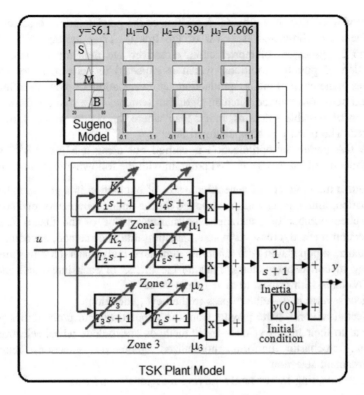

Fig. 1.6 TSK plant model for the air temperature in a laboratory dryer

between experimentally recorded real plant output y_e and simulation model output y_{TSK} with respect to real plant output:

$$\mathbf{F_p} = \int [E(t)/y_e(t)]^2 \cdot dt, \qquad (1.1)$$

where $E(t) = y_{TSK}(t) - y_e(t)$ and both $y_e(t)$ and $y_{TSK}(t)$ are responses to one and the same pattern of input $u(t)$. The plant output and input data is collected from an empirically designed FLC closed loop system in real time operation at sample period $dt = 0.3$ s. The data $[u(t), y_e(t)]$ is filtered if necessary and reduced in number to avoid correlation, here by taking every 4th point (the sampling becomes 1.2 s). The new sample period is selected to preserve the character of the signals and reduce the adjacent samples which do not differ. The pattern of input signal $u(t)$ is controlled by the reference steps to satisfy the requirements to be rich in magnitudes and frequencies and to cover the whole range of the input signal, so that the TSK plant model can learn the real plant nonlinearity.

Genetic algorithm mimics the evolution of populations. First, different possible solutions to a problem are generated. They are tested for their performance, that is, how good a solution they provide. A fraction of the good solutions is selected, and the others are eliminated (survival of the fittest). Then the selected solutions undergo the processes of reproduction, crossover and mutation to create a new generation of possible solutions, which is expected to perform better than the previous generation. Finally, production and evaluation of new generations is repeated until convergence. Such an algorithm searches for a solution from a broad spectrum of possible solutions, rather than where the results would normally be expected. The penalty is computational intensity.

The GA parameter optimization is carried out using a MATLAB^{TM} genetic algorithm embedded in a developed program with the following algorithm [18, 57].

1. **Input data**—number of generations, size of population in a generation, fitness function, initial upper and lower bound for the tuning parameters, end condition (reached number of generations or minimal value of the fitness function), selection method, crossover points, mutation rate and method, encoding type
2. Initialize a population with randomly generated individuals (chromosomes—an array of coded parameters), evaluate the fitness of each individual and rank individuals with respect to their fitness value
3. Select several best—elite to pass to the new generation
4. Select survivors—mate parents from the population with probabilities proportional to their fitness values—basic methods are roulette wheel selection, uniform, stochastic uniform, normalized geometric selection, remainder, tournament selection
5. Create offspring by randomly varying individuals—parents

    5.a. Apply crossover over parents with a probability equal to the crossover rate —basic crossover methods are single point, multipoint, uniform, scattered
    5.b. Apply mutation with a probability equal to the mutation rate—mutation methods—Gaussian, uniform, adaptive feasible (in one or more bits)

6. Evaluate the fitness function and accept the offspring in the new generation if better than the parents, else repeat from step 4 to step 6.
7. Repeat from step 4 to step 6 until enough members are generated to form the next generation.
8. Repeat from step 3 till the end condition is met—the number of generations or the desired accuracy (minimum) is reached.

Each parameter set is first encoded for instance into a concatenated bit string representation making a chromosome for specific parameters values. Each parameter in the chromosome is a gene. After a population is created the fitness (objective) function is computed for each member (chromosome). Then parents are selected with probability proportional to their fitness value for producing off-springs for the new generation. The idea is to let members with above-average fitness reproduce and replace members with below-average fitness. Crossover generates new chromosomes that are expected to retain the good features of the previous generation. In a single point crossover, the point is selected at random and the parent chromosomes swap their bit strings to the right of this point. Then mutation takes place by flipping a bit. Its aim is to prevent the population from converging towards a local minimum. The mutation rate is low in order to preserve good chromosomes.

This algorithm is repeated several times—each time expanding the upper and lower bounds for the parameters, till a desired minimum of the fitness function Eq. (1.1) is reached. In this way the subjective assignment of these bounds is avoided. The fitness function is evaluated after running a Simulink model of the TSK plant model from Fig. 1.6 with current values for the tuning parameters and the same inputs to the real plant—the input to the PWM that controls the relative duration of the connection of the electrical heater to the mains voltage.

For the laboratory dryer example the GA optimization reaches minimum of $\mathbf{F_p} = 2.9$ for the optimal computed parameters $\mathbf{q_{TSK}} = [K_{1TSK} = 13.72,$ $K_{2TSK} = 12.2,$ $K_{3TSK} = 8.3,$ $T_{1TSK} = 57.3,$ $T_{2TSK} = 96.5,$ $T_{3TSK} = 49.1,$ $T_{4TSK} = 11.6, T_{5TSK} = 15.4, T_{6TSK} = 2.5]$. The parameters of the GA are: population size = 20; number of generations = 20; elite = 2; crossover rate = 0.8 and method—single point; mutation operator–adapt feasible; fitness scaling—rank based; selection—roulette wheel; binary coding.

The ready TSK plant model is validated comparing its response to the real plant experimental response in real time operation for input signals, different from the used in modeling. Usually the TSK plant model is connected in parallel to the real plant that is on-line real time controlled in a closed loop. The TSK plant model is fed with the same control action and its output is compared with the real plant output. If the two outputs are closed, i.e. satisfy the criterion for small integral squared relative error, the TSK plant model is accurate and can substitute the plant for the range of input and output it is constructed. Here the validation is carried out comparing the real plant step responses in real time operation and the TSK plant model responses to the same step inputs—Fig. 1.7.

In case of a MIMO plant individual channels input-output are TSK modeled and then unite. For example for a Heating, Ventilation and Air-Conditioning (HVAC) plant the output controlled variables are the room air relative humidity $y_1 = RH,\%$ and temperature $y_2 = T, °C$, controlled by the duty ratio of switching of humidifier u_1 and heater u_2 to the nets voltage supply. The plant is TITO (two-input and two-output) with cross impact—the humidifier affects besides RH also T and the heater influences T, which changes also RH. So, the plant is multivariable, and can be represented with the block diagram in Fig. 1.8a. Similar TSK models to the given in Fig. 1.6 for each of the four main and cross channels are built and separately or jointly GA optimized to yield the compound TITO TSK plant model in Fig. 1.8b, where Hum1j denotes the channels with output 1—RH—main channel for input u_1 and output component y_{11}, and cross channel for u_2 and output component y_{12}, T2j denotes the channels with output 2—T and inputs u_1 for cross channel y_{21} and u_2 for main channel y_{22}.

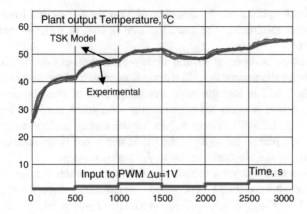

Fig. 1.7 TSK plant model output versus dryer's air temperature from real time experiments

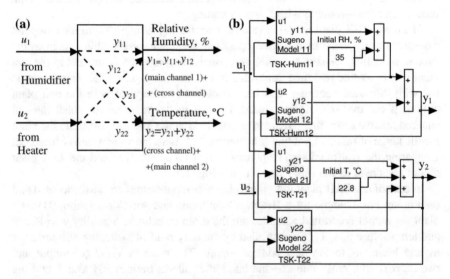

Fig. 1.8 TITO TSK plant model: **a** a general block diagram of a HVAC TITO plant; **b** a compound TITO TSK plant model of the TSK models for each channel

1.5 GA Optimization of the FLS-Based AFLC

The GAs can be applied to optimize off-line the FLS-based AFLC parameters. They provide a proper optimization approach because of the effective parallel within a generation random gradient-free search technique which ensures fast computing of the global extremum of a complex nonlinear non-analytically defined multi-modal fitness function of many parameters and constraints.

The tuning parameters of the fuzzy two-level controller are the parameters of the main controller $\mathbf{q}_{main}$ and the parameters of the FLS $\mathbf{q}_{FLS}$ − $\mathbf{q} = [\mathbf{q}_{main}\mathbf{q}_{FLS}]$. The initial empirically tuned controllers' parameters are used to define the required by the GAs upper and lower bounds for the search of the optimal parameters. The fitness function is computed off-line after simulations of the closed loop system using the TSK plant model. It may unite several requirements for the performance of the closed loop system with the FLS-based AFLC and then a multi-objective optimization is carried out. Here, the following multi-objective fitness function is proposed:

$$F = \frac{\max(y_m)}{D_r} + \frac{1}{N}\sum_{n=1}^{N}|e_n| + \frac{1}{N}\sum_{n=1}^{N}|u_n| + \max(\frac{y_m}{y_r D_r}) \to \min_{\mathbf{q}}, \qquad (1.2)$$

where N is the total number of samples for a preset final time of the simulation experiments. In Eq. (1.2) four objectives are combined—minimization of: (1) relative maximal deviation from reference y_m with respect to the range of the step changes in reference D_r; (2) mean absolute error; (3) mean absolute control effort—an estimate for energy efficiency; (4) maximal relative output with respect to reference y_r and D_r—an estimate for the maximal overshoot $\sigma = y/y_r - 1$.

The parameter optimization based on simulations is adequate and realistic if the computation of Eq. (1.2) uses a sample of the nonlinear closed loop system responses to step reference changes with a variety of random magnitudes and durations within the ranges of the real system operation in order to reflect all possible operation conditions.

1.6   Case Studies for FLS-Based AFLC Design and GA Optimization

Case study 1—FLS-based AFLC with main linear controller
    The block diagram of a FLS-based AFLC with a main PID linear controller and a FLS of one 2I FU is shown in Fig. 1.9. The parameters optimized by the GA tuner are $\mathbf{q}_{FLS} = [k_p\ k_i\ k_d]$ and $\mathbf{q}_{PID}^o = [K_p^o\ T_i^o\ KD^o]$ and the fitness function is Eq. (1.2). Then these parameters are continuously on-line adapted by the FLS at each sample period $t_n = n \cdot dt$:

$$K_p^n = k_p^n\ K_{FLS}[|e|, (y/y_r)]\ K_p^o, T_i^n = k_i^n\ K_{FLS}[|e|, (y/y_r)]\ T_i^o, KD^n$$
$$= k_d^n\ K_{FLS}[|e|, (y/y_r)]\ KD^o.$$

The FLS-based AFLC from Fig. 1.9 is designed for the control of the temperature of the air in the laboratory dryer. Experimental step responses of temperature of the designed closed loop systems from real time control are depicted in Fig. 1.10. The control action is presented in Fig. 1.11. The overshoot and the settling time are

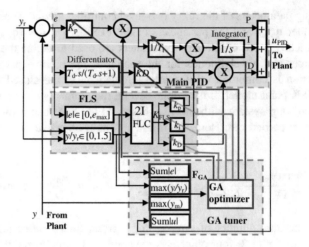

Fig. 1.9 FLS-based AFLC with GA tuner, main PID controller and FLS on 2I FU

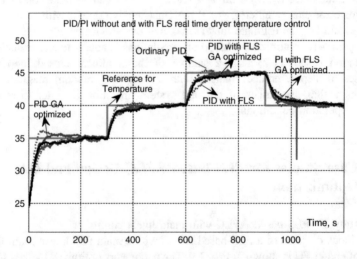

Fig. 1.10 Temperature step responses in different operation points (references) of designed systems based on FLS, GA tuner and main PID/PI linear controllers

reduced after GA optimization of the tuning parameters. The PI/PID-FLS with GA optimization is the best solution. The GA optimized PID-FLS ensures the fastest step responses. Either of the two intelligent techniques—GA optimization only or FLS only, leads to a significant reduction of system settling time and overshoot. With respect to economic control and safety and reliability of the final control elements the best is the controller that ensures the good system performance measures at low control action with less oscillations and reduced noise effect— among all studied PI/PID–FLS AFLCs the GA optimized PI-FLS is the best [21],

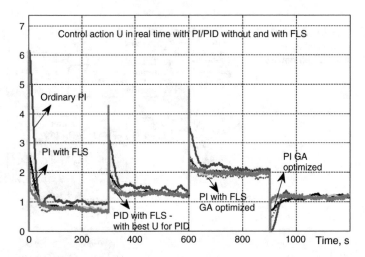

Fig. 1.11 Step responses of control action U in different operation points (references) of designed systems based on FLS, GA tuner and main PID/PI linear controllers

followed by the PID-FLS with empiric tuning. The improvement due only to GA optimization is similar to the improvement due only to the introduction of a FLS. The effect of GA, however, is dependent on a reliable plant model, needs off-line simulations, a rich sample of data collected and a proper fitness function, while the FLS operates on-line, considering all current moment system peculiarities and needs scanty information about the plant.

Case study 2—FLS-based AFLC with main PDC

The PDC approach is attractive in designing of simple FLCs for nonlinear plants taking advantage of the well-developed linear control theory. The nonlinear plant is represented by a TSK plant model with local linear plant models in the conclusions of the fuzzy rules, each described in the state space. The rules premises of the TSK plant model and of the PDC are identical and are based on variables that recognize the belonging to the defined local linearization areas. The PDC is a TSK controller model with linear controllers described in the state space in the rules conclusions. Each linear controller is designed for the corresponding linear plant model. Thus each controller's rule compensates the corresponding plant model rule. The design of the local linear controllers is based on the linear control theory to ensure local closed loop system stability, robustness, desired performance, compensation of disturbances, etc. [11, 15]. After or parallel to the local linear controllers design the stability of the designed global nonlinear system is analyzed using Lyapunov indirect method. The nonlinear PDC system stability inequalities may consider different constraints over signals, model uncertainties, performance indicators and they are solved applying the LMI numerical technique.

In [19] the FLS-based AFLC design and optimization methodology is applied when the main controller is a PDC from Fig. 1.2b. The local linear controllers are PI and a TSK plant model is available. The plant is the air temperature in the

Fig. 1.12 FLS-based AFLC
with GA tuner, main PI-PDC
controller and FLS on one SI
FU

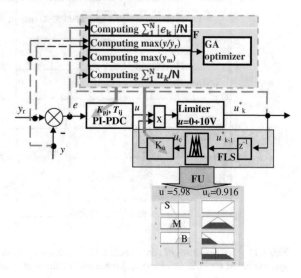

laboratory dryer. In the operation range it exhibits different responses which can be classified in three linearization areas.

The block diagram of the PI-PDC-FLS AFLC is given in Fig. 1.12. The parameters $\mathbf{q} = [\mathbf{q}_{PI}\ \mathbf{q}_{FLS}]$ to be GA optimized are the parameters of the three PI controllers $\mathbf{q}_{PI} = [K_{p1}, T_{i1}; K_{p2}, T_{i2}; K_{p3}, T_{i3}]$ and the FLS output ScF $\mathbf{q}_{FLS} = [K_u]$. The fitness function is Eq. (1.2). The FLS operates on-line in real time and can compensate for different signals, disturbances and changes in plant behaviour, etc., which are not included in the sample used in the GA optimization. The suggested FLS has an input the bounded control u_{k-1}^* in previous time moment t_{k-1} and output u_c which provides a multiplicative correction of the PDC-FLC output u according to Fig. 1.12. The correction is assumed to vary in the range [0.3, 1.7] not to influence system stability but also for the purpose of its tuning a scaling gain K_u is introduced. The input and output MFs of the FLS fuzzy block and the rules are selected standard and are shown also in Fig. 1.12. The meaning of this embedding of a FLS is to keep the control action in the defined by the MFs norm thus ensuring energy efficiency in control. The bigger the control—the longer the period when the dryer and the fan of the dryer are connected to the nets supply, so the greater the consumed electrical energy. Oscillations with great peaks of control action are not good for the heater durability and reliability, and also show energy inefficiency—too much heating followed by a necessary cooling.

The step responses of temperature in systems with 3PI-PDCs with and without FLS and a linear PI controller in real time control are depicted in Fig. 1.13. The FLS does not significantly improve the system performance as it is designed to reduce the control effort. The GA parameter optimization minimizes the fitness function (Eq. 1.2) that unites several criteria—three related to system performance (mean absolute error, overshoot σ, maximal deviation from reference y_m), and one

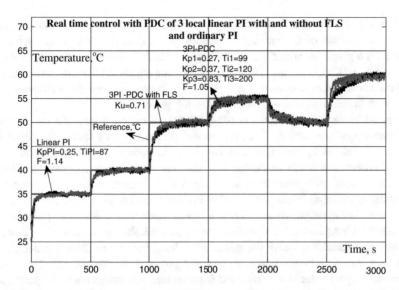

Fig. 1.13 Temperature step responses in different operation points (references) of designed systems based on FLS, GA tuner and main 3PI-PDC

—to the control action (mean control effort). The PDC systems with and without the FLS have close settling time, only for the last reference the FLS system has a slower response. The FLS, however, contributes to reduction of the control effort as seen in Fig. 1.14, so the control is more economic and smooth (small peaks and oscillations) and the system—more energy efficient.

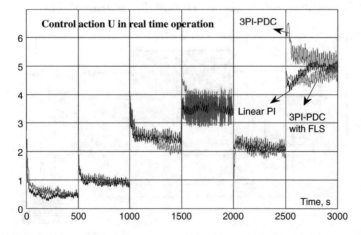

Fig. 1.14 Step responses of control action U in different operation points (references) of designed systems based on FLS, GA tuner and main 3PI-PDC

The parameters of a PDC-FLS-based AFLC in Fig. 1.3e can be on-line tuned using an estimate of the current plant gain as input to the FLS $K_k = \Delta y_k / \Delta u_{k-d}$, where for a given Δt sample period $d \cdot \Delta t$ represents the a-priory estimated plant time delay, $\Delta y_k = y_k - y_{k-n}$, $\Delta u_{k-d} = u_{k-d} - u_{k-d-n}$ are computed from measured values at time instants $k.\Delta t$ and $n.\Delta t - n = 3 \div 5 < d$ is chosen from the estimated plant inertia in order to escape too small or too big gains.

The adaptive controller aims at compensation of the plant variance via changing the controllers' parameters. The plant model parameters vary due to the following main reasons. First, the operation point moves along a nonlinear plant static characteristic as a result of changes of different plant inputs (control and disturbances). Plants with smooth nonlinearity can be represented by a number of identical in structure linear plant models but with variable parameters—each for an identified linearization zone around a given operation point. So, plant model parameters change with the operation point. Second, the parameters are a function of auxiliary variables. Third, the parameters change with time (time variant plants) due to aging, pollution, change of load and physical parameters, etc. The easiest plant model parameter to be estimated from on-line measurements is the plant gain. Also it is the most important for the great impact on the closed loop system stability, robustness and performance. Therefore, the changes of the plant gain can be compensated by adaptation of the controllers' parameters which will ensure the closed loop system stability and its desired performance despite changes of the operation point along nonlinear static characteristic or changes in the auxiliary variable or with time.

A FLS-based AFLC structure using estimation of current plant gain is shown in Fig. 1.15. The main controller can be of various types. Here two variants are assumed: (1) a PDC, consisting of j local PI linear controllers with tuning parameters gains and time constants (K_{pj}, T_{ij}) and a Sugeno model for fuzzy blending of the local controllers' outputs u_{kj} at each sample time $k - u_{kPDC} = \sum_{j=1}^{N} \mu_{kj} u_{kj}$ according to the defined linearization zones; (2) a Mamdani PI incremental FLC with tuning

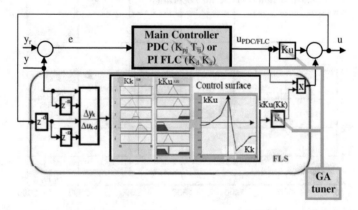

Fig. 1.15 FLS-based AFLC with GA tuner, main PI-PDC controller and FLS on one SI FU with input—the estimate of the current plant gain

parameters the differentiator's gain K_d for the de/dt and the denormalization gain T_a for Δu before the integral post processing. The FLS estimates the current plant gain K_k from the rate of change of its output Δy_k and its delayed input Δu_{k-d} and adapts on-line the control action $u_{k+1}(K_k) = u_{kPDC/FLC} \cdot [K_u + K_s k K_u(K_k)]$. GA tuner is used to optimize the parameters of the FLS and of the main controller.

The FLS-based AFLC concept on system performance indicator—an estimate of the current plant gain in Fig. 1.15 is applied for the design of the control of the following complex plants with available approximate nonlinear models: (1) the product concentration in a chemical reactor with a PDC main controller; (2) the liquid levels in a two-tank TITO plant with a main Mamdani PI incremental FLC in two identical structures; (3) the biogas production rate with a main Mamdani PI incremental FLC. By simulation is proven that the FLS in the three applications reduces both the overshoot and the settling time in all step responses in different operation points and simulated changes of plant parameters. In the TITO system the coupling is reduced as well. The systems improve their robustness preserving their performance despite the plant parameter variations due to nonlinearity and specific chemical or biochemical reactions. In Fig. 1.16 the step responses of the biogas production rate in various simulated control systems are depicted at different reference changes and disturbances on the initial organic waste. In Fig. 1.17 the corresponding control actions—the dilution rates D, are presented.

The comparison shows that the PI FLC-FLS based on an estimate of the current plant gain has similar step responses at different references and disturbances on the initial organics, characterized by the smallest overshoot and the shortest settling time, and a smooth fast control with few oscillations and the smallest peaks. This

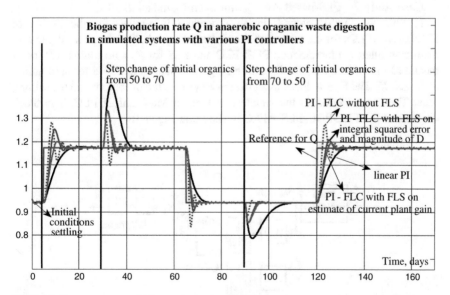

Fig. 1.16 Step responses of biogas production rate in simulated control systems with various controllers

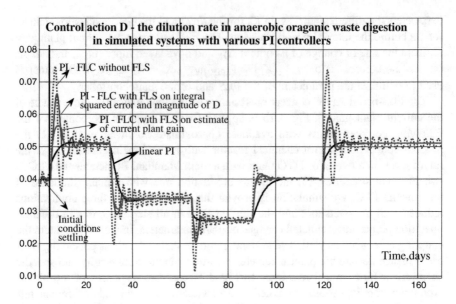

Fig. 1.17 Control actions–dilution rates in simulated control systems with various controllers

observation proves the improved system performance and robustness at reduced control effort (energy—economic control). Next rated is the system with FLS based on integral squared error estimation and the magnitude of the control, followed by the FLC system without a FLS and the worst—the linear PI control system.

Case study 3—FLS-based AFLC with main Mamdani 2I FLC

In [20, 48] the developed design procedure is applied for the design of FLS-based AFLCs for the control of the air temperature in the laboratory dryer with main controller: (1) incremental PI 2I FLC and FLS for K_u auto-tuning; (2) incremental PI PDC and a FLS for K_u auto-tuning; (3) linear PI and FLS for auto-tuning of K_p and K_i. The FLS reduces in all the three systems the overshoot and the settling time. The control effort is the smallest with main Mamdani 2I FLC. The block diagram of the FLC/PDC-FLS AFLC is shown in Fig. 1.18.

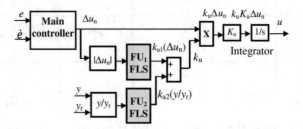

Fig. 1.18 FLS-based AFLC with GA tuner, main controller and FLS on SI FU_1 and 2I FU_2

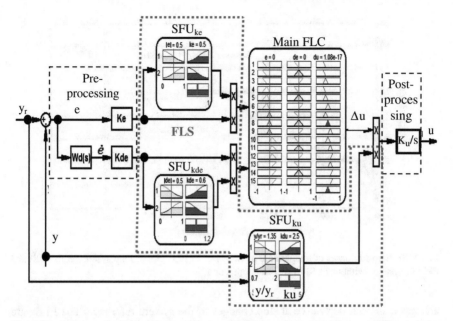

Fig. 1.19 FLS-based AFLC with main incremental PI-2I FLC and FLS on two SI FU_1—SFU_1 and SFU_2, and one 2I FU_2—SFU_3

In [11] the FLS of the AFLC is built of three FUs—SFU in order to easily design, where the subscripts denote the SFU output which scales the corresponding ScF—Fig. 1.19. The MFs of the SFU are small in number and in absolute universes of discourse. The main controller is PI-2I FLC designed from robust performance criterion [11]. The FLS improves the system performance measures and energy efficiency.

In [47] a FLS-based AFLC is designed for the same plant using the same main controller as in [11]. The auto-tuning of K_e and K_{de} is also the same but the FLS structure for the adaptation of K_u—the signal Δu before the post-processing is modified by two FUs to yield Δu $\{1 + [k_{u1}(y/y_r) + k_{u2}(|\Delta u|)]$, where $k_{u1}(y/y_r)$ and $k_{u2}(|\Delta u|)$, are the outputs of the two FUs, which on-line change with the change of the accepted performance measures—y/y_r—the estimate for σ, and $|\Delta u|$—the estimate for energy efficiency. The improvement in the system performance is seen in Fig. 1.20, where the real time step responses of the temperature in the laboratory dryer of the two-level PI FLC system (with FLS) and the PI-FLC system (without FLS) are overlaid to ease comparison—the FLS reduces significantly the settling time. By further GA optimization the overshoot that appeared in the first two step responses can be additionally reduced.

In [23] a TITO FLS-based AFLC is designed for the control of the air humidity and temperature in a HVAC system, which consists of two identical structures for each channel. The main controllers are PI position 2I FLCs with input the channel system error e_i and the derivative of the plant output dy_i/dt, which changes smoothly

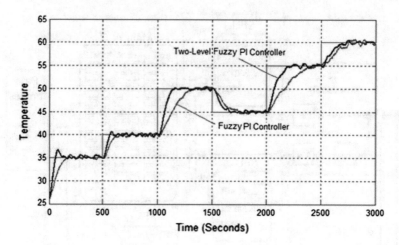

Fig. 1.20 Step responses of temperature of two-level PI FLC closed loop system (with FLS) and PI-FLC system (without FLS) in real time operation

and has a smooth derivative at step changes of the system reference. The FLSs are of three FUs—each for tuning of ScFs K_e, K_{dy} for the inputs and K_p in the post-processing. The FUs inputs are the performance indicators for dynamic accuracy $|e_i|$ and $|dy_i/dt|$ and economic control effort $|\Delta u_i|$ at the output of the FU of the i main controller.

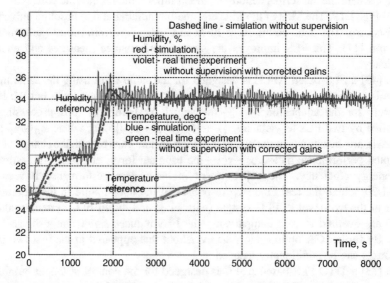

Fig. 1.21 Step responses of humidity and temperature of a TITO FLS-based AFLC closed loop system from simulation—with on-line FLS adaptation (*solid line*), without the FLS (*dashed line*), and from real time experimentation with corrected gains (measurement noise effect observed on the graphs)

Using the GA optimized model in Fig. 1.8 off-line adaptation is simulated. The mean values of the adapted ScFs are then fixed as corrected gains for the ScFs and the closed loop system tested in real time experimentation. In Fig. 1.21 the system step responses are depicted from simulation with the on-line FLS adaptation and without the FLS, and from the real time control with fixed to mean values correcion gains (ScFs). Simulations and real time control are carried out in MATLAB^{TM} environment [55, 56]. The general improvement of 46 % for AFLC system in real time control with respect to the system without FLS is estimated from introduced functional that includes the sum of three components: (1) the accumulated relative control effort for the two channels and the four step response; (2) the sum of relative settling times; (3) the sum of relative maximal deviations of outputs from the corresponding references.

1.7 Approximation of Adaptive FLS-Based FLC with PDC of Simple Structure for the Purpose of PLC Implementation and System Stability Analysis

1.7.1 Approximation TSK Modeling of the FLS-Based AFLC

The next stage in the design of a FLS-based AFLC is related to the embedding of the controller in an industrial PLC for wide engineering applications. The ease and the transparency of the design and optimization of the FLS-based AFLC leads to the use of several FUs which complicates or even makes impossible its PLC completion as the required resources of memory and processing time surpass the limits set for real time operation in industrial environment [6–9]. This is the reason to search for approximation of the designed FLS-based AFLC to a simpler structure which is equivalent in functionality.

There are various approaches to approximation of a designed FLS-based AFLC from experimental data, collected from real time operation of the closed loop system in MATLAB^{TM} real time. In [11, 13] a neuro-fuzzy Sugeno structure of one FU is trained to equivalently replace the designed AFLC with many FUs. The collected experimental data should contain signals from real time operation in all possible operating points and conditions, the data should be pre-processed—filtered from noise and correlation, normalized or standardized. Training is long and after it generalization and validation tests with independent input data are required.

Here a procedure for approximation of a designed FLS-based AFLC to a simple PDC equivalent is illustrated. It is based on available TSK plant model and GA optimization techniques. The TSK plant model determines the structure of the PDC like the given in Fig. 1.2b, where the Sugeno model is the same in structure and in function as the used in the TSK plant model in Fig. 1.6. Local linear PI/PID controllers of the PDC are selected and their parameters are computed by GA minimization of the error between the outputs of the closed loop system with the FLS-based AFLC $y_{AFLC}(t)$ and the closed loop system with its PDC equivalent $y_{PDC}(t)$. The accepted fitness function is:

$$\mathbf{F} = \int [E(t)/y_{AFLC}(t)]^2 \, dt, \tag{1.3}$$

where $E(t) = y_{AFLC}(t) - y_{PDC}(t)$.

Both systems are simulated using the TSK plant model when subjected to the same random in magnitude and duration reference changes in order to imitate all possible operating conditions, so that the PDC equivalent learns the FLS-based AFLC nonlinearity and ability to adapt. The collected in simulation data is big in size and correlated. Therefore a sample of it is selected for the GA optimization with a multiple sample period $dt_m = m \ dt \ (m > 2)$, preserving the basic pattern of the random reference signal and reducing both size and correlation effect.

The simulation is fast, more realistic, with full control on the experiment, safe for the real world plant. It allows various inputs—reference changes and disturbances with different magnitude and frequency from the realistic industrial environment range. However, a reliable plant model is required for the whole range of operation conditions.

The simplified PDC substitute for the designed FLS-based AFLC is validated in parallel work and open loop during the real time control of the plant via the FLS-AFLC. It is connected in the closed loop system with the FLS-AFLC that controls the plant to receive the same error signal and its output control action $u_{PDC}(t)$ is compared with the control of the FLS-AFLC $u_{AFLC}(t)$. For small relative error between $u_{PDC}(t)$ and $u_{AFLC}(t)$ the simplified PDC replaces the FLS-based AFLC.

The approximation procedure is applied to a designed FLS-AFLC controller for the air temperature in a laboratory dryer. In Fig. 1.22 the step responses of the air temperature in a laboratory dryer and the corresponding control are seen for the adaptive real time control with the initial PI-FLS-AFLC and the simplified PI-PDC-AFLC. The settling time and the overshoot in the two systems are close, the control action of the PDC is on average smaller in magnitude and without peaks and oscillations which is both more economic and safe for the final control elements. This simplification with a PDC equivalent is easy to apply to a great number of nonlinear plants and plants with variable parameters from industry when their operation range that includes all operation points, modes and conditions, can be split into a few overlapping zones in which the plant can be represented by average linear plant models. This allows to build the Sugeno model in the structure of the TSK plant model in Fig. 1.6 or Fig. 1.8 and to obtain its parameters via GA optimization.

Fig. 1.22 Step responses of
temperature and control of a
FLS-based AFLC closed loop
system and a PDC AFLC
system from real time control

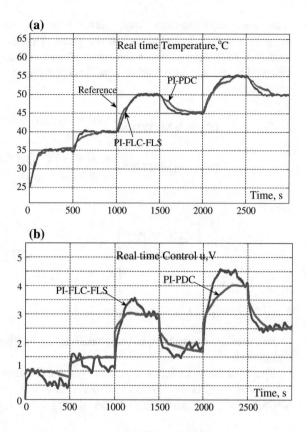

1.7.2   Lyapunov Stability Analysis of System with PDC-AFLC

The PDC representation of the FLS-based AFLC together with the TSK plant
model facilitates the study of the closed loop system stability using Lyapunov
indirect method and the LMI numerical technique [11, 15, 16].

The classical approach to a PDC design is selection on the basis of the TSK plant
model of a PDC structure and local linear controllers and tuning their parameters
using the linear control theory methods for ensuring of local linear systems stability
and desired performance. Then stability analysis of the global fuzzy nonlinear
system on the basis of the TSK models of the PDC and the plant via the Lyapunov
indirect method is carried out. In this investigation the PDC is tuned to substitute
the complex FLS-AFLC. This makes the problem of global system stability very
important especially considering the adaptive properties of the system.

The closed loop system fuzzy rules are derived for a TSK plant model with local
ZN linear models and a TSK-PDC with local PI linear controllers. First the PI
controllers are turned to incremental controllers to eliminate integration in their

state space representation. Their output becomes the control rate which needs integration before being passed to the plant. This integration is included in the modified TSK description of the plant. Thus the fuzzy rules of the closed loop system for the ith linearization zone become respectively for the plant and the controller:

$$
\textbf{IF } \mathbf{y}(t) \text{ is } M_i \textbf{ THEN } \left| \begin{array}{l} \dot{x}_i(t) = \mathbf{A}_{io} x_i(t) + \mathbf{B}_{id} u_i(t - \tau_i) \\ y_i(t) = \mathbf{C}_i x_i(t) \end{array} \right.
$$

$$
\textbf{IF } \mathbf{y}(t) \text{ is } M_i \textbf{ THEN } \left| \begin{array}{l} \dot{u}_i(t) = -\mathbf{F}_i x_i(t) + \mathbf{G}_i x_r \\ \text{or } \dot{u}_i(t) = K_{pi}\dot{e}(t) + (K_{pi}/T_{ii})e(t) \end{array} \right. ,
\tag{1.4}
$$

where

$$
x_i(t) = \begin{bmatrix} x_{i1}(t) = y(t) \\ x_{i2}(t) = \dot{x}_{i1}(t) \end{bmatrix}, \ \mathbf{A}_{io} = \begin{bmatrix} 0 & 1 \\ 0 & -1/T_i \end{bmatrix}, \ \mathbf{B}_{id} = \begin{bmatrix} 0 \\ K_i/T_i \end{bmatrix},
$$

$$
\mathbf{C}_i = \begin{bmatrix} 1 & 0 \end{bmatrix} \text{ and } x_r = \begin{bmatrix} x_{r1} = y_r \\ x_{r2} = 0 \end{bmatrix};
$$

$$
\mathbf{F}_i = \begin{bmatrix} K_{pi}/T_{ii} & K_{pi} \end{bmatrix}, \ \mathbf{G}_i = \begin{bmatrix} K_{pi}/T_{ii} & 0 \end{bmatrix}
$$

In a similar way fuzzy rules can be derived for any other local linear plant model and controller.

The Lyapunov sufficient condition for quadratic stability of the closed loop system Eq. (1.4) is the existence of matrices $\mathbf{P} > 0$, and $\mathbf{Q} > 0$ such that the following matrix inequalities are satisfied for $i, j = 1 \div r, j > i$ [11, 15]:

$$
\left| \begin{array}{l} \mathbf{PA}_{io} + \mathbf{A}_{io}^T \mathbf{P} + \mathbf{PB}_{id}\mathbf{F}_i\mathbf{Q}^{-1}\mathbf{F}_i^T\mathbf{B}_{id}^T\mathbf{P} + \mathbf{Q} < 0 \\ 0.5\mathbf{P}(\mathbf{A}_{io} + \mathbf{A}_{jo}) + [0.5(\mathbf{A}_{io} + \mathbf{A}_{jo})]^T\mathbf{P} + 0.5(\mathbf{B}_{id}\mathbf{F}_j\mathbf{Q}^{-1}\mathbf{F}_j^T\mathbf{B}_{id}^T + \mathbf{B}_{jd}\mathbf{F}_i\mathbf{Q}^{-1}\mathbf{F}_i^T\mathbf{B}_{jd}^T) + \mathbf{Q} \le 0 \end{array} \right.
\tag{1.5}
$$

The solution of Eq. (1.5) is searched by using the LMIs numerical computation technique for solving optimization problems of the mathematical programming under convex restrictions [11, 15, 60, 61]. For the transformation of the nonlinear inequalities (1.5) into linear the Shur decomposition is applied:

$$
\begin{bmatrix} \mathbf{PA}_{io} + \mathbf{A}_{io}^T\mathbf{P} + \mathbf{Q} & \mathbf{PB}_{id}\mathbf{F}_i \\ \mathbf{F}_i^T\mathbf{B}_{id}^T\mathbf{P} & -\mathbf{Q} \end{bmatrix} < 0
\tag{1.6}
$$

Then LMIs, related to the requirements for positively determined solution, are added. A harder condition is introduced for the searched matrices

$$
-\mathbf{P} < -\mathbf{O} \text{ and } -\mathbf{Q} < -\mathbf{O},
$$

where $\mathbf{O}$ is accepted to be a small positively determined matrix $\mathbf{O}$ (e.g. $\mathbf{O} = 10^{-3}\mathbf{I}$) instead of a zero matrix. This aims at ensuring of a LMIs solution of positively determined matrices $\mathbf{P}$ and $\mathbf{Q}$ in case of a critically stable open loop system due to the PI-PDC which contains an integrator. The solutions $\mathbf{P}$ and $\mathbf{Q}$ should also guarantee a low sensitivity to parameter variations, inaccurate data and computational errors, so they should have small condition numbers.

1.8 Conclusion

In this chapter, an approach for the design of adaptive controllers based on a two level structure of a main controller and a fuzzy logic supervisor is presented. The design considers simple means for on-line real time continuous adaptation by FLS auto-tuning of gains and scaling factors of the main controller based on assessed from current measurements performance indicators. The approach is built on easy and transparent procedures that integrate TSK modeling on experimental data and optimization using genetic algorithms. The GA optimization is used for tuning of the parameters of the TSK plant/controller model and of the FLS-based AFLC. The two techniques are applied to modeling of the designed FLS-based AFLC for the purpose of its approximation with a simple structure of one fuzzy unit. An approximation to a simple PDC functional equivalent of the FLS-AFLC is demonstrated in order to facilitate the PLC implementation of the AFLC and to enable the stability analysis of the adaptive closed loop system presented. The case studies show the application of the approach for the design of various FLS-based AFLCs with different main controllers—PI/PID linear, PI/PID FLC, PI/PID PDC, and FLS structures for the control of temperature, levels, concentration, humidity and temperature. The simulation and real time investigations prove the improved closed loop system performance and energy efficiency—reduced settling time, overshoot, control action and increased robustness, in the whole range of operation and of plant parameters variations. The comparison is with ordinary FLC and linear control.

The future research will focus on PLC implementations and industrial applications.

Acknowledgments The author would like to express her gratitude to Prof. Lakhmi C. Jain from the University of Canberra, Australia for his invaluable assistance and the many discussions that helped improving the presentation of the research, to the reviewers for their competent comments and recommendations and to the Bulgarian diploma and Ph.D. students who were involved in many of the experiments and the investigations, contained in this chapter.

References

1. Driankov, D., Hellendorn, H., Reinfrank, M.: An Introduction to Fuzzy Control. Springer (1993)
2. Jang, J.-S.R., Sun, C.-T., Mizutani, E.: Neuro-fuzzy and soft computing. In: A computational Approach to Learning and Machine Intelligence. Matlab Curriculum Series, Prentice-Hall Inc., NJ (1997)

3. Ross, T.J.: Fuzzy Logic with Engineering Applications. McGraw Hill, Inc (1995)
4. Yager, R.R., Filev, D.P.: Essentials of Fuzzy Modelling and Control. Wiley Inc., NY (1994)
5. Jantzen, J.: Foundations of Fuzzy Control. Wiley (2007)
6. SIMATIC S7 Fuzzy Control. User's Manual, Siemens AG
7. Auto-tune PID Temperature Controller. Fuzzy Logic plus PID Microprocessor-based Controller Series. BRAINCHILD Electronic Co. Ltd (2003)
8. Auto-tune Fuzzy PID Process Temperature Controller, Manual TEC-220/920 Revision 12/2012—TEC-220 and TEC-920, TEMPCO Electrical Heater Corporation
9. Future Design Controls, Inc. http://www.futuredesigncontrols.com/
10. Georgieva, O., Sekoulov, I., Behrendt, J.: Intelligent control design of activated sludge wastewater treatment plants. Int. J. Knowl. Based Intell. Electron. Syst. **4**, 110–117 (2000)
11. Yordanova, S.: Methods for Design of Fuzzy Logic Controllers for Robust Process Control. KING, Sofia (2011). (in Bulgarian)
12. Yordanova, S.: Fuzzy two-level control for anaerobic wastewater treatment. In: Yager, R., Sgurev, V. (ed.) Proceedings of the 2nd International IEEE Conference on Intelligent Systems I, Varna, Bulgaria, pp. 348–352 (2004)
13. Yordanova, S., Petrova, R., Mladenov, V.: Neuro-fuzzy control for anaerobic wastewater treatment. WSEAS Trans. Syst. **2**(3), 724–729 (2004)
14. Yordanova, S., Tsekova, R., Tabakova, B., Mladenov, V.: MATLAB Real-time two-level fuzzy control of nonlinear plant. In: Proceedings of the 11th WSEAS International Conference on Systems, Crete, Greece, pp. 183–188 (2007)
15. Tanaka, K., Wang, H.O.: Fuzzy Control Systems Design and Analysis: A Linear Matrix Inequality Approach. Wiley, New York (2001)
16. Yordanova, S.: Lyapunov stability and robustness of fuzzy process control system with parallel distributed compensation. Inform. Technol. Control **4**, 38–48 (2009)
17. Angelov, P.: Automomous Learning Systems. From Data Streams to Knowledge in Real-time. Wiley, UK (2013)
18. Yordanova, S., Georgieva, A.: Genetic algorithm based optimization of fuzzy controllers tuning in level control. J. Electrotechnica Electronica E+E **48**(9–10), 45–51 (2013)
19. Yordanova, S., Sivchev, Y.: Design and tuning of parallel distributed compensation-based fuzzy logic controller for temperature. J. Autom. Control **2**(3), 79–85 (2014)
20. Yordanova, S., Yankov, V.: Fuzzy supervisor for nonlinear self-tuning of controllers in real time operation. Proc. Tech. Univ. Sofia **64**(1), 131–140 (2014)
21. Yordanova, S.: Intelligent approaches for linear controllers tuning with application to temperature control. J. Intell. Fuzzy Syst. **27**(6), 2809–2820 (2014)
22. Yordanova, S., Petrova, R., Mastorakis, N., Mladenov, V.: Sugeno predictive neuro-fuzzy controller for control of nonlinear plant under uncertainties. J. WSEAS Trans. Syst. **5**(8), 1814–1821 (2006)
23. Yordanova, S., Merazchiev, D., Jain, L.: A two-variable fuzzy control design with application to an air-conditioning system. IEEE Trans. Fuzzy Syst. (2014). doi: 10.1109/TFUZZ.2014. 2312979 (Early Access)
24. Astrom, K.J., Wittenmark, B.J.: Adaptive Control. Addison-Wesley, CA (1989)
25. Yamazaki, T., Mamdani, E.H.: On the performance of a rule-based self-organising controller. In: Proceedings of IEEE Conference on Applications of Adaptive and Multivariable Control, Hull, England, pp. 50–55 (1982)
26. Palma, L., Moreira, J., Gil, P., Coito, F.: Hybrid approach for control loop performance assessment. In: Proceedings 5th International Conference on Intelligent Decision Technologies —Frontiers of Artificial Intelligence and Applications (FAIA), KES IDT, IOS Press, pp. 235–244 (2013)
27. Yamashita, Y., Matsumoto, S., Suzuki, M.: Start-up of a catalytic reactor by fuzzy controller. Chem. Eng. Japan **21**, 277–281 (1988)
28. Cho, Y.-W., Seo, K.-S., Lee, H.-J.: A direct adaptive fuzzy control of nonlinear systems with application to robot manipulator tracking control. Int. J. Control Autom. Syst. **5**(6), 630–642 (2007)

29. Ram, A.G., Lincoln, S.A.: A model reference-based fuzzy adaptive PI controller for non-linear level process system. Int. J. Res. Rev. Appl. Sci. 14(2), 477–486 (2013)
30. Passino, K., Yurkovich, S.: Fuzzy Control. Addison-Wesley Longman Inc, CA (1998)
31. Angelov, P., Skrjanc, I., Blazic, S.: Robust evolving cloud-based controller for a hydraulic plant. In: Proceedings of IEEE international conference on evolving and adaptive intelligent systems EAIS-2013, Singapore, 16–19 April 2013, pp. 1–8 (2013)
32. Costa, B., Skrjanc, I., Blazic, S., Angelov, P.: A practical implementation of self-evolving cloud-based control of a pilot plant. In: IEEE International Conference on Cybernetics CYBCONF-2013, Lausanne, Switzerland, 13–15 June, pp. 7–12 (2013)
33. Procyk, T.J., Mamdani, E.H.: A linguistic self-organising process controller. Automatica 15 (1), 15–30 (1979)
34. Treesatayapun, C.: Balancing control energy and tracking error for fuzzy rule emulated adaptive controller. Appl. Intell. 40(4), 639–648 (2014)
35. Kaminskas, V., Liutkevicius, R.: Adaptive fuzzy control of nonlinear plant with changing dynamics. Informatica 13(3), 287–298 (2002)
36. Rojas, I., Pomares, H., Gonzalez, J., Herrera, L., Guillen, A., Rojas, F., Valenzuela, O.: Adaptive fuzzy controller: application to the control of the temperature of a dynamic room in real time. Fuzzy Sets Syst. 157, 2241–2258 (2006)
37. Qing, L., Mahfouf, M.: A model-free self-organizing fuzzy logic control system using a dynamic performance index table. Trans. Inst. Measur. Control 32(1), 51–72 (2010)
38. Kazemian, H.K.: Comparative study of a learning fuzzy PID controller and a self-tuning controller. ISA Trans. 40, 245–253 (2001)
39. Sharma, K., Ayyub, M., Saroha, S., Faras, A.: Advanced controllers using fuzzy logic controller for performance improvement. Int. Electr. Eng. J. 5(6), 1452–1458 (2014)
40. Karasakal, O., Yesil, E., Guzelkaya, M., Eksin, I.: Implementation of a new self-tuning fuzzy PID controller on PLC. Turk. J. Electr. Eng. 13(2), 277–286 (2005)
41. Soyguder, S., Karakose, M., Alli, H.: Design and simulation of self-tuning PID-type fuzzy adaptive control for an expert HVAC system. Expert Syst. Appl. 36(3), 4566–4573 (2009)
42. Zulfatman, Rahmat M.: Application of self-tuning fuzzy PID controller on industrial hydraulic actuator using system identification approach. Int. J. Smart Sens. Intell. Syst. 2(2), 246–261 (2009)
43. Li, H.-X., Tso, S.: A fuzzy PLC with gain-scheduling control resolution for a thermal process —a case study. Control Eng. Pract. 7, 523–529 (1999)
44. Zhao, Z.-Y., Tomizuka, M., Isaka, S.: Fuzzy gain scheduling of PID controllers. IEEE Trans. Syst. Man Cybern. 23(5), 1392–1398 (1993)
45. Akkizidis, I.S., Roberts, G.N., Ridao, P., Batlle, J.: Designing a fuzzy-like PD controller for an underwater robot. Control Eng. Pract. 11, 471–480 (2003)
46. Mudi, R.K., Pal, N.R.: A self-tuning fuzzy PI controller. Fuzzy Sets Syst. 115(2), 327–338 (2000)
47. Yordanova, S., Yankov, V.: Design and implementation of fuzzy two-level control. Proc. Tech. Univ. Sofia 63(1), 79–88 (2013)
48. Visioli, A.: Fuzzy logic based set-point weight tuning of PID controllers. IEEE Trans. Syst. Man Cybern.—Part A 29(6), 587–592 (1999)
49. He, S.-Z., Tan, S., Xu, F.-L.: Fuzzy self-tuning of PID controllers. Fuzzy Sets Syst. 2, 37–46 (1993)
50. Rohmanuddin, M., The, H., Itoh, K., Ohtani, T., Ahmad, A., Nazaruddin, Y., Sitompul, J.: Fuzzy high-level control of a nonlinear MIMO plant in a PID system environment and its parameter tuning using genetic algorithm. In: Proceedings of the 4th World CSCC, Athens (2000)
51. Zheng, J., Zhao, S., Wei, S.: Application of self-tuning fuzzy PID controller for a SRM direct drive volume control hydraulic press. Control Eng. Pract. 17, 1398–1404 (2009)
52. Ghaffari, A., Shamekhi, A.H., Saki, A., Kamrani, E.: Adaptive fuzzy control for air-fuel ratio of automobile spark ignition engine. Proc. World Academy Sci. Eng. Technol. 48, 284–292 (2008)

53. Bandyopadhyay, R., Patranabis, D.: A new auto-tuning algorithm for PID controllers using dead-beat format. J. ISA Trans. **40**, 255–266 (2001)
54. Wu, C.-J., Ko, C.-N., Fu, Y.-Y., Tseng, C.-H.: A genetic-based design of auto-tuning fuzzy PID controllers. Int. J. Fuzzy Syst. **11**(1), 49–58 (2009)
55. Fuzzy Logic Toolbox. User's guide for use with MATLAB. The, MathWorks Inc (1998)
56. MATLAB—Real-time workshop. User's Guide. The MathWorks, Inc (1992)
57. MATLAB—Genetic algorithm and direct search toolbox. User's Guide. The MathWorks, Inc (2004)
58. Yordanova, S.: Fuzzy smith predictor for nonlinear plants with time delay based on parallel distributed compensation. Proc. Tech. Univ. Sofia **62**(2), 15–24 (2012)
59. Yordanova, S., Tashev, T.: Fuzzy internal model control of nonlinear plants with time delay based on parallel distributed compensation. WSEAS Trans. Circuits Syst. **11**(2), 56–64 (2012)
60. Lam, H., Leung, F.: Fuzzy controller with stability and performance rules for nonlinear systems. J. Fuzzy Sets Syst. **158**(2), 147–163 (2007)
61. Balas, G., Chiang, R., Packard, A., Safonov, M.: Robust Control Toolbox™. The MathWorks Inc, User's Guide (2009)

Chapter 2
Intelligent Carpooling System

A Case Study for Bacău Metropolitan Area

Elena Nechita, Gloria-Cerasela Crişan, Sergiu-Mădălin Obreja
and Constantin-Sebastian Damian

Abstract Mobility is one of the most basic features of our modern society. People and goods move around the entire Earth in a continuous and broad attempt to fulfill economic, safety and environmental goals. The Mobility Management or Transportation Demand Management is a collection of strategies for encouraging more efficient traffic patterns towards achieving specific planning objectives. For example, people can choose to switch from peak hours to non-peak time, or to cycle instead of using the car. Administrative regulations could introduce incentives or reimbursements when alternative commuting modes are used. Governmental policies could include fuel tax increases or pay-as-you-drive freeway taxes or car insurances. The goal of this chapter is to present several alternative travel modes, their social impact and their utility. As an example, we present two applications for shared-use mobility in the metropolitan area of Bacău, Romania. The applications integrate diverse computing languages with platforms, standards and technologies. The experimental results are encouraging, allowing us to consider that seamless integration of hybrid management systems for transportation could have tremendous economic and social impact at global scale.

Keywords Traffic management · Carpooling · Transportation on demand · GPS data sets

2.1 Introduction

Gathering people into common trips leads to individual and social efficiency. At a personal level, it reduces the total travelling cost and the driving stress as well. Although it is less comfortable than using the personal car and people usually need more time for performing the travel, the broad acceptance of the shared-use

E. Nechita (✉) · G.-C. Crişan · S.-M. Obreja · C.-S. Damian
Department of Mathematics, Informatics and Educational Sciences, "Vasile Alecsandri"
University of Bacău, Calea Mărăşeşti 157, 600115 Bacău, Romania
e-mail: enechita@ub.ro

© Springer International Publishing Switzerland 2016   43
K. Nakamatsu and R. Kountchev (eds.), *New Approaches in Intelligent Control*,
Intelligent Systems Reference Library 107, DOI 10.1007/978-3-319-32168-4_2

mobility shows its viability, with practical developments not entirely explored. From the social point of view, less fuel consumption, less CO_2 emission, less traffic congestion and more social interaction are the benefits of sharing cars or vans. By saving natural resources like oil, by producing less pollution and by cutting the travel expense, using a car for many persons is an advance to sustainable transportation systems. As an alternative to public transportation and taxi services, sharing a car combines the benefits of a shared cost with the flexibility of a taxi ride. The main idea of shared-use mobility is to share the transportation cost between multiple participants.

There are many ways people could use the same car or van. Car sharing means reserving and using a car for a short period of time (usually for several hours) and returning it before the reservation expires. It is a very convenient option for people needing rare and short rides: common vehicle costs show that this is the most efficient decision for less than 10.000 km/year. With an annual fee, and paying by hour when needing a ride, one could avoid buying a car and paying for its registration, insurance, maintenance, etc. Carpooling means using the same car for simultaneous transport of several people from a common starting point to a common end point. The most usual case is when neighbours work at the same facility and agree to travel using only one car. Modern systems use the internet and communication technologies in order to dynamically organise optimal trips. Such examples are the smartphone apps like *Lift*, *Sidecar* or *Uber*.

2.2 Shared-Use Mobility—Historical Evolution

With more than 70 years of development, the shared-use mobility had different materializations, addressing specific needs, and using the available resources at that time:

- During the World War II, the American companies, churches and social associations were encouraged to "…organise state and local transportation committees and car sharing clubs" [1] in order to preserve resources for the war (Fig. 2.1a). After the war, the model of the modern family and the increasing quality of life brought less concern in sharing cars under institutional frameworks, but instead self-organised "fampools" (family and friends) became natural.
- The energy crises from the 70s spiked up the gasoline price, so people again turned to the idea of grouping for common travels. Governments supported such initiatives through High-Occupancy Vehicle (HOV) lanes, park-and-ride facilities (Fig. 2.1b), or by sponsoring ridesharing projects. A unique project still in use is the Morgantown Personal Rapid Transit (PRT) system from the West Virginia University, USA [2]. Small cars are dispatched only at request, but in crowded hours the system switches to a classic public transportation one. This is a hybrid system, but is worth mentioned as it is running for almost 40 years, with a reliability rate of 98.5 %.

(a)   (b)

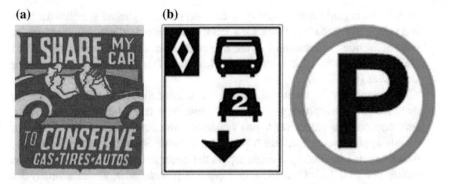

Fig. 2.1 a Car-sharing advertising (World War II). *Source* Ames Historical Society's World War II Veterans Project. b HOV lane sign and free carpool parking sign in Ontario, Canada. *Source* http://www.mto.gov.on.ca

- Until the end of the last century, several early systems were organised either by large-scale employers, or by transportation associations, in order to mitigate the traffic congestion, to lower the air pollution and to reduce the parking lot strain. The matches were made by-hand, after collecting data on the employees' addresses. Local transportation firms supplied vans for longer commutes when around ten employees came from the same residential area [3] (Fig. 2.2a). For one-time carpooling, telephone-based ride matching systems were set-up. The transition to the next systems is made by enhancing the systems through Personal Digital Assistants (PDAs) and Geographic Information Systems (GIS) capabilities.
- After 2000, the new communication technologies and the internet broad facilities had great impact on the reliability and on the responsiveness of the ridesharing systems. The clients manage to post online their commands, and connected services are now offered to the interested public. For example, the San Francisco 511 platform offers complex traffic information in the SF Bay

(a)   (b)

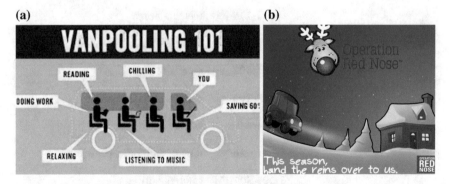

Fig. 2.2 a Vanpool advertising in Vermont, USA. *Source* Vermont Agency of Transportation. b Red Nose 2014 advertising in Canada. *Source* http://www.mto.gov.on.ca

area, including rideshare options [4]. *Ecolutis* offers aggregate services for firms and public in France [5]. There are special occasion ride offers, organised by the local authorities. An example is the *Red Nose* operation in Canada: the drivers are suggested to call for a volunteer who can safely drive them home during December (Fig. 2.2b).

- The current developments of mobile communications provide real-time or dynamic ridesharing today. The drivers post their trip while they are driving, and the potential riders warn just before the intentioned departure time. The mobile application notifies each part's smartphone about the available pairing, which can be accepted by a single tap on the smartphone screen. *Villefluide* is an online platform for ridesharing in France [6], *Ridefinder* works globally for Europe [7]. Successful smartphone apps are *Carma*, *Lift*, and *Sidecar*. Complimentary rewards are offered to people choosing to use these services. For example, *NuRide* computes and reports all its green activities in a specific area [8] (Fig. 2.3).

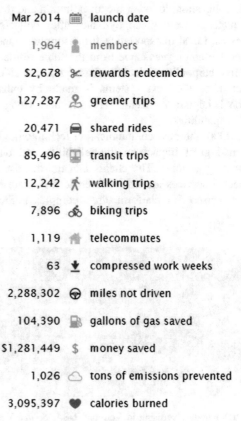

Activity in Rhode Island

Mar 2014	📅	launch date
1,964	👤	members
$2,678	✂	rewards redeemed
127,287	👣	greener trips
20,471	🚗	shared rides
85,496	🚋	transit trips
12,242	🚶	walking trips
7,896	🚲	biking trips
1,119	🏠	telecommutes
63	⬇	compressed work weeks
2,288,302	🌐	miles not driven
104,390	⛽	gallons of gas saved
$1,281,449	$	money saved
1,026	☁	tons of emissions prevented
3,095,397	♥	calories burned

Fig. 2.3 *NuRide* Rhode Island report on December 1st, 2014. *Source* www.ripta.com

Dynamic carpooling, also known as casual carpooling because it arranges shared rides on very short notice, arises supplementary challenges regarding the matching of drivers and travelers in real time. Adding this request to the flexibility required in routes leads to complicated algorithms, possibly leading to combinatorial explosion [9]. A review on the available automatic and heuristic data processing routines to support efficient matching in carpool schemes is presented in [10]. An application for dynamic carpooling also needs safety authentication of the rider and driver before making the match. Feedback or reputation systems may be useful in providing information about who to trust, as inter-personal constraints (such as punctuality, smoking versus non-smoking, male versus female preferences) could also interfere [11]. A state of the art on dynamic carpooling and an identification of the issues against the adoption of carpooling systems together with some suggestions to solve them are presented in [12].

For a comprehensive ridesharing classification scheme and for an evolution of the concept the reader can access the paper of Chan and Shaheen [13].

The following examples furthermore give some details on the functional vehicle sharing solutions, illustrating the potential of these initiatives, with implications in traffic flow improvement, public safety and optimisation of movement of people and goods.

- *Car2Go* [14] is the Smart division of Daimler AG, offering services in European and Nord-American cities. The company provides *Smart Fortwo* highly energy-efficient vehicles, available through booking online with applications for mobile devices. Moreover, the system is based on a "one-way" model, which makes it very flexible. The clients are charged per minute, hour or day. The rates include renting, fuel costs, insurance, maintenance and parking charges. Sometimes, a small annual fee may be applied. In May 2014, the company owned over 10,000 "car2gos" in eight countries, distributed in 26 cities and servicing more than 700,000 clients. According to [15] the service is providing electric cars exclusively in USA since November 2011.
- *Zipcar* [16] is the largest car sharing service, available in USA, Canada, Great Britain, Spain and Austria. The fleet is diverse, the cars may be driven by hour or day but they have to be returned to the same reserved parking spot. The driver pays a registration fee, an annual fee and reservation fees for the regular usage. The fuel costs, parking, and insurance are included, and facilities are provided for business and for universities.
- *CityCarClub* [17] is a UK car sharing network, available in 15 cities. The charge includes insurance, tax and fuel. The fleet is composed of low emission cars, experimenting alternative technologies such as biodiesel, hybrid vehicles and Stop-Start technology [18]. The network members use smart cards to access the vehicles.

2.3 Conceptual Models for Shared Mobility Systems

The current transportation domain encompasses many stakeholders with hetero-geneous needs and goals. The Intelligent Transportation System (ITS) must operate with a large number of concepts, must balance the requests with the available resources, and are supposed to manage a wide variety of models and implemen-tations. We expect that the ITSs will bring major social and economic benefits, due to the greater efficiency of the transport system and increase security. One direction in ITS research is to design specific ontologies, representing dictionaries of relevant concepts from traffic domains organised in a hierarchical structure of classes and taxonomy.

Examples of top-level ontologies that can derive transportation concepts are: Suggested Upper Merged Ontology (SUMO) [19], OpenCyc [20]. ISO 14825:2011 specifies the data model and physical encoding formats for geographic databases for ITS applications and services [21]. Semantic web for Earth and Environmental Terminology (SWEET) is a collection of dedicated top-level ontologies, maintained by NASA Jet Propulsion Laboratory [22].

By simultaneously treating Intelligent Vehicles and Intelligent Infrastructure, the low-level ontologies allow comprehensive and targeted approaches to Transportation problems. The Open Geospatial Consortium [23] offers geospatial and location standards, providing community-specific OGC Geography Markup Language (GML) application schemas (AIXM for aviation, GeoSciML for geo-sciences, etc.). Transportation simulations using ontologies are presented in [24, 25], and specific carpooling simulations based on ontologies are published in [26].

The complexity of the current transportation situations and the need of rapid actions when sensitive events occur (i.e. bottlenecks, accident, and sudden road closures) request decision support systems to assist traffic engineers. The knowledge-based systems are designed to enhance the dynamicity and respon-siveness, to mitigate the effects of the traffic incidents, and to correctly predict them [27]. Data analysis or data mining can be used for real-time suggestions for human operators. One such example is the opening of new lanes when the sensors register heavy traffic. Another approach is to install mobile applications which can rec-ommend to the driver new paths for completing his/her daily trip from home to work.

Conceptual models and simulations are used to get insights on the strengths and weaknesses of the carpooling systems for a given region. A simulation-based methodology which emphasizes the construction process of a logic flow diagram that translates the proposed methodology is presented in [28]. Multiple, heteroge-neous models have been integrated using ontology techniques [26], especially for predicting the demand. User-centred research has also been developed. This approach is helpful for assessing the potential of the carpooling concept in general, as well as to improve the interaction of the users with the existing web-based platforms [29]. Statistical testing shows diverging requirements between different groups; age, gender and individual economic situation (employment, for example) are factors that influence carpooling acceptance.

Culture-specificity is also very important. A case study for Switzerland is presented in [30] and an analysis on Vermont carpool data is available in [31]. The paper includes some considerations on the perceptions of carpooling, which have been found to play a larger role than cost or convenience. For Portugal, the work of [32] reveals the results of a *stated preference* experiment meant to assess the enhancement possibilities for carpooling and carpool clubs. For China, where the private car ownership had exceeded 88 million in 2012, a design of a commute carpooling system is given in [33] and an investigation of human mobility patterns, for predicting owners' driving trajectories and real-time position for carpooling, is presented in [34]. As taxicab systems play a prominent role in transportation models, recommendation systems for carpoolable taxicabs come from China, too. In [35, 36] the authors design such systems, based on a data driven approach.

2.4 Modelling a Carpooling System for Bacău, Romania

This section presents two implementations for carpooling services in the Bacau area, Romania. The models are based on multi-agent interactions, with specific characteristics and use multiple communication and representation technologies.

2.4.1 Country Context

Romania is located at the crossroads of many routes connecting Eastern to Western Europe and Northern to Southern Europe. Of the total National road network, 5,868 km (37.3 %) are classified as European roads [37]. Moreover, Romania's location on the transit axes connecting Europe to Asia generated numerous studies regarding the infrastructure and transport capabilities. The Romanian Government identified [38] several key transport-related issues:

- Domestic transport, although diversified, has insufficient capacity for transporting freight and passengers;
- The transport infrastructure is insufficiently developed and requires significant investment in order to meet European standards;
- Access to the West-European corridors, as well as to the Eastern and Southern Europe ones, is limited;
- The access roads from national roads to town centres and cities are inadequate and most towns located on National and European roads lack bypasses.

At the end of 2013, more than 5 million privately-owned vehicles have been recorded in Romania [39]. Nevertheless, the number of new cars has registered a significant decrease during the last 5 years. This is due mainly to the poor purchasing power of the population and to the fluctuations of the pollution tax. Meanwhile, the market of the second-hand vehicles, imported from the Western Europe, has increased. Under these circumstances, the pollution could also increase.

Romanian law no 3/2001 ratifies the Kyoto Protocol of the United Nations Framework Convention on Climate Change. According to it, Romania is obliged to reduce its emissions of greenhouse gases with 8 %, considering the year 1989 as baseline. If we refer to the emissions from road transport, the average carbon dioxide emissions from new passenger cars was, in 2013, 132.1 gCO_2/km (decreasing with 5.22 % compared to 2012) in Romania, while the EU Member States' average was 126.7 gCO_2/km (decreasing with 4.1 % compared to 2012) [40]. The Romanian government encouraged the rolling stock renewal through national policies, in order to eliminate the vehicles with significant pollutant capacity. However, this initiative should be complemented with "soft" measures, such as intelligent traffic management and transport systems.

The following subsection presents an overview of the data which are relevant for Bacău, influencing the transportation conditions and the transit system as they stand at the moment of our study.

2.4.2 City Context

In order to implement a sharing mobility system that is functional and beneficial for a certain area, specific factors also have to be taken into account. This is the approach that we have considered for Bacău metropolitan area. Bacău, the 15th largest city in Romania, is the residence city of Bacău County, part of the historical region Moldavia and of the North-East Development Region (according to the EU regional policy [41]). The city is situated at about 300 km from Bucharest, the capital of Romania. Two European routes cross Bacău: E85 links the South area with the Northern cities Iaşi and Suceava, while E574 provides access to the South-Western part, with Braşov, Piteşti and Craiova (the last two are important centres of the European automotive industry). There is no beltway for Bacău; therefore the transit through the city often faces congestion, consequently leading to increased travel time, vehicle operating costs, accidents and environmental damage.

Current data (June, 2014, [42]) show a population slightly over 170,000 inhabitants, distributed within the city limits of 43.1 km^2, while the whole county has more than 706,000 inhabitants. Bacău metropolitan area [43] integrates Bacău and the neighbouring communities, with a population of around 250,000.

The transportation needs are high and have increased during the last 5 years, along with the development of SMEs around Bacău. Several divisions of important international companies are hosted in or in the vicinity of Bacău as well. The rolling stock has increased. Consequently, the traffic has reached in Bacău a level where the key performance indicators (especially the indices for traffic efficiency and for pollution reduction [44]) are alarming.

For Bacău County, the car ownership has been reported at 111,500 at the end of 2013 [42]. The public transport is based on three major median busways and seven secondary ones in the city. In addition there are regular buses for the neighbouring communities. This system is far from providing comfortable connections between the rural area and Bacău, where most of the employers are concentrated. Thus there is an important flow of vehicles to and from Bacău, especially during the working days. An important share of them uses E85 regularly, therefore generating agglomeration with all its consequences.

The development plan of Bacău [45] follows an integrated approach to the transport and land use. Key directions include improvement of infrastructure, management and control of land use citywide and traffic decongestion of the city centre, but important investments (which are not scheduled in the near future) are needed. In the central area of the city, roadside parking is limited in location and duration and is well enforced. The number of off-street parkings is very small.

As far as we know, no shared-use mobility study has been performed for the city of Bacău. The two implementations which are presented in Chap. 5 aim at encouraging the authorities to consider new policies regarding the local transportation means and to raise awareness among the local community towards the green solutions.

2.4.3 Information Requirements for an Efficient Model

In order to incentive the traveller's choice, a carpooling system has to provide several features [46], among which we mention:

- *Accessibility*. The stops need to be located near residential neighbourhoods and/or in the business area of the city.
- *Affordability*. The rates should be reasonable; it is important that the system is available for short trips.
- *Convenience*. If the vehicles are easy to check in and out at any time, this would highly encourage their use.
- *Reliability*. It is recommendable that the vehicles are available and have minimal failures.

Besides the above listed aspects, the targeted users in Bacău would certainly require to access the service on smart devices, hence the necessity of a modern, adequate implementation.

Considering these characteristics, the model we propose focuses mainly on fuel economy and reduction of transport-related greenhouse gas emissions through an improved degree of occupancy of the vehicles. The mobility of users in the metropolitan area and the connectivity to the public transportation systems have been simulated through the flexibility of the implementations. Therefore, for the target area we have studied:

- the main transport routes and the most important congestion points
- the most important, necessary stops
- the zones with maximum passengers flow
- traffic monitoring (traffic volume, daily travel data)
- sample data on travel time and speed
- socioeconomic information (workplaces, schools, public institutions and number of their attendees)
- public transport and parking surveys
- consumers' satisfaction and perception data on the public transport systems
- environmental impact.

The corresponding data have been provided by the Committee for urbanism, territorial planning and environmental protection of Bacău County Council [47].

For an efficient carpooling system, the characteristics of the vehicle fleet are also essential. In order to score improvements resulting from its implementation, fuel consumption and emissions of CO_2 must be reduced. This can be done gradually, introducing alternative vehicles (such as hybrid electric or fully electric) and/or fuels (bio- and coal-based, compressed natural gas or liquefied petroleum gas). As these facilities are scarcely present in Bacău County, we only simulate their influence by means of two parameters: the average fuel consumption and the average CO_2 emission per car, for the system fleet. As a conclusion, our model strengthens the role of shifting from privately-owned vehicles to carpooling, and assesses the reliability of the solution considering the measures suggested in [48, 49].

According to these goals, the followings represent input data for the model:

- number of main stops
- number of secondary stops
- number of routes
- the set of routes
- number of clients in each main stop, on time intervals
- average fuel consumption
- average greenhouse gas emissions for the vehicles involved in the simulation.

For a given number of clients, the carpooling simulation algorithm allows us to load a car with 1, 2, 3, 4 or with a random number (between 1 and 4) of clients. For the chosen routes and stops, the calculations provide as output data:

- number of needed cars (for each of the five variants of loading)
- total fuel consumption
- total greenhouse gas emissions.

2.5   Applications and Experiments

The applications presented in this section integrate diverse computing languages with platforms, standards and technologies. They have been used to perform a preliminary experiment regarding the appropriateness of implementing a carpooling service in the metropolitan area of Bacău.

2.5.1   SplitCar

The *SplitCar* application implements the model with Microsoft solutions: the *Visual Studio 2013* suite, the *.NET* Framework 4.5 package, *C#* and *XML*, the components *Windows Presentation Foundation* (WPF) for controlling the map delivered by *Bing*, and the *REST* service for locations and routes, offered by *Bing Maps*.

Given that the majority of the handled data represent locations, distances and routes, we have chosen as support the *Bing Maps* tool. In order to use maps, the *Map* control will be placed in a WPF window. Besides the *Map* control, an active Microsoft account and a *Bing* key are needed for the use of the map-related services: locations finding, distance calculator, route finding. After setting the Microsoft account and getting the key access to the *Bing* services, this key is assigned to the *Map* control *CredentialsProvider* attribute:

```
<WPF:Map CredentialsProvider=''key''/>
```

The map is initialized using the attributes *Mode* (*RoadMode*, *AerialMode*, *AerialModeWithLabels*), *ZoomLevel*, *AnimationLevel*, *Heading*, which can be modified at any time through the controller, using *C#*. For a complete control over the map, several functionalities have been implemented in the main window of the application, as Fig. 2.4 presents.

Several child controls may be added on the *Bing* map with the command

```
<map_name>.Children.Add(<control>);
```

These controls could represent locations, routes (*MapPolyline* objects), surfaces (*MapLayer* objects) or other objects. Only the first two are compulsory for our model implementation.

The locations are defined through *Pushpin* type controls, whose attributes allow the customization desired by the user. In Fig. 2.4, the main stops are coloured in green and the secondary ones appear in blue. Both types are created with a left-click of the mouse in the *Map* control. Right click on a location removes it from the map.

In the main window of the application, several items are designed to offer an improved control over the map. The user can set the maximum numbers of main stops (between 2 and 4) and secondary or via-stops (between 0 and 9).

Routes can be defined on the map between any two points, with a *MapPolyline* control, according to the following conventions: any route starts in a main stop and

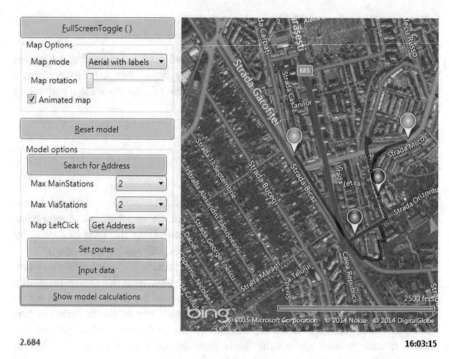

Fig. 2.4 The main window of *SplitCar*. Routes and stops

ends with a main stop (different from the origin); any route can pass through zero, one or more secondary stops; routes are displayed on the map with different colours, automatically computed according to the number of routes; the routes take into account the one-way roads; if a location (part of one or more routes) is deleted, the user implicitly agrees to delete all the routes that pass through that location/stop. Any number of routes can be created, and the definition of a route can be cancelled at any time.

The routes are created when activating the *Set routes* button (which opens the dialog window only if at least two main stop are already defined). For the accuracy of the implementation, we have used the services *REST* from *Bing Maps*. These services receive information on locations and routes, interpret them and display them on the map. The distance between two points on the map is computed in kilometres. In Fig. 2.4, the length of the routes appears in the low, right corner.

The implementation of a class that finds a certain location on the map is simple. The button *Search for address* opens a window where the user can introduce an address. After providing it, the class *LocationFromAddress* returns the corresponding location (latitude and longitude coordinates) on the map:

```
class LocationFromAddress
{
  private string BaseUrl =
  "http://dev.virtualearth.net/REST/v1/Locations?query=";
  private string BingAdditional =
  "&output=xml&include=queryParse";
  private string BingMapsKey = "&key=my_private_key";
  private string BingQuery = string.Empty;
  private Location point;
  public LocationFromAddress();
  public void SetQuery(string);
  public Location Run();
  private Uri AssembleLink();
  private string NormalizeQuery(string);
}
```

Conversely, the class *AddressFromLocation* returns an address if the location is given in latitude and longitude:

```
class AddressFromLocation
{
  private string BaseUrl
     "http://dev.virtualearth.net/REST/v1/Locations/";
  private string BingAdditional =
     "?output=xml&includeEntityTypes=
     Address&includeNeighborhood=1";
  private string BingMapsKey = "&key=my_private_key";
  private Location point;
  public AddressFromLocation();
  public void SetPoint(Location);
  public string Run();
  private Uri AssembleLink();
}
```

The next step is to compute the optimal routes and their lengths. The route between two main stops (with or without intermediary stops) is generated with the class *RouteBetweenTwoPoints*, whose components are as follows:

```
class RouteBetweenTwoPoints
{
  public enum MapOptions { Start = 0, End,
   ViaPoint, Optimize,  TravelMode, Points, Tolerance };
  public enum OptimizeRoute { Time = 0, Distance,
   TimeAvoidClosure, TimeWithTraffic } ;
  public enum TravelMode { Driving = 0, Walking, Transit };
  private LocationCollection viaPoints;
  private Location Start, End;
  private List<string> Options = new List<string>();
  private string BaseUrl =
     "http://dev.virtualearth.net/REST/v1/Routes?o=xml";
  private string BingMapsKey = "&key=my_private_key";
  public RouteBetweenTwoPoints();
  public string SetRequestParameter(MapOptions, object);
  public MapPolyline Run();
  public string GetLink();
  private Uri AssembleLink();
}
```

We notice that the function *SetRequestParameter* receives as first parameter an object of *MapOptions* type and holds the information on what needs to be established, while the second parameter is an object which depends on the type of the first parameter. For example, if *RouteBetweenTwoPoints* is used to define a route, the function *SetRequestParameter* sets the following:

- the start and stop points (main stops)
- the intermediary points of the route (the secondary stops, if any)
- the optimisation type (there are four possibilities, to optimise upon: time, distance, traffic jams avoidance, time and traffic data)
- the travel type (with three options: driving, walking, transit)
- the type of the desired output, indicating if we need the points returned by the *REST* service to create the route)
- the tolerance allowed for the points to determine the route.

If all the data have been appropriately setup with *SetRequestParameter*, the route is received in a *MapPolyline* object, with the function *Run*. Afterwards, the computations are made with the class *DistanceBetweenTwoPoints*.

For each route and four time intervals (6:00–10:00, 10:00–14:00, 14:00–18:00, 18:00–22:00), a number of persons requiring the carpooling services can be introduced when activating the button *Input data*, as in Fig. 2.5. Moreover, the average fuel consumption and the average greenhouse gas emissions are required, per private, regular car and per carpooling vehicle.

When all the input data are available, the module *RouteData* performs the computations of the output information (total fuel consumption and greenhouse gas emissions) for a certain route. The data are presented in several formats: text, tables and histograms, according to the users' choice, when activating the button *Show model calculations*. An example is given in Fig. 2.6, where the value "5" on the

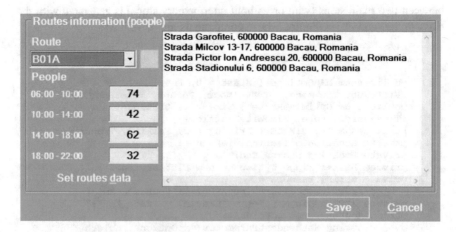

Fig. 2.5 Data acquisition window for *SplitCar*

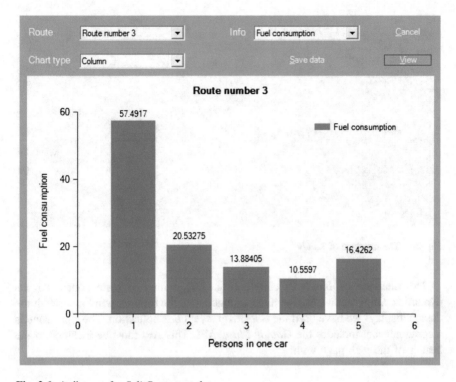

Fig. 2.6 A diagram for *SplitCar* output data

horizontal axis encodes what was denoted as "random setting" in Table 2.2, namely a *random* number (between 1 and 4) of occupants in a car. The total fuel consumption is measured in litres.

2.5.2 Buddy

The *Buddy* application is a web-based solution. We have chosen to use *ASP .NET* for the server-side, *JavaScript*, *JQuery*, *Ajax* as programming languages, *HTML*, *CSS* and other applications for improving the design of the web interface (*JQueryUI*, *Bootstrap*, *jChartFX*), all of them embedded in *Visual Studio*. We also used the API provided by Google for the well-known mapping service *Google Maps*. The API also allows the use of maps stored on sites of third parties, and includes a locator for organisations (as well as other objectives) placed on the territories of numerous countries all over the world.

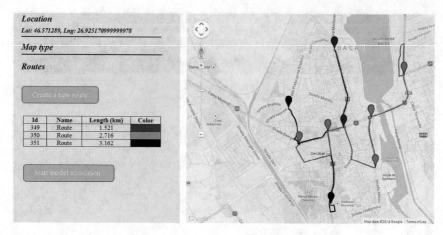

Fig. 2.7 The interface of *Buddy*

The interface of *Buddy* is simple (Fig. 2.7), with two main zones. For its dynamic configuration (resizing in accordance with the browser window or with the device display) we have used the *Bootstrap* style class collection. The right zone is a container that includes the *Google Maps* API. This API must be included in the header of the web page with:

```
<script
  type="text/javascript"
  src="https://maps.googleapis.com/maps/
  api/js?sensor=false">
</script>
```

and an object map is defined with:

```
var mapOptions =
{
  mapTypeControl: false,
  mapTypeId: google.maps.MapTypeId.ROADMAP,
  center: new
  google.maps.LatLng(46.571289, 26.925170999999978),
  zoom: 14 //zoom 14 times on Bacau
};
map = new google.maps.Map(
  document.getElementById("#map-canvas"),
  mapOptions);
geocoder = new google.maps.Geocoder();
```

The left zone of the interface uses the *accordion* widget from the *JQuery UI* package and offers access to three tabs: *Location*, *Map type*, and *Routes*.

The packages *jQuery* and *jQuery UI* have to be included in the header of the web page with:

```
<head runat="server">
  <script src="jquery-1.9.0.min.js" type="text/javascript">
  </script>
  <script src="jquery-ui-1.10.4.js" type="text/javascript">
  </script>
  <link href="jquery.ui.accordion.css"
  rel="stylesheet" type="text/css" />
</head>
```

As the widgets in *jQuery UI* use the *jQuery* classes, it is compulsory that the order of the lines in the above sequence code is maintained.

The user can search for a new location, either introducing its name or its coordinates. When searching for an address upon its name, the function *findAdress()* gets the input text and (if this is consistent) uses the *geocode* method, as follows:

```
geocoder.geocode(
   { 'address': address }, function (results, status)
   {
      if (status == google.maps.GeocoderStatus.OK)
      {
         map.setCenter(results[0].geometry.location);
         $("#location-coordinates").html
         "Lat: " + map.getCenter().lat() + ", Lng: " +
         map.getCenter().lng());
      }
      else
         alert('Geocode was not successful
         for the following reason: ' + status);
   });
```

If the corresponding search is successful and the function returns a list of locations, the map is centred in the first result, which is considered to be the most significant. When searching for an address upon its coordinates, the map is centred in the point defined by the specified coordinates (if they are correct) with

```
map.setCenter(new google.maps.LatLng(lat, lng));
```

Fig. 2.8 A hybrid view over Bacău, with *Google Maps*

With *Map type*, the user changes the view, choosing among: *Roadmap* (as in Fig. 2.7), *Satellite*, *Hybrid* (as in Fig. 2.8) or *Terrain*.

The tab *Routes* is the most complex. It provides functionalities that allow to create and to modify routes on the *Google* map. When the web application is launched, an *Ajax* function calls an ASP.NET function from the server. It returns (if any) the routes previously saved in the database. The results provided by this function appear to the user as in Fig. 2.7, and are stored in a JavaScript object of JSON type. For the routes displayed in Fig. 2.7, the JSON object is as follows:

```
46.58299327827027-26.912510097026825/46.56989553861151-
26.91491335630417;
46.57880470562833-26.930191218852997/46.57160668424229-
26.92186564207077/46.56328646931478
26.922981441020966/46.56912845580439-
26.910965144634247/46.56706317894082-26.903841197490692;
46.57302276393836-
26.90118044614792/46.556971695640904-26.911823451519012;
```

The function *parseExistingRoutes* receives the above string and decomposes it in values representing the coordinates of the routes:

```
function parseExistingRoutes(string)
  {
    string = string.substring(0, string.length - 1);
    var routes_strings = string.split(';');
    for (var i = 0; i < routes_strings.length; i++)
    {
      var route_string = routes_strings[i].split('/');
      var lcs = new Array();
      for (var j = 0; j < route_string.length; j++)
      {
        var locations_strings =
        route_string[j].split('-');
        lcs[j] = new Object();
        lcs[j].lat = locations_strings[0];
        lcs[j].lng = locations_strings[1];
      }
      r++;
      showRoute(lcs, function (distance) {});
    }
  }
```

For each location, an object such as

```
{ lat: 46.57302276393836, lng: 26.90118044614792 }
```

is created. For each route, the application generates a vector of objects. This is further sent to the function *showRoute()* which recomposes the route. The first element of the vector always becomes the start point and the last one is seen as the destination. If the route is made of more than three locations, the intermediary ones become via-stops:

```
var start = new google.maps.LatLng(ls[0].lat, ls[0].lng);
var end = new google.maps.LatLng(ls[len - 1].lat,
ls[len - 1].lng);
var waypoints = [];
for (var i = 1; i < len-1; i++)
waypoints.push({location: new google.maps.LatLng
ls[i].lat, ls[i].lng), stopover: true});
```

where the function *google.maps.LatLng(x,y)* creates *Google* objects of type *point*.

To display a route, the algorithm finds the shortest path that connects the starting point with the destination, preserving the position and the order of the intermediary locations. This is done with a *request* object, which is further used as a parameter for the functions in the *Google* API:

```
var request =
  {
    origin: start,
    destination: end,
    waypoints: waypoints,
    optimizeWaypoints: true,
    travelMode: google.maps.TravelMode.DRIVING
  };
```

Each point is identified with a *Marker* type object on the Google map. Ten images (for the marker objects) in ten different colours are stored in the resources file of the system, to enhance the routes readability on the map. The marker objects are initialised with

```
var icon = "Resources/marker" + r%10 + ".png";
for (var i = 0; i < len; i++)
  {
    var marker = new google.maps.Marker({
    position: new google.maps.LatLng(ls[i].lat, ls[i].lng),
    map: map,
    icon: new google.maps.MarkerImage(icon)});
    newmarkers.push(marker);
  }
```

where r is the index of the route.

The computations and the route display are done through a call of the *Google* API:

```
var directionsDisplay =
new google.maps.DirectionsRenderer({
polylineOptions:
  {strokeColor: colors[(r - 1)%10] },
  suppressMarkers: true, preserveViewport: true });
directionsDisplay.setMap(map);
```

which specifies the desired colour for the route, cancels the standard markers display and indicates no zoom over the created route.

If the call is successful, the route is displayed (with the customized markers) and its length (in kilometres) is computed:

```
if (status == google.maps.DirectionsStatus.OK)
{
  directionsDisplay.setDirections(response);
  for (var i = 0; i < markers.length; i++)
  if (markers[i]) markers[i].setMap(null);
  var thisroute = response.routes[0];
  var distance = 0;
  for (var i = 0; i < thisroute.legs.length; i++)
  distance += thisroute.legs[i].distance.value;
  callback(distance / 1000);
  callback(response.routes[0].legs[0].distance.value/1000);
}
```

The application offers to the user the possibility of defining his/her own routes, when activating *Create new route*. This option calls the function *createRoute()*, whose main functionalities are to place markers in the desired locations and to reuse the already existing markers (both with right-click of the mouse). Instances of *Listener* type objects of the API are instantiated to activate these facilities:

```
listener1 = google.maps.event.addListener
map, 'rightclick', function (event)
{
var marker=
new google.maps.Marker
({position:event.latLng,map:map,title:'Marker-'+(l+1)});
markers[l] = marker;
var loc = new Object();
loc.lat = event.latLng.lat().toString();
loc.lng = event.latLng.lng().toString();
locations.push(loc);
...

for (var i = 0; i < newmarkers.length; i++)
newmarkers_listeners[i]=
google.maps.event.addListenerOnce
newmarkers[i],'rightclick',
function ()
{
var marker =
new google.maps.Marker
({ position: this.getPosition(), map: map, title:
'Marker-' + (l + 1) });
markers[l] = marker;
var loc = new Object();
loc.lat = this.getPosition().lat().toString();
loc.lng = this.getPosition().lng().toString();
locations.push(loc);
...
```

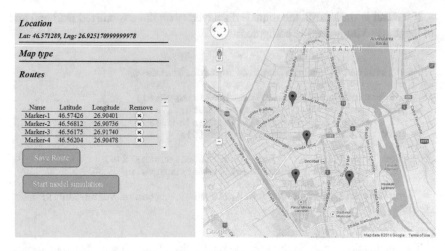

Fig. 2.9 Defining a route to visit the locations defined by user markers

The chosen locations are stored in an array and their coordinates are displayed in a temporary table, as in Fig. 2.9. The user can remove some locations, if these are misplaced. When the configuration is final, the route can be saved.

The system performs an *ajax* call of the function *saveRoute( )* on the server:

```
$.ajax({
  type: "POST",
  url: "Default.aspx/saveRoute",
  data: "{jsonstring:'" + JSON.stringify(locations) + "'}",
  contentType: "application/json; charset=utf-8",
  dataType: "json",
  success: function (msg)
    {
    ...
```

The array *locations* is sequenced and sent as a JSON object. After saving the route, the right-click events detection over the markers and the over the map is stopped. Moreover, the route length (returned by the function *showRoute( )* above) is also saved (Fig. 2.10) through an *ajax* call of the function *saveRouteDistance( )* on the server:

```
$.ajax({
  type: "POST",
  url: "Default.aspx/saveRouteDistance",
  data: "{jsonstring:'" + route_id + "-" + distance + "'}",
  contentType: "application/json; charset=utf-8",
  dataType: "json",
  success: function (msg)
    {
    ...
```

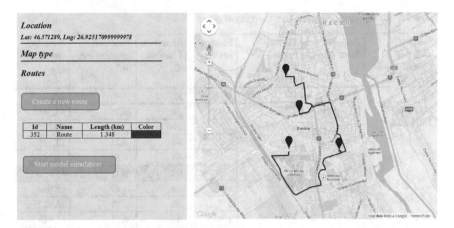

Fig. 2.10 The definition of the route is completed

The simulation starts when activating the button *Start model simulation*, which opens the windows where the user presents:

- the parameters (the same as in the first application: number of passengers, the average fuel consumption and the average greenhouse gas emissions per privately-owned car and per carpooling vehicle);
- data for each route: number of travellers on that route, per day and time intervals.

Both applications presented here are available at [50], where we have collected a series of applications designed to illustrate the role of Informatics in developing intelligent systems to support the sustainable development of our society. The results presented in the following Section were obtained on a desktop PC with a 2 GHz quad-core processor and 2 GB RAM capacity.

2.6 Results and Discussion

The application *Buddy* generates, for each route, a table (see Table 2.1) where the user can compare the figures related to each type of mobility setting, defined by the column *Persons per car* (4). The lines shadowed in grey correspond to the mean value of the vehicles loading, determined as the average of the sequence of random numbers (between 1 and 4) whose sum cover the travel request. The values in column (7), computed as the product between the route lengths, the number of cars and the quantity of CO_2 emissions per car and per kilometre, have been rounded to the nearest integer.

Table 2.1 Comparison between five mobility choices

Route ID/Route name/Route length (km)	Time intervals	Travel request (number of persons)	Persons per car	Number of needed cars	Total fuel consumption (l)	Total CO_2 emissions (g)
(1)	(2)	(3)	(4)	(5)	(6)	(7)
363/Route 363/1.614	6:00–10:00	250	1	250	28.25	64,561
			2	125	13.11	20,174
			3	84	8.11	13,556
			4	63	6.61	10,167
			2.58	97	10.18	15,654
	10:00–14:00	190	1	190	21.47	49,066
			2	95	9.97	15,332
			3	64	6.71	10,328
			4	48	5.04	7748
			2.79	68	7.13	10,976
	14:00–18:00	100	1	100	11.30	25,823
			2	50	5.25	8071
			3	34	3.57	5488
			4	25	2.62	4036
			2.56	39	4.09	6293
	18:00–22:00	80	1	80	9.04	20,670
			2	40	4.20	6457
			3	27	2.83	4358
			4	20	2.10	3229
			2.76	29	3.04	4681

The data in columns (6) and (7) can also be seen in charts, built with *jChartFX* support. The package *jChartFX* (which requires the presence of *jQuery* tools) must be included in the header of the web page:

```
<script src=
  "jchartfx.advanced.js" type="text/javascript"></script>
<script src=
  "jchartfx.system.js" type="text/javascript"></script>
<script src=
  "jchartfx.coreVector.js" type="text/javascript"></script>
```

The simulation has been run for a flow of 10,000 persons (which represents an estimation for the number of the commuters in a working day for Bacău city) and for an average route of 12 km length per car and per day. We used various routes

Table 2.2 Sample output data

	Number of needed cars	Total fuel consumption (l)	Total CO_2 emissions (kg)	Fuel economy and CO_2 emissions decrease (%)
(a) Personal ownership (2 persons per car)	5,000	4,200	7,800	–
(b) Carpooling (1 to 4 persons per car; random setting) *Set 1* of parameters	3,900	3,276	6,084	22.00
(c) Carpooling (1 to 4 persons per car; random setting) *Set 2* of parameters	3,900	2,340	4,680	44.28

and two sets of parameters. *Set 1* models the case when the cars used for carpooling have the same characteristics as the regular ones. As most of the passenger cars used in Romania complies with the *Euro 4* emission standard, we have set the fuel consumption per 100 km at 7 L (of gasoline) and the CO_2 emissions at 130 g/km [39]. *Set 2* simulates a modern carpooling fleet, using vehicles with superior performances, for which the two parameters were set to 5 L/100 km and 100 g/km, respectively.

Table 2.2 presents a sample of output data, which display substantial fuel savings and important CO_2 emissions decrease. Although the applications allow the user to simulate a vehicle loading with any number (from 1 to 4) of passengers, the table includes only the lines for two persons per car and for the random loading, which is the most plausible situation to happen in real life. For the computation in the last column, the reference line is (a), for both (b) and (c).

If we consider an average price of 1.1 euros per gasoline litre in 2014 [40], we get a daily saving of 1,016 euros for case (b) and 2,046 euros for case (c). According to the SWOT analysis performed by Bacău County Council [45], the investments of the inhabitants in health-care and education decreased during the last years. The savings could be directed towards these domains, which could improve the quality of life, both at personal and community level. Moreover, the impact of the shared mobility services on the environment must not be disregarded. The results of the simulations are represented in Figs. 2.11 and 2.12.

Despite these encouraging results, we should question the attitude of the public in Bacău metropolitan area towards the carpooling option. A study on this issue is part of our future work, but preliminary research show that there are both psychological barriers and lack of involvement from the local stakeholders regarding this mobility solution.

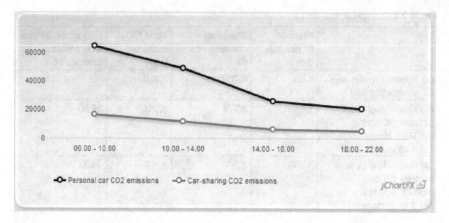

Fig. 2.11 The total CO_2 emissions for single-driver case (*indigo*) and for random number of passengers case (*grey*)

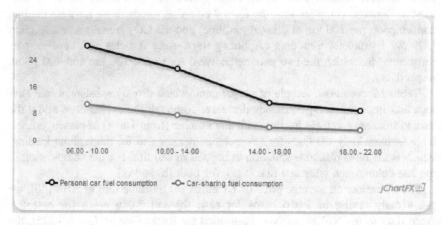

Fig. 2.12 The total fuel consumption for single-driver case (*indigo*) and for random number of passengers' case (*grey*)

2.7 Conclusions and Further Work

Transportation activities have major adverse impact on most of the urban agglomerations, as the traffic is more and more busy and the pollution becomes difficult to mitigate.

The cars cannot be eliminated from our lives but the unwanted effects of their extensive use could be reduced through intelligent strategies. There are several solutions for that: ecological car design, efficient metropolitan networks, multi-modal planning etc. Among them, *car-sharing* and *carpooling* are two modern concepts and should be put forth; specific solutions and facilities should be developed, as both are cost-effective and social rewarding.

Car-sharing and carpooling become more effective and more attractive when seen as an alternative means of transport that can be used together with the personal vehicles and/or with the public transport systems. Practically, all the shared-mobility companies started by manually managing their services: the users place a booking of a vehicle to a human operator, get the key and register the personal data in a form stored in the car. As the programmes have extended their fleet rapidly, the manual-operated systems became expensive and inconvenient, generating errors. Automated reservations, key management and billing offer make the share-used solutions more attractive and effective.

In Europe, car-sharing and carpooling have already proved to be effective in several countries, including: Austria, Germany, Switzerland and Netherlands. In Romania, the participation of several municipalities in European projects (such as Bucharest in the *CIVITAS TELLUS—Transport and Environment Alliance for Urban Sustainability* [51], Sibiu and Timişoara in *TRANSPOWER—Supervised Implementation of Sustainable Urban Transport Concepts* [52], Suceava, in *MIDAS—Measures to Influence transport Demand to Achieve Sustainability* [53], Brăila in *PILOT—Planning Integrated Local Transport* [54]) and organisations (such as *The Romanian Public Transport Association* in *LINK—The European forum on intermodal passenger travel* [55]) started to implement several forms of ICT-based intelligent mobility solutions.

There is a ride sharing community and several initiatives that Romania joined, such as [56, 57]. EkoRoad [58] and 4 in masina [59] are two Romania-based car-sharing/carpooling services available on mobile devices. The former has expanded its area at European level [60].

In Romania, the market share of electric and hybrid vehicles is now less than 0.01 % [61]. A single public charging point is available in Bucharest (however, it is estimated [62] that in 2020, the market share will reach over 5 %). Fuel prices have increased in recent years and their evolution is now uncertain. As the development of infrastructure is not expected soon, the city of Bacău could significantly benefit from the implementation of several "soft" measures for improving its transportation capacity, efficiency and comfort (see [63] for a review on Clean Transport Systems and related issues). Bike-sharing (which will be implemented in the future) and car-sharing/carpooling are such measures which may have a high benefit-cost ratio.

Our future work will focus on developing a carpooling application for mobile devices. The already implemented features (such as routes definition, or operation on maps) will be preserved and some more will be added, such as finding the nearest vehicle available on the way, therefore approaching a dynamical perspective designed to fit the users' needs and to optimise the efficiency of the shared system. Another research direction that we will investigate refers to the use of ACO metaheuristic [64, 65] for solving the *Daily Car-Pooling Problem* and to the assessment of this metaheuristic performance when compared to the results in [66].

Acknowledgments This research was supported by the project "Bacău and Lugano—Teaching Informatics for a Sustainable Society", co-financed by Switzerland through the Swiss-Romanian Cooperation Programme to reduce economic and social disparities within the enlarged European Union.

References

1. Coates, A.: Guide to victory. Popular Gov. **9**(1–4), (1943). https://archive.org/details/populargovernmen914inst. Accessed Nov 2014
2. Transportation and Parking, West Virginia University. http://transportation.wvu.edu/prt. Accessed Nov 2014
3. DART's RideShare program site. http://www.ridedart.com/services/rideshare. Accessed Nov 2014
4. San Francisco Bay 511 program site. http://www.511.org. Accessed Nov 2014
5. Covouiturage Ecolutis site. http://www.ecolutis.com. Accessed Nov 2014
6. Villefluide project site. http://www.villefluide.fr. Accessed Nov 2014
7. Carpooling and Ridesharing across Europe site. http://www.ridefinder.eu. Accessed Nov 2014
8. NuRide project site. http://www.nuride.com. Accessed Nov 2014
9. Knapen, L., Yasar, A., Cho, S., Keren, D., Dbai, A.A., Bellmans, T., Janssens, D., Wets, G., Shuster, A., Sharfman, I., Bhaduri, K.: Exploiting graph-theoretic tools for matching in carpooling applications. J. Ambient Intell. Humaniz. Comput. **5**(3), 393–407 (2014)
10. Ferrari, E., Manzini, R., Pareschi, A., Persona, A., Regattieri, A.: The car pooling problem: Heuristic algorithms based on saving functions. J. Adv. Trans. **37**, 243–272 (2003)
11. Nagare, D.B., Moore, K.L., Tanwar, N.S., Kulkarni, S.S., Gunda, K.C.: Dynamic carpooling application development on Android platform. Int. J. Innov. Technol. Explor. Eng. **2**(3), 136–139 (2013)
12. Graziotin, D.: An analysis of issues against the adoption of dynamic carpooling. http://arxiv.org/abs/1306.0361 (2010)
13. Chan, N.D., Shaheen, S.A.: Ridesharing in North America: past, present and future. Transp. Rev. **32**(1), 93–112 (2012)
14. Austin car2go site. https://www.car2go.com. Accessed Nov 2014
15. U-T San Diego News site. http://www.utsandiego.com. Accessed Nov 2014
16. Zipcar car sharing project site. http://www.zipcar.com. Accessed Nov 2014
17. City Hire Club site. http://www.citycarclub.co.uk/. Accessed Nov 2014
18. DeBruin, L.A.: Energy and feasibility analysis of gasoline engine start/stop technology. Undergraduate Honors Thesis, Ohio State University (2013). https://kb.osu.edu/dspace/bitstream/handle/1811/54555/Start-Stop_Thesis.pdf. Accessed Nov 2014
19. Suggested Upper Merged Ontology site. http://www.adampease.org/OP/
20. Open Cyc Platform site. http://www.cyc.com/platform
21. ISO 14825:2011 site. http://www.iso.org/iso/iso_catalogue/catalogue_tc/catalogue_detail.htm?csnumber=54610
22. SWEET Overview site. http://sweet.jpl.nasa.gov/
23. Open Geospatial Consortium site. http://www.opengeospatial.org/
24. Becker, M., Smith, S.: An ontology for multi-modal transportation planning and scheduling. Tech. report CMU-RI-TR-98-15, Robotics Institute, Carnegie Mellon University (1997)
25. Houda, M., Khemaja, M., Oliveria, K., Abed, M.: A public transportation ontology to support user travel planning. In: Loucopoulos P., Cavavero J.L. (eds.) Proceedings of the International Conference on Research Challenges in Information Science, RCIS'2010 19-21, Nice, France (2010)

26. Cho, S., Kang, J.I., Yasar, A.U.H., Knapen, L., Bellemans, T., Janssens, D., Wets, G., Hwang, C.S.: An activity-based carpooling microsimulation using ontology. procedia computer science. In: The 4th International Conference on Ambient Systems, Networks and Technologies (ANT 2013), The 3rd International Conference on Sustainable Energy Information Technology (SEIT-2013), vol. 19, pp. 48–55 (2013)
27. Dimitrakopoulos, G., Bravos, G., Nikolaidou, M., Anagnostopoulos, D.: Proactive, knowledge-based intelligent transportation system based on vehicular sensor networks. Intell. Trans. Syst. 7(4), 454–463 (2013)
28. Correia, G., Viegas, J.M.: A conceptual model for carpooling systems simulation. J. Simul. 3, 61–68 (2009)
29. Wilkowska, W., Farrokhikhiavi, R., Ziefle, M., Vallée, D.: Mobility requirements for the use of carpooling among different user groups. In: Ahram, T., Karwowski, W., Marek, T. (eds.) Proceedings of the 5th International Conference on Applied Human Factors and Ergonomics (AHFE 2014) (2014)
30. Ciari, F.: Why do people carpool? Results from a Swiss survey. Swiss Transport Research (2012). http://www.strc.ch/conferences/2012/Ciari.pdf
31. Watts, R., Belz, N., Fraker, J., Gandrud, L., Kenyon, J., Meece, M.: Increasing carpooling in Vermont: opportunities and obstacles. University of Vermont Transportation Research Center (2010). http://www.uvm.edu/~transctr/trc_reports/UVM-TRC-10-010.pdf
32. Correia, G., Viegas, J.M.: Carpooling and carpool clubs: clarifying concepts and assessing value enhancement possibilities through a stated preference web survey in Lisbon, Portugal. Trans. Res. Part A 45, 81–90 (2011)
33. Zhou, G., Huang, K., Mao, L.: Design of commute carpooling based on fixed time and routes. Int. J. Veh. Technol. (2014). http://www.hindawi.com/journals/ijvt/2014/634926/
34. Liu, N., Feng, Y., Wang, F., Liu, B., Tang, J.: Mobility crowdsourcing: toward zero-effort carpooling on individual smartphone. Int. J. Distrib. Sens. Netw. (2013). http://www.hindawi.com/journals/ijdsn/2013/615282/
35. Zhang, D., He, T., Liu, Y., Stankovic, J.A.: CallCab: a unified recommendation system for carpooling and regular taxicab services. In: Proceedings of the 2013 IEEE International Conference on Big Data, pp. 439–447 (2013)
36. Zhang, D., Li, Y., Zhang, F., Lu, M., Liu, Y., He, T.: coRide: carpool service with a win-win fare model for large-scale taxicab networks (2013). http://www-users.cs.umn.edu/~tianhe/Papers/coRide.pdf
37. Asakaite, I., Celik, B.: Mapping of logistics infrastructure of central and Eastern Europe for automotive Industry. Master Thesis, Göteborg University (2006). https://gupea.ub.gu.se/bitstream/2077/9603/1/2006_72.pdf
38. Government of Romania. Ministry of transport (2007) Sectorial operational programme transport 2007–2013. http://www.fonduri-ue.ro/res/filepicker_users/cd25a597fd-62/Doc_prog/prog_op/5_POST/2_POST_Eng.pdf. Accessed Nov 2014
39. Romania—Open datasets site. http://date.gov.ro/. Accessed Nov 2014
40. European Environment Agency (2014) monitoring CO_2 emissions from passenger cars and vans in 2013. EEA Technical report 19/2014. http://www.eea.europa.eu/publications/monitoring-co2-emissions-from-passenger. Accessed Nov 2014
41. Inforegio EU Regional Policy site. http://ec.europa.eu/regional_policy/index_ro.cfm. Accessed Nov 2014
42. National Institute of Statistics (2014). http://www.insse.ro. Accessed Oct 2014
43. European Statistics site. http://ec.europa.eu/eurostat. Accessed Oct 2014
44. Kaparias, I., Bell, M.G.H.: Key performance indicators for traffic management and intelligent transport systems. 7th framework programme, contract no. 218636 (2011). http://www.polisnetwork.eu/publicdocuments/download/516/document/d-3-5—key-performance-indicators-for-traffic-management-and-its-final.pdf. Accessed Nov 2014
45. Bacău County Council (2014) strategy for sustainable development of Bacău county for the period 2010–2030. http://www.bids-see.ro/. Accessed Nov 2014

46. European Commission, DG Enterprise & Industry (2009) The potential of intelligent transport systems for reducing rod transport related greenhouse gas emissions. A Sectorial e-Business watch study by SE Consult, Final Report. http://ec.europa.eu/enterprise/archives/e-business-watch/studies/special_topics/2009/documents/SR02-2009_ITS.pdf. Accessed Nov 2014
47. Bacău County Council site. http://www.csjbacau.ro. Accessed Nov 2014
48. Millard-Ball, A., Murray, G., Schure, J., Fox, C., Burkhardt, J.: TCRP Report 108: Car-sharing: Where and How it Succeeds. Transportation Research Board of the National Academies, Washington DC (2005)
49. Shaheen, S.A., Wright, J., Sperling, D.: California's zero-emission vehicle mandate: linking clean-fuel cars, car sharing, and station car strategies. Trans. Res. Rec. **1791**, 113–120 (2001). TRB Paper 02-3587, Institute of Transportation Studies, UC Davis
50. BLISS project site. http://www.bliss.ub.ro. Accessed Nov 2014
51. Transport and Environment Alliance for Urban Sustainability (CIVITAS TELLUS project). http://www.civitas.eu/content/tellus. Accessed Nov 2014
52. Transport Research and Innovation portal. http://www.transport-research.info/web/projects/project_details.cfm?id=11355. Accessed Nov 2014
53. Midas Europe project site. http://www.midas-europe.com. Accessed Nov 2014
54. PILOT—Planning Integrated Local Transport. http://www.pilot-transport.org. Accessed Nov 2014
55. TRANSPOWER—Supervised Implementation of Sustainable Urban Transport Concepts. http://www.transport-research.info/web/projects/project_details.cfm?id=35675. Accessed Nov 2014
56. Roadsharing services project site. http://www.roadsharing.com. Accessed Nov 2014
57. Carpool Rideshare project site. http://www.carpoolworld.com. Accessed Nov 2014
58. Ekoroad travel project for Romania site. http://www.ecodrum.ro. Accessed Nov 2014
59. 4 in masina project site. http://4inmasina.ro. Accessed Nov 2014
60. Ekoroad travel project site. http://www.ekoroad.com. Accessed Nov 2014
61. Varga, B.O.: Electric vehicles, primary energy sources and CO_2 emissions: Romanian case study. Energy **49**, 61–70 (2013)
62. APIA (2014) Romanian association of the vehicle producers and importers. http://apia.ro. Accessed Nov 2014
63. European Commission, Directorate-General for Mobility and Transport (2011). Study on clean transport systems. Project 2010 final report. http://ec.europa.eu/transport/themes/urban/studies/doc/2011-11-clean-transport-systems.pdf. Accessed Nov 2014
64. Crişan, G.C., Nechita, E.: Solving TSP with Ant algorithms. J. Comput. Commun. Control **3**, 228–231 (2008)
65. Pintea, C.M., Crişan, G.C., Chira, C.: Hybrid Ant models with a transition policy for solving a complex problem. Log. J. IGPL **20**(3), 560–569 (2012)
66. Swan, J., Drake, J., Özcan, E., Goulding, J., Woodward, J.: A comparison of acceptance criteria for the daily car-pooling problem. Comput. Inf. Sci. **III**, 477–483 (2013)

Chapter 3
Naval Intelligent Authentication and Support Through Randomization and Transformative Search

Stuart H. Rubin and Thouraya Bouabana-Tebibel

Abstract The problem addressed, in this chapter, pertains to how to represent and apply knowledge to best facilitate its extension and use in problem solving. Unlike deductive logics (e.g., the predicate calculus), an inherent degree of error is allowed for so as to greatly enlarge the inferential space. This allowance, in turn, implies the application of heuristics (e.g., multiple analogies) to problem solving as well as their indirect use in inferring the heuristics themselves. This chapter is motivated by the science of inductive inference. Examples of state-space search, linguistic applications, and a focus methodology for generating novel knowledge (components) for wartime engagement for countering (cyber) threats (WAMS) are provided.

3.1 Introduction

Deductive methods, such as the predicate calculus, have met with failure in the Japanese Fifth Generation project [1]. The follows because there was no way to heuristically drive the back-cut mechanism to limit the combinatoric explosion. Rule-based systems are limited by the need to know the rules of engagement, which is practically unobtainable. Evolutionary methods require suitable representations for knowledge (as do neural networks to maintain tractability). These are not supplied a priori because again the rules of engagement are not well understood; hence, their representations are not well understood either. Inferential methods, like ID3, require decision points, which essentially defines a post-mortem battle management dump and analysis. Given the degrees of freedom inherent to each battle,

S.H. Rubin (✉)
SSC-PAC 71730, San Diego, CA 92152-5001, USA
e-mail: stuart.rubin@navy.mil

T. Bouabana-Tebibel
LCSI Laboratory, École Nationale Supérieure d'Informatique,
Algiers, Algeria
e-mail: t_tebibel@esi.dz

© Springer International Publishing Switzerland 2016
K. Nakamatsu and R. Kountchev (eds.), *New Approaches in Intelligent Control*,
Intelligent Systems Reference Library 107, DOI 10.1007/978-3-319-32168-4_3

such an analysis is not forthcoming either. Hence, an inductive case-based approach is the remaining methodology of choice for battle management.

This contribution is motivated by the science of inductive inference. A focus methodology for generating novel knowledge (components) for wartime engagement for countering (cyber) threats (WAMS) will be provided. More specifically, an area of military concern pertains to the generation of maximally effective plans for conventional and asymmetric battles [2]. Such plans include strategic as well as tactical components. A problem is that it is not practical to plan for every contingency. Thus, a methodology will be provided, which can extract and extend the applicability of naval (wartime) knowledge to contemporary situations.

3.1.1   Context of the Study

Randomization is currently very common in CBR processing. Many works show that hybridization of CBR and fuzzy methods leads to compromise solutions with respect to accuracy and computational complexity, since the accuracy of CBR grows as the number of cases increases. More cases mean added computational complexity in terms of space and time. Randomization is used for case retrieval in CBR. In [3–5] fuzzy similarity measures are applied to range and retrieve similar historical case(s).

Randomization is also used in the reuse phase of CBR to make a decision on whether or not the most similar case should be reused. In [6], a random function is applied on a special category of users that have made similar requests to compute their mean opinion.

Similarly, in [7], decision making on knowledge handling within a CBR system is based on fuzzy rules—thus providing consistent and systematic quality assurance with improvement in customer satisfaction levels and a reduction in the defect rate. In [8], a scoring model based on fuzzy reasoning over identified cases is developed. It is applied to the context of cross-document identification for document summarization.

However, recent work in Case-Based Reasoning [9, 10] has evidenced that while cases provide excellent representations for the capture and replay of military plans, the successful generalization of those plans requires far more contextual knowledge than economically sound planning will allow. For example, a generalization of gasoline is fuel. However, only a proper subset of automobiles can burn fuels other than gasoline. Knowledge of which automobiles burn what fuels is not coincidently obtained through the generalization process. In practice, this problem, which arises out of context, is referred to as, overgeneralization. Rather, the approach to randomization, proposed herein, enables the dynamic extraction of transformation cases, which permits the inference of novel cases through the application of set/sequence context—along with a relative estimate of case validity.

3.1.2   Problem Description

Using inductive case-based reasoning, an inherent degree of error is allowed for so as to greatly enlarge the inferential space. This allowance, in turn, implies the application of heuristics (e.g., multiple analogies) to problem solving as well as their indirect use in inferring the heuristics themselves. Indeed, the origins of case-based reasoning as an episodic memory support its use for analogic reasoning. This is because a most-similar case is sought; rather than a proper sequence of applicable rules.

Consider the following problem, where the assigned task is the lossless randomization of a sequence of integers [11]. A randomization is given here by $n_{i+1} \leftarrow 2n_i + i$.

$$\text{Randomize} \quad n: \quad 0 \quad 0 \quad 1 \quad 4 \quad 11 \quad 26 \quad 57$$
$$i: \quad 0 \quad 1 \quad 2 \quad 3 \quad 4 \quad 5 \quad 6$$

We say that this randomization is lossless because the associated error-metric (e.g., the 2-norm) is zero. Randomizations may or may not exist given the operator, operand set, and the set error-metric bounds. Furthermore, even in cases where randomizations exist, they may not be discoverable in the allocated search time on a particular processor(s) [12]. In view of this, the general problem of randomization is inherently heuristic.

Note that a slightly more complex (real-world) task would be to randomize a similar sequence of integers, where the error-metric need not be zero, but is always bounded. Such sequences arise in the need for all manner of prediction (e.g., from the path of an incoming missile to the movement of storm tracks). This abstraction underpins the novel aspects of the WAMS battle (cyber) management systems.

Clearly, there is no logic that can solve the inductive inference problem [12]. Rather, one needs to define a search space such that the search operators are informed. The more informed the search operators, the less search that is required (i.e., successful or not). Here is one possible schema to delimit the search space in this problem:

$$n_{i+1} \leftarrow M \; \{*, /, +, -, **\} \; n_i \{*, /, +, -\} \{i, n_{i-1}\}$$

Partially assigning mnemonics, this schema can be described as follows:

$$n_{i+1} \leftarrow \text{int extended-ops } n_i \text{ ops } \{i, \; n_{i-1}\}$$

But, even here, it is apparently ambiguous as to how such a schema might be found. To answer this question, consider the randomization of the even sequence, $2n$, and the odd sequence, $2n + 1$. The randomization of these two sequence definitions is given by $2n + j, j \in \{0, 1\}$. Next, note that "+" $\subset$ ops $\subset$ extended-ops. Each replacement, at the right, represents a level of generalization. Generalizations are made to randomize two or more instances. For example, if the odd sequence were defined by $2n - 1$, then a first-level randomization (i.e., based on the given mnemonics) of $2n + 1$ and $2n - 1$ is given by 2n ops 1. Clearly, having multiple mnemonics can greatly enlarge the search space and result in intractable solutions. An evolutionary approach to reducing the implied search time is to perform a gradient search outward from known valid points. Here, search reduction is obtained by commensurately reducing search diversity. It is claimed that this process is what enables most of us to solve inferential randomization problems such as this one, most of the time. The dual constraints of available search time versus the generality of the candidate solution space serves to dynamically contract or expand the search space.

Notice that the process of randomization not only captures existing instances in a more compact form, but in so doing embodies similar instances, which may or may not be valid. The point is that by limiting the degree of generalization, one tightens the degree of analogy and in so doing, increases the chance of a valid inference. The inferences found to be valid are fed back to the randomization process. This results in a more delimited search space and provides for multiple analogies—increasing the subsequent chance for valid inferences. Moreover, according to Solomonoff [13–15] the inference of grammars more general than regular grammars is inherently heuristic. The context-free grammar (CFG) is the lowest-level such grammar. All non deterministic grammars may be statistically augmented—resulting in stochastic grammars [16]. Furthermore, where heuristics serve in the generation of new knowledge and that knowledge serves in the generation of new heuristics, the amplification of knowledge occurs by way of self-reference [17]! Allowing for the (self-referential) application of knowledge bases, any practical methodology serving in the discovery of these heuristics must be domain-general. The transformative search for randomization is the most general such methodology because it extracts self-referential knowledge from situational as well as action knowledge in context [11, 12, 18].

In the remainder on this paper, Sect. 2 presents the theoretical aspect of the approach, where an illustrative development of the proposed technique is provided. Section 3 shows the appropriateness of the approach for some sensitive domains of applications. In Sect. 4, a methodology of implementation is developed. Section 5 discusses the approach and provides some concluding remarks.

3.2   Technical Approach

Phrase structure (i.e., Type 0) grammatical inference will be applied to induce novel case knowledge. The number of derivational paths, necessarily including the time for derivation, provides a heuristic estimate of the derivational error (Fig. 3.1). Higher-level descriptive mnemonics are acquired through usage—eliminating the need for query here. Such query does not work well in general because most associations are novel and hierarchical (i.e., not colloquial).

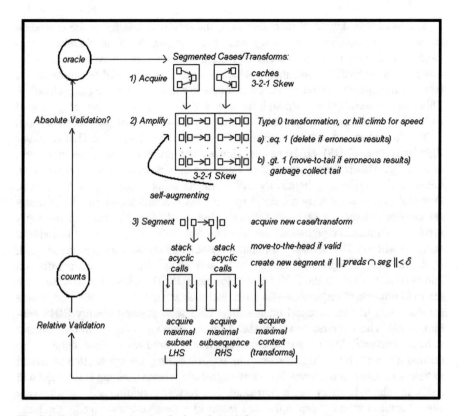

Fig. 3.1 Dynamic representation for knowledge inference

The more constrained the search for knowledge, not only the faster that knowledge may be induced, but the faster heuristics aiding in the induction of that knowledge may be obtained as well. This knowledge may appear to be rules, but it is actually cases. This is because rule applications may be chained, which prevents immediate feedback on their validity. On the other hand, validity can always be ascertained after a single case application. Also, for this reason, case bases are easier to maintain and can be grown far larger. Derivational paths are characterized as transformational sequences for the synthesis of individual cases. Situations and/or actions may be transformed to create a candidate case. Cases found to be in error are deleted from their containing grammatical segment(s). Only chains of length one may be deleted if an erroneous case should result. When longer transformational chains are used and result in an erroneous case, the involved transforms are logically moved-to-the-tail of their segment in their order of application. Thus, these transforms will be among the first to fall off of their tails when space is reclaimed. Convergence upon correct cases and thus correct transforms is assured. This methodology for knowledge inference is depicted in Fig. 3.1.

The attainable degree of randomization, the resultant validity of the embodied cases, and the rapidity of discovery will always be greater if the domain is segmented to facilitate parallel processing. Segmentation also reduces the need for superfluous comparison operations, which insures that they will be tractable. Of course, one segment can refer to another if care is exercised to avoid daisy chaining. This may be accomplished through the use of a non-redundant stack of generated calls. The assignment of segments itself may properly be under case-based control.

The invocation of distinct segments for (partial) left-hand side (LHS) and/or right-hand side (RHS) predicates maximizes the reuse of case-based/transform knowledge. This is a primary tenet of randomization theory. Thus, case/transform LHSs and/or RHSs may iteratively call segments using subsets of LHS contexts. Potential cycles are readily managed by the LHS case/transform predicates themselves. For example, a segment for predicting what crops to sow will no doubt invoke a segment for predicting the weather. Similarly, a segment for planning a vacation will likewise invoke a segment for predicting the weather (i.e., reuse) [19].

A grammatical segment is defined to be domain specific. Segments are divided into two parts—that for the LHS and that for the RHS. The LHS consists of sets and the RHS consists of sequences—for cases and their transformations alike. Segments acquire cases having maximal intersection (LHS sets) or commonality (RHS subsequences). These metrics are normalized for the number of entries in the segment to insure compatibility. Each case/transformation is stored twice. New segments are created if there is no intersection or commonality. Setting the intersection threshold to zero will result in relatively few large segments; whereas, setting it too high will result in relatively many small segments. The former condition will be relatively slow, capture too many analogies, and result in excessive errors; while, the latter condition will be relatively fast, not capture enough analogies, and result in too little creativity. Thus, the intersection threshold may be dynamically adjusted for a given domain. Referenced segments are logically moved to the head of a containing list. Unreferenced segments are subject to tail deletion. Cases/transforms may be cached for maximal speed of retrieval. There are two basic types of class equivalence, introduced by Rubin [18], and made realizable in this chapter. Here, { } denotes sets, while ( ) denotes phrase sequences.

1. Case pairs of the form: $A \rightarrow C$; $B \rightarrow C$. This implies that {A, B} may be substituted for one another as conditional predicates, where C is a subsequence or a super-sequence of a RHS immediately implied by the substitution. Substitutions are initially hill-climbed (i.e., non contracting transformations—getting closer to a match with each substitution), given a matching action subsequence or super-sequence, to most rapidly enable a situation to be covered by a supplied context. Upon failure of such search, substitutions are made in random order, until interrupt, to enable Type 0 generality. The first most-specific case situation to be covered, and/or having the most instances of derivation, is selected. This metric is inherently heuristic.

2. Non deterministic case pairs of the form: $A \rightarrow C$; $A \rightarrow D$. This implies that (C D) may be substituted for one another as non deterministic actions, where A is a subset

or a superset of a LHS immediately implied by the substitution. Substitutions are initially hill-climbed, given a matching situational subset or super-set, to most rapidly enable a (partial) action substitution. Upon failure of such search, substitutions are made in random order, until interrupt, to enable Type 0 generality. The first most-specific transformation to be applied, and/or having the most instances of derivation, is selected. This metric is inherently heuristic.

3.2.1   Illustrative Developments

Consider the following case-based knowledge segment. The use of frames will be elaborated upon at a subsequent point so as not to obscure the basic concepts.

$$C1: \quad \{A, B, C\} \rightarrow (X)$$
$$C2: \quad \{A, E\} \rightarrow (X)$$
$$C3: \quad \{B, D\} \rightarrow (X)$$
$$C4: \quad \{A, B, C\} \rightarrow (Y)$$

Clearly, C1 and C4 are non deterministic. Moreover, C1, C2, and C3 are conceptually (i.e., if not literally) equivalent in the context of the RHS, X. Case pairings are made on the basis of the most-recently fired first. Such pairings are made using the 3-2-1 skew to favor the just in time induction of new cases in a domain that has shown itself to be most-recently needed (see Appendix I). Also, the acquisition of new cases leads to the acquisition of new transforms. Note that if the system sits idle for long, it enters dream mode via the 3-2-1 skew. That is, it progressively incorporates less recently acquired/fired cases.

Thus, where our case-based segment reflects a logical move-to-the-front ordering, C1 and C2 are most-likely paired. This induces a context-sensitive case (i.e., which is also limited in applicability to the segment from which it was derived) for the transformation of LHS representations as follows.

T1 [C1, C2]: $\{A, B, C\} \longleftrightarrow \{A, E\} \mid (X)$

Next, consider the acquisition of a fifth valid case:

C5: $\{A, B, C, D\} \rightarrow (Z)$, where $X \in Z$, or $Z \in X$, read as subsequence.

The mean probability of validity, for the inferred case, is given by:

$$\frac{\max(|X||X \in Z, |Z||Z \in X)}{\max(|X|, |Z|)}.$$

Applying T1 to C5 yields:

C6 = T1 [C5]: $\{A, D, E\} \rightarrow (Z)$

If C6 is validated, T1 is moved to the logical head of a validated LHS trans-formational segment. Otherwise, T1 and C6 are deleted. Notice that the acquisition of C6 may result in an iterative propagation of synthesized cases and associated acquisitions. It is important to provide for immediate feedback from a single case to squelch such possible explosions of bad knowledge. Moreover, suppose the LHS transformational segment includes T2 as follows.

T2 [,C2]:{A, D} $\leftarrow \rightarrow$ {A, E} | (W), where $X \in W$, or $W \in X$.

The mean probability of validity, for the inferred case, is given by:

$$\frac{\max(|X| \mid X \in W, \ |W| \mid W \in X)}{\max(|X|, |W|)}.$$

Applying T2 to T1 induces the further transformational case:

T3 [C1,]: {A, B, C} $\leftarrow \rightarrow$ {A, D} | $(X \mid X \in W, W \mid W \in X)$

Transformations are adjudicated as soon as feedback can be had on the validity of an induced case. Furthermore, the synthesis of redundant cases serves to provide predictive evidence of validity. Distinct case pairings are selected at random, where the selection space is gradually increased downward, using the segmented list head (see Appendix I). Again, this tends to favor the synthesis of knowledge in pro-portion to its recent utility. Transformations, like cases, are logically moved within their segment when acquired (i.e., at the head) or fired.

Suppose that C1 and C4 are the next cases to be paired. This induces a segment-specific context-sensitive case for the transformation of the RHS repre-sentations as follows.

T4 [C1, C4]: (X) $\leftarrow \rightarrow$ (Y) | {A, B, C}

The mean probability of validity, for the inferred case, is given by:

$$\frac{|\{A, B, C\} \cap \{A, B, C\}|}{|\{A, B, C\} \cup \{A, B, C\}|} = 1.0.$$

Next, consider the acquisition of a seventh valid case:

C7: {A, C} $\rightarrow$ (Y Z), where {A, C} $\subseteq$ {A, B, C} (or vice versa), where this is read as subsets.

The mean probability of validity, for the inferred case, is given by:

$$\frac{|\{A, C\} \cap \{A, B, C\}|}{|\{A, C\} \cup \{A, B, C\}|} = 0.67.$$

Applying T4 to C7 yields:

C8 = T4 [C7]: {A, C} $\rightarrow$ (X Z) with stochastic probability of 1.0 (the probability for T4) * 0.67 (the probability for C7) = 0.67

Similarly, the joint probability of LHS and RHS transformations is multiplicative. If C8 is validated, T4 is moved to the logical head of the appropriate validated RHS transformational segment. Otherwise, T4 and C8 are deleted.

Notice that the acquisition of C8 may result in an iterative propagation of symmetric cases and associated acquisitions. It is again important to provide for immediate feedback from a single case to squelch such possible explosions of bad knowledge. Moreover, suppose the RHS transformational segment includes T5 as follows, where Q is the LHS of whatever the (resultant) transform is applied to (i.e., transform or case).

T5 [,C4]: (W) $\leftarrow \rightarrow$ (Y) | {A, B, C}, where $Q \subseteq$ {A, B, C}, or {A, B, C} $\subseteq Q$.

The mean probability of validity, for the inferred case, is given by:

$$\frac{|Q \cap \{A, B, C\}|}{|Q \cup \{A, B, C\}|}.$$

Applying T4 to T5 induces the further transformational case:

T6 [C1,]: (W) $\leftarrow \rightarrow$ (X) | {A, B, C}, where $Q \subseteq$ {A, B, C}, or {A, B, C} $\subseteq Q$.

The mean probability of validity, for the inferred case, is given by:

$$p(T4), \frac{|\{A, B, C\} \cap \{A, B, C\}|}{|\{A, B, C\} \cup \{A, B, C\}|} * p(T5), \frac{|Q \cap \{A, B, C\}|}{|Q \cup \{A, B, C\}|} = \frac{|Q \cap \{A, B, C\}|}{|Q \cup \{A, B, C\}|}.$$

The transformational cases are segmented and logically ordered to facilitate pattern matching as well as for determining the relative time of last use. Cases are selected for firing using a most-specific inference engine. Unreferenced cases, as well as cases having validated generalizations (i.e., subsumption) may drop off of the logical tail of their segments.

3.2.2 Stochastic Validation

The process of grammatical randomization is not deterministic. Thus, if the order of randomizations is determined at chance, then distinct transformational paths may result. The relative validity of an induced case bears proportion to the number of paths by which it may be derived by transformational process. Unfortunately, there is no formula for ascertaining the probability of such derivations in an absolute sense. This is because the quality of the cases comprising transformations as well as the applicability of the transformations themselves cannot be known a priori.

Stochastic validation is ideally suited to parallel processing. It is to be used where feedback is long in coming and thus individual cases need to be tagged with a relative predictive validity. The use of stochastic validation does not preempt the need for immediate feedback because that which can be learned is limited by the availability of feedback.

3.3   Domains of Application

We propose in what follows to apply the proposed approach to some situations well-qualified to enhance the utility of use of such a technique and illustrate its functioning.

3.3.1   State-Space Prediction for Autonomic Computing

The task of this example is to show how randomization can be used for the abstract solution of transforming one representation of knowledge, or model, into a similar (i.e., symmetric) one. The application here is to predict a proper operator sequence, as applied to a state space, in a plan to reach a goal. Note that the methodology can supply heuristics to accelerate state-space search. The approach employed here learns similar maps, whose value (i.e., faster heuristic search) and validity converge with scale. To begin, since robot problems are simple and intuitive, examples from this domain are used to illustrate the major ideas embodied in this chapter. Programming a robot involves integrating many functions—incorporating percep-tion, the formulation of plans of action, and monitoring the execution of these plans [10]. Here, the problem is to synthesize a sequence of robot actions that map the robot from some initial state to some goal state.

Robot actions are modeled by a production system, where state descriptions are constructed from predicate calculus well-formed formulas (*wffs*) [20]. The task here is to stack three blocks (i.e., A, B, C). The formula CLEAR (B) means that block B has a clear top with no block on top of it. The ON predicate is used to describe which blocks are (directly) on other blocks. The predicate HANDEMPTY has value T just in case the robot (placement) hand is empty. The use of other predicates is similar. Suppose the initial state description has blocks A and B on the table and block C on top of block A. Further suppose that the goal state has block C on the table, block B on top of block C, and block A on top of block B. Let's begin with a common initial state (0):

(0) CLEAR (A), ONTABLE (A); CLEAR (B), ONTABLE (B); CLEAR (C), ONTABLE (C); HANDEMPTY

Next, the following two cases are acquired in the first segment.

(1) pickup (C) → CLEAR (A), CLEAR (B), HOLDING (C), ONTABLE (A), ONTABLE (B)
(1) pickup (A) → CLEAR (B), CLEAR (C), HOLDING (A), ONTABLE (B), ONTABLE (C)

The general theory has us looking to equate the LHSs here. This is possible by way of an adjustment for the use of wffs through the use of parametric fusion;

although, a pair of matching LHS instances does not result, which precludes equating the LHSs here:

(1) pickup (U) → CLEAR (V), CLEAR (W), HOLDING (U), ONTABLE (V), ONTABLE (W)

Similarly, (1) pickup (B) induces the instantiation:

(1) pickup (B) → CLEAR (V), CLEAR (W), HOLDING (B), ONTABLE (V), ONTABLE (W)

Here, it is discovered through search that B ≠ V ≠ W and {B, V, W} = {A, B, C}. Taking V = A and W = C yields:

(1) pickup (B) → CLEAR (A), CLEAR (C), HOLDING (B), ONTABLE (A), ONTABLE (C)

Next, the following five cases are acquired in the second segment.

(2) stack (B, A) → CLEAR (C), ON (B, A), CLEAR (B), ONTABLE (A), ONTABLE (C), HANDEMPTY

(2) stack (C, B) → CLEAR (A), ON (C, B), CLEAR (C), ONTABLE (A), ONTABLE (B), HANDEMPTY

(2) stack (C, A) → CLEAR (B), ON (C, A), CLEAR (C), ONTABLE (A), ONTABLE (B), HANDEMPTY

(2) stack (A, C) → CLEAR (B), ON (A, C), CLEAR (A), ONTABLE (B), ONTABLE (C), HANDEMPTY

(2) stack (A, B) → CLEAR (C), ON (A, B), CLEAR (A), ONTABLE (B), ONTABLE (C), HANDEMPTY

Parametric fusion here yields (ONTABLE is commutative):

(2) stack (U, V) → CLEAR (C), ON (U, V), CLEAR (U), ONTABLE (V), ONTABLE (C), HANDEMPTY

Similarly, (2) stack (B, C) induces the instantiation:

(2) stack (B, C) → CLEAR (?), ON (B, C), CLEAR (B), ONTABLE (C), ONTABLE (?), HANDEMPTY

Here, it is discovered through search that ? ≠ U ≠ V and {?, U, V} = {A, B, C}. Taking ? = A yields:

(2) stack (B, C) → CLEAR (A), ON (B, C), CLEAR (B), ONTABLE (C), ONTABLE (A), HANDEMPTY

Again, a pair of matching LHS instances does not result, which precludes equating the LHSs here.

Next, the following five cases are acquired in the third segment. These build on the results found in the second segment.

(3) pickup (C) → ON (B, A), CLEAR (B), HOLDING (C), ONTABLE (A)
(3) pickup (A) → ON (C, B), CLEAR (C), HOLDING (A), ONTABLE (B)
(3) pickup (B) → ON (C, A), CLEAR (C), HOLDING (B), ONTABLE (A)
(3) pickup (B) → ON (A, C), CLEAR (A), HOLDING (B), ONTABLE (C)
(3) pickup (C) → ON (A, B), CLEAR (A), HOLDING (C), ONTABLE (B)

Parametric fusion here yields:

(3) pickup (U) → ON (V, W), CLEAR (V), HOLDING (U), ONTABLE (W)

Similarly, (3) pickup (A) induces the instantiation:

(3) pickup (A) → ON (V, W), CLEAR (V), HOLDING (A), ONTABLE (W)

Here, there are two pairs of cases having common LHSs in the third segment. They are equated as follows.

ON (B, A), CLEAR (B), HOLDING (C), ONTABLE (A) →
ON (A, B), CLEAR (A), HOLDING (C), ONTABLE (B) | pickup (C)

ON (C, A), CLEAR (C), HOLDING (B), ONTABLE (A) →
ON (A, C), CLEAR (A), HOLDING (B), ONTABLE (C) | pickup (B)

These are parametrically fused, through search, as follows.

ON (U, V), CLEAR (U), HOLDING (W), ONTABLE (V) →
ON (V, U), CLEAR (V), HOLDING (W), ONTABLE (U) | pickup (W)

Instantiating A for W (satisfying the context) yields:

ON (U, V), CLEAR (U), HOLDING (A), ONTABLE (V) →
ON (V, U), CLEAR (V), HOLDING (A), ONTABLE (U)

Thus, (3) pickup (A) → ON (B, C), CLEAR (B), HOLDING (A), ONTABLE (C), where if this were not predicted valid (e.g., using multiple derivations), which it is →
ON (C, B), CLEAR (C), HOLDING (A), ONTABLE (B) as a second candidate (or vice versa)

Next, the following five cases are acquired in the fourth segment. These build on the results found in the third segment.

(4) stack (C, B) → CLEAR (C), ON (C, B), ON (B, A), ONTABLE (A), HANDEMPTY

(4) stack (A, C) → CLEAR (A), ON (A, C), ON (C, B), ONTABLE (B), HANDEMPTY

(4) stack (B, C) → CLEAR (B), ON (B, C), ON (C, A), ONTABLE (A), HANDEMPTY

(4) stack (B, A) → CLEAR (B), ON (B, A), ON (A, C), ONTABLE (C), HANDEMPTY

(4) stack (C, A) → CLEAR (C), ON (C, A), ON (A, B), ONTABLE (B), HANDEMPTY

Parametric fusion here yields:

(4) stack (U, V) → CLEAR (U), ON (U, V), ON (V, W), ONTABLE (W), HANDEMPTY

Similarly, (4) stack (A, B) induces the instantiation:

(4) stack (A, B) → CLEAR (A), ON (A, B), ON (B, W), ONTABLE (W), HANDEMPTY

Here, it is discovered through search that W = C, which yields:

(4) stack (A, B) → CLEAR (A), ON (A, B), ON (B, C), ONTABLE (C), HANDEMPTY

This is readily verified to be a goal state (e.g., as predicted by multiple derivations), as desired. It may appear that simple parametric substitution, as illustrated here, is the key to randomization. Thus, the question arises as to why the need for conformant LHSs and/or RHSs for knowledge amplification. The answer is that in general, equivalence transformations can be arbitrarily complex. Unless the domain is well understood, it is not possible to catalog most equivalence transformations a priori. It follows that the proposed methodology, which is knowledge-based rather than algorithmic, is needed for randomization in the general case. Furthermore, this methodology provides for a composition of transforms, which serves to extend the initial transformational space. This represents a deepest form of reuse.

3.3.2 WAMS: Wartime Associate for Military Strategy

An area of military concern pertains to the generation of maximally effective plans for conventional and asymmetric battles [2, 21]. Such plans include strategic as well as tactical components. A problem is that it is not practical to plan for every contingency. Thus, a methodology is needed, which can extract and extend the applicability of naval (wartime) knowledge to contemporary situations.

The approach to randomization, proposed herein, enables the dynamic extraction of transformation cases, which permits the inference of novel cases through the application of set/sequence context—along with a relative estimate of case validity.

All cases and transforms are segmented. Supplied basis cases may be mined from actual wartime data (i.e., as opposed to military war games) and/or supplied by a military expert. The use of cases here, as opposed to rules, means that the military strategist need not know precise causality. Thus, far more basis knowledge can be supplied. The methodology will find, extract, apply, and to some level validate the causal knowledge prior to amplification. As a result, wartime decision support systems can be delivered, which strategize solutions to novel, but not totally unexpected situations as they develop. The resulting System of Systems, or Wartime Associate for Military Strategy (WAMS), amplifies existing naval knowledge for wartime engagement.

Situations and actions are given a frame format to facilitate pattern matching as well as descriptive content and are of the form.

Context-Free Category: context-sensitive phrase instance.

For example, Contact Method: shortwave radio transmission. The use of distinct frames and/or instances implies distinct cases. A transform, or a sequence of transforms, can map a case so that it is covered by a context. In what follows, Fi is frame i followed by a phrase instance. Searches specific to a segment(s) utilize segment name(s).

Inherent sequence in frames, on the LHS, may be specified through the use of *join* frames. For example, the frames, SOCKS-ON and SHOES-ON might include the join frame, SOCKS-ON-BEFORE-SHOES-ON. Similarly, a relaxation of the sequence requirement, on the RHS, may be specified through the use of *generalized* frames. For example, STEP-LEFT-FOOT, then STEP-RIGHT-FOOT may be replaced by STEP-FOOT.

Situational frames are marked with a "#" on the LHS to indicate that the order of those so marked must be in the relative order specified (i.e., at least two # frames for effect). Action frames are marked with a "#" on the RHS to indicate that the relative order of those so marked is not constrained (i.e., at least two # frames for effect). A transform {Fj:, Fi:} matches {#Fi, #Fj}. Similarly, a transform (Fj: Fi:) matches (#Fi #Fj). The "#"s are immutable as this is case-specific information (e.g., a sequence for assembly may not match a sequence for disassembly), carried by the context. A sample case pairing follows.

$$
\begin{array}{ll}
\text{C9:} & \text{C10:} \\
\{\#F1:a, & \{F2:b, \\
\#F2:b\} & F1:a\} \\
\rightarrow & \rightarrow \\
(F3:c & (\#F3:c \\
F4:d) & \#F4:f)
\end{array}
$$

These cases induce the non deterministic case transform:

T7 [C9, C10]: (F3:c F4:d) $\leftarrow \rightarrow$ (#F3:c #F4:f) | {F1:a, F2:b}

Notice that this transform can reverse the order of RHS frames, while changing the phrase supplied to F4 from d to f, or vice versa.

Similarly, the case pairing:

C11:   C12:
{#F1:a,   {F2:b,
#F2:b}   F1:a}
$\rightarrow$ $\rightarrow$
(F3:c   (#F4:d
F4:d)   #F3:c)

induces the case transform:

T8 [C11, C12]: {#F1:a, #F2:b} $\leftarrow \rightarrow$ {F2:b, F1:a} | (#F3:c #F4:d)

Notice that this transform can reverse the order of LHS frames, or serve as a generalized id transform. RHS and LHS transformations may co-occur, where the most frequently synthesized conformant cases are selected for firing.

A transformational protocol is necessary to insure that all realizations function in an identical manner. This is best defined, for the programmer, by way of examples.

Given: {A, B, C, D}, transform {A, D} to {E}, yields {E, B, C}

Given: {#A, #B, C, #D}, transform {#B, #D} to {#F, #E}, yields {#A, #F, C, #E}

Given: {#A, #B, C, #D}, transform {#A, #D} to {#E, #F, #G}, yields {#E, #B, C, #F, #G}

Given (A, B, C, D), transform (A, D) to (E), yields (E B C)

Given (#A #B C #D), transform (#B #D) to (#F #E), yields (#A #F C #E)

Given (#A #B C #D), transform (#A #D) to (#E #F #G), yields (#E #B C #F #G)

The transformational protocol could involve hierarchy, as with say:

{A, #{B, C}, #{#E, F, #G}(H #I #J)} $\rightarrow$ (#K #(L M) (#N O #P #{Q, #R, #S}));

but, not only is this specification methodology increasingly complex and unnatural—it is unnecessary. Rather, non monotonic cases can use a global blackboard to build up a vocabulary of case definitions. It is scientifically proper to subdivide complex cases into non monotonic ones to enable reuse (a tenet of randomization theory) [12]. For example:

{#Laces: Pull laces, #Tie: Make bow} $\rightarrow$ (Foot-ware: Shoes are on)

is simple, easy to understand, and reusable. The global blackboard utilizes posting and retraction protocols to change the context. For example, the previous non monotonic case might be specified as:

{#Laces: Pull laces, #Tie: Make bow} → RETRACT: (Foot-ware: Shoes are off); POST: (Foot-ware: Shoes are on)

The order of retract and post actions is critical and is preserved in any action transformation.

To continue, here is a simple example of naval wartime knowledge expressed in the form of stochastic cases. Transformations are made within context as scale is impossible to simulate in this example:

C1 (<u>basis</u> or n-step induced):
Situation:
{
Occupation: Marines holding enemy-occupied island
Terrain: Many large boulders on island
Situation: Marines under attack by enemy bombers
Situation: Marines under attack by machine gun fire from enemy fortifications
Situation: Marines under attack by enemy mortars
CASREP: Marine casualties high
Ammo: Marine ammunition low
Rockets: Marine rockets medium
Reinforcements: 3-5 hours out
}

Action (p = 0.4):
(
Quick Action: Move for cover
Psyops: Create decoys
#Rifle Instructions: Fire to suppress enemy fire
#Rocket Instructions: Random fire rockets at enemy fortifications
)

Predicted Success: 88%
Predicted (Relative) Validity: 99%

C1 is a basis (i.e., user-supplied) case. n-step includes LHS + RHS transformations. The stochastic probability of this non deterministic case is 0.4. The rifle and rocket instructions may be issued at any time relative to the other instructions embodied by this case. The order of Quick Action and Psyops is immutable.

Next, second and third similar cases are supplied in preparation to illustrate the induction of LHS and RHS transformations. Thus, let

C2 (<u>basis</u> or n-step induced):
Situation:
{
Occupation: Marines holding enemy-occupied building
Situation: Marines under attack by snipers
Situation: Marines under attack by machine gun fire from enemy placements
Reinforcements: less than 10 minutes out
Terrain: There are surrounding buildings
CASREP: Marine casualties medium
Ammo: Marine ammunition medium
Rockets: Marine rockets medium
}

Action (p = 1.0):
(
#Rocket Instructions: Random fire rockets at enemy fortifications
Quick Action: Move for cover
Psyops: Create decoys
#Rifle Instructions: Fire to suppress enemy fire
)

Predicted Success: 95%
Predicted Validity: 99%

C2 is a basis case. This case is deterministic, since p = 1.0. The rocket and rifle instructions may be issued at any time relative to the other instructions embodied by this case action. The order of Quick Action and Psyops is immutable.

C3 (<u>basis</u> or n-step induced):
Situation:
{
Reinforcements: 3-5 hours out
Occupation: Marines holding enemy-occupied island
Terrain: Many large boulders on island
Situation: Marines under attack by enemy bombers
Situation: Marines under attack by machine gun fire from enemy fortifications
Situation: Marines under attack by enemy mortars
CASREP: Marine casualties high
Ammo: Marine ammunition low
Rockets: Marine rockets medium
}

Action (p = 0.6):
(
Action: Minimal movement for immediate cover
Action: Radio for air support
#Mortar Instructions: Launch mortars at enemy fortifications
#Rifle Instructions: Fire to suppress enemy fire
)

Predicted Success: 90%
Predicted Validity: 99%

C3 is a basis case. The stochastic probability of this non deterministic case is 0.6. The mortar and rifle instructions may be issued at any time relative to the other instructions embodied by this case. The order of Action instructions is immutable.

Next, a LHS and a RHS transformation case will be formed and acquired by the proper segments. Situations and actions may be logically stored to save space. The search for a LHS/RHS match is made within a segment upon the acquisition of a new case by the segment, the movement of a fired case within the segment, or when attempting to match a context. First, a LHS transformation (ST1) is given:

C1.LHS ←→ C2.LHS | C1 RHS = C2 RHS:
ST1:
Situation C1:
{
Occupation: Marines holding enemy-occupied island
Terrain: Many large boulders on island
Situation: Marines under attack by enemy bombers
Situation: Marines under attack by machine gun fire from enemy fortifications
Situation: Marines under attack by enemy mortars
CASREP: Marine casualties high
Ammo: Marine ammunition low
Rockets: Marine rockets medium
Reinforcements: 3-5 hours out
}
        ←→
Situation C2:
{
Occupation: Marines holding enemy-occupied building
Situation: Marines under attack by snipers
Marines under attack by machine gun fire from enemy placements
Reinforcements: less than 10 minutes out
Terrain: There are surrounding buildings
CASREP: Marine casualties medium
Ammo: Marine ammunition medium
Rockets: Marine rockets medium
}

in the context of :
Action min (0.4, 1.0) = 0.4:
(
Quick Action: Move for cover
Psyops: Create decoys
#Rifle Instructions: Fire to suppress enemy fire
#Rocket Instructions: Random fire rockets at enemy fortifications
)

Predicted Success: min (88%, 95%) = 88%
Predicted Validity: min (99%, 99%) = 99%

This transform case is acquired by the segment(s) with which the RHS has a maximal normalized (i.e., for the number of entries in the segment) number of subsequences and super-sequences in common. A new segment is created if this intersection is empty, or below threshold. Referenced segments are logically moved to the head of a containing list. Unreferenced segments are subject to tail deletion.

Here, a new segment is created at the outset. Nevertheless, an example of a RHS, which is a subsequence of this one that adds three to the commonality count for the segment, which is divided by the total number of RHSs in the segment (i.e., normalized) to insure that the number of entries does not skew the result, may be given by:

Action ():
(
#Rifle Instructions: Fire to suppress enemy fire
Quick Action: Move for cover
Psyops: Create decoys
)

Then, the following situation may be pattern matched by the situation for C2 and transformed into the situation for C1. That is, for example:

Situation:
{
1. Occupation: Marines holding enemy-occupied building
2. Situation: Marines under attack by snipers
Air Support: Helicopter gunships
3. Situation: Marines under attack by machine gun fire from enemy placements
4. Reinforcements: less than 10 minutes out
5. Terrain: There are surrounding buildings
6. CASREP: Marine casualties medium
7. Ammo: Marine ammunition medium
8. Rockets: Marine rockets medium
}

is transformed into:

Situation:

{
1. Occupation: Marines holding enemy-occupied island
5. Terrain: Many large boulders on island
Air Support: Helicopter gunships
2. Situation: Marines under attack by enemy bombers
3. Situation: Marines under attack by machine gun fire from enemy fortifications
Situation: Marines under attack by enemy mortars
6. CASREP: Marine casualties high
7. Ammo: Marine ammunition low
8. Rockets: Marine rockets medium
4. Reinforcements: 3-5 hours out
}

with the associated
Action = 0.4:
(
Quick Action: Move for cover
Psyops: Create decoys
#Rifle Instructions: Fire to suppress enemy fire
)

Predicted Success = 88%
Predicted Validity = 99%

Frame names and their associated phrases are attempted to be matched as one descriptive entity. The above numbering scheme shows a candidate association. This association is not necessarily unique.

Next, a RHS transformation (AT1) is given:

C1.RHS ←→ C3.RHS | C1 LHS = C3 LHS:
AT1:
Action C1 (p = 0.4):
(
Quick Action: Move for cover
Psyops: Create decoys
#Rifle Instructions: Fire to suppress enemy fire
#Rocket Instructions: Random fire rockets at enemy fortifications
)

Predicted Success: 88%
Predicted Validity: 99%
Action C3 (p = 0.6):
(
Action: Minimal movement for immediate cover
Action: Radio for air support
#Mortar Instructions: Launch mortars at enemy fortifications
#Rifle Instructions: Fire to suppress enemy fire
)

Predicted Success: 90%
Predicted Validity: 99%

in the context of:
Situation:
{
Reinforcements: 3-5 hours out
Occupation: Marines holding enemy-occupied island
Terrain: Many large boulders on island
Situation: Marines under attack by enemy bombers
Situation: Marines under attack by machine gun fire from enemy fortifications
Situation: Marines under attack by enemy mortars
CASREP: Marine casualties high
Ammo: Marine ammunition low
Rockets: Marine rockets medium
}

This transform case is acquired by the segment with which the LHS has a maximal normalized (i.e., for the number of entries in the segment) number of subsets and supersets in common. A new segment is created if this intersection is empty, or below threshold. Again, referenced segments are logically moved to the head of a containing list. Unreferenced segments are subject to tail deletion.

Here, a new segment is created, since this is a situation and situations and actions lie in distinct segments, by definition. An example of a LHS, which is a subset of this one that adds three to the commonality count for the segment, which is divided by the total number of LHSs in the segment (i.e., normalized) to insure that the number of entries does not skew the result, may be given by:

Situation:
{
CASREP: Marine casualties high
Ammo: Marine ammunition low
Rockets: Marine rockets medium
}

Next, let the following case be present in the assumedly single situational segment.

C4 (basis or n-step induced):
Situation:
{
Light: It is dark outside
Occupation: Marines holding enemy-occupied island
Terrain: Many large boulders on island
Situation: Marines under attack by enemy bombers
Situation: Marines under attack by machine gun fire from enemy fortifications
Situation: Marines under attack by enemy mortars
CASREP: Marine casualties high
Ammo: Marine ammunition low
Rockets: Marine rockets medium
Reinforcements: 3-5 hours out
}

Action (p = 0.7):
(
Quick Action: Move for cover
Psyops: Create decoys
#Rifle Instructions: Fire to suppress enemy fire
#Mortar Instructions: Launch mortars at enemy fortifications
#Rocket Instructions: Random fire rockets at enemy fortifications
)

Predicted Success: 92%
Predicted Validity: 99%

Now, suppose that the following is the supplied context.

{
Weather: It is raining
Light: It is dark outside
Occupation: Marines holding enemy-occupied building
Situation: Marines under attack by snipers
Situation: Marines under attack by machine gun fire from enemy placements
Reinforcements: less than 10 minutes out
Terrain: There are surrounding buildings
CASREP: Marine casualties medium
Ammo: Marine ammunition medium
Rockets: Marine rockets medium
}

Clearly, no case will be triggered. However, hill climbing followed by Type 0 search, until success or interrupt may enable a LHS transformation to result in an applicable case. Here, the RHS of ST1 is a subsequence of the RHS of C4. Moreover, the LHS of C1 is a subset of the LHS of C4. Thus, according to ST1, the LHS of C2 may be substituted for the portion of the LHS of C4 that matches the LHS of C1. The resulting candidate case, C5, follows.

C5 (n-step induced):
Situation:
{
Light: It is dark outside
Occupation: Marines holding enemy-occupied building
Situation: Marines under attack by snipers
Situation: Marines under attack by machine gun fire from enemy placements
Reinforcements: less than 10 minutes out
Terrain: There are surrounding buildings
CASREP: Marine casualties medium
Ammo: Marine ammunition medium
Rockets: Marine rockets medium
}

Action (p = 0.7):
(
Quick Action: Move for cover
Psyops: Create decoys
#Rifle Instructions: Fire to suppress enemy fire
#Rocket Instructions: Random fire rockets at enemy fortifications
)

Predicted Success: 92%
Predicted Validity: 99%

Notice that the LHS of C5 is now covered by the context (i.e., a most-specific covering). A transformational search is conducted for the first transformed case to be covered. The case RHS is next adjudicated for validity. This can be through an oracle or relative frequency counts for the number of syntheses paths.

Suppose that the probability of C5's action being correct is below squelch. Then, a sequence of RHS transformation(s) (AT22) is sought until one successive transformation brings C5 into validity. Here is one such RHS transformation for the sake of argument:

Ci.RHS ←→ Cj.RHS | Ci LHS = Cj LHS:
AT2:
Action Ci (p = 0.7):
(
Quick Action: Move for cover
Psyops: Create decoys
#Rocket Instructions: Random fire rockets at enemy fortifications
#Rifle Instructions: Fire to suppress enemy fire
)

Predicted Success: 92%
Predicted Validity: 99%

Action Cj (p = 0.5):
(
Action: Minimal movement for immediate cover
Action: Radio for air support
#Mortar Instructions: Launch mortars at enemy fortifications
#Rifle Instructions: Fire to suppress enemy fire
)

Predicted Success: 88%
Predicted Validity: 99%

in the context of:
Situation:
{
Light: It is dark outside
Occupation: Marines holding enemy-occupied building
Situation: Marines under attack by snipers
Terrain: There are surrounding buildings
CASREP: Marine casualties medium
}

This LHS is a subset of the situation for C5. Also, the first action for AT2 is a subsequence of the action for C5. Thus, the second action for AT2 may be substituted for the matching first action in C5 with the result:

Action (p = 0.7):
(
Action: Minimal movement for immediate cover
Action: Radio for air support
#Mortar Instructions: Launch mortars at enemy fortifications
#Rifle Instructions: Fire to suppress enemy fire
)

Predicted Success: min (88%, 92%) = 88%

Predicted Validity: min (99%, 99%) = 99%

This case is covered by the context. Again, if the validity is below squelch, another synthesis (chain) is attempted until such relative validation and/or absolute validation is achieved or abandoned. Subsequences and their substitutions are always determinable.

Key to the success of this methodology is to have intersection thresholds set for optimal creativity for the domain(s). The methodology is linearly scalable using parallel processors.

3.3.3   Cyber WAMS

An adaptation of WAMS will demonstrate the efficacy of the concept. Let frames be synonymous with components and their associated phrases be mapped onto component parameters. Next, introduce distinct situational and action components, which compute the same functions as a set of base components. Clearly, this will enable the system to form many situational and action transformations as has been previously illustrated. Thus, a given situation may be mapped by many equivalent combinations of software components to different action components—all of which lead to the same desired actions; albeit, using distinct code realizations. A (dynamic) lexicon is needed to map the semantics of the desired functionality onto a set of semantically equivalent components and their parameters. This allows for a contextual covering to be obtained. Note that such a covering and such action substitutions are not necessarily context free. The attendant benefits of our transformational approach follows.

For example, the situation pattern-matching search known as backtracking can iteratively expand the leftmost node, or the rightmost node on Open [20]. Results here are not identical, but are statistically equivalent. If one component is provided with one expansion search parameter, the other component must be provided with the same search parameter, or the resultant dual-component search will have some breadth-first, rather than strictly depth-first characteristics. This will change the semantics resulting from the use of large search spaces. Clearly, components need to be instantiated with due regard for subtle context to preserve their aggregate semantics.

Similarly, the OS action functions, move-to-the-head and move-to-the-tail can be implemented logically or physically (e.g., for small lists). But, one cannot substitute for one component parameter without substituting for the other, if they share a global data structure; since, logical and physical data movement is based on distinct data structures and operators. Such knowledge is dynamic and context sensitive and thus cannot be properly captured by a diversification engine (multi-compiler), such as that proposed by Mark Bilinski (UCI) [22]. Our transformative methodology will automatically capture valid context-sensitive knowledge, such as illustrated here.

Let the (dynamic) lexicon of Table 3.1 be used for the specification of component-based code and its parameters. There are many ways to specify and index such a lexicon (e.g., using a variant of keyword search). The guiding purpose

Table 3.1 Lexicon for the specification of component-based code and its parameters

No	Logical link	Concept	Component/parameter(s)
1 \<START\>	2	Retrieve the maximum or minimum of a list	**Find**—operates on a list with a parameter to return the maximum or minimum
2	3	Retrieve and destructively remove the specified value from a list of values. This value is used to form a new list in the order of retrieval	**Select**—takes a list and returns the specified value and removes that value from the list. The removed value is concatenated at the head of the list
3	4	Iteratively apply a sequence of arguments to a list until that list is null	**Iterate**—takes a list argument and iterates all contiguous operations in sequence, which affect that list
4	5	Apply an efficient sort technique for use on a serial processor where the elements are not in sorted order and can be relatively large in number	**Quicksort**—efficiently sort a list of at least 21 unsorted elements on a serial processor
5	6	Compute statistical information pertaining to a numerical list	**Statistics**—computes the mean, median, variance, std. deviation, Durbin-Watson statistic, etc. for a supplied numerical list. Must precede any printing of statistics
6	7	Send information to the TTY	**Print**—will print a list, statistics, or defined statements —depending upon the supplied parameter
7	\<END\>	Compute all important statistics and flag the meaning of those included in the generated report	**Stat-Pack**—takes a numeric list and computes the mean, median, variance, std. deviation, Durbin-Watson statistic, etc. Unlike Statistics, provides explanations of the meaning for all generated statistics

is to facilitate the translation of domain-specific concepts onto a space of existing components and their parameters to permit the desired computation to occur. A logical move-to-the-head ordering upon reference may be used to facilitate search based on *temporal locality* [23].

Consider the three cases:

C13:	C14:	C15:
{#Find (list, minimum),	{Quicksort (list)}	{Quicksort (list)}
#Select (list, minimum),		
#Iterate (list)}		
→	→	→
(Statistics (list)	(Statistics (list)	(Print (Stat-Pack
(list))) #Print (statistics)	#Print (list)	
#Print (list))	#Print (statistics))	

C14 and C15 are non deterministic. They induce the RHS non deterministic case transform:

T9 [C14, C15]: (Statistics (list); #Print (list); #Print (statistics)) ← → (Print (Stat-Pack (list)))

C16: T9 (C13): {#Find (list, minimum); #Select (list, minimum); #Iterate (list)} → (Print (Stat-Pack (list)))

C15 and C16 induce the LHS case transform:

T10 [C15, C16]: {#Find (list, minimum); #Select (list, minimum); #Iterate (list)} ← → {Quicksort (list)}

C17: T10 (C13): {Quicksort (list)} → (Statistics (list); #Print (statistics); #Print (list))

Notice that T9 (C17): C15. This would increase the relative validity of C15 and its associated non deterministic action (see below). For the cyber domain, relative validity can be interpreted to mean the preferred method of computation among all *non-compromised* solutions found. The methodology replays and creates dynamic case actions, which are implied by a dynamic situational context. These actions, while syntactically distinct, may or may not be semantically distinct. In either case, the overarching plan of action is preserved. This makes for greater security than syntactic variation alone could achieve. The reason is that the number of candidate attack methods need grow only linearly to counter changes in syntax, but exponentially to counter changes in the underlying semantics—though I/O behavior is preserved. Thus, cyber WAMS can potentially provide exponentially more security than can a multi-compiler by finding multiple paths from start to goal states [20, 22]. Under syntactic differentiation, achieving the same results implies computing the same component semantics. Under transformational equivalence, one need not compute the same component semantics—only ones that achieve the same results in context. Given sufficiently large problem spaces and sufficient computational power, "exponential" cyber security can be had even in the absence of absolute validation!

3.4   The System of Systems Randomization Methodology

To this point, the theory of dynamic representation, based on randomization, has been discussed and exemplified. However, when it comes to implementation, there are some simplifying assumptions, which can be made as follows. The embodied heuristics, which follow, derive from a corner point analysis of scalability and are noted. The resulting System of Systems methodology for randomization is depicted in Fig. 3.2.

Assumption 1   Case pairs of the form, A → C; B → C imply that situational pairs {A, B} may be substituted for one another *without* regard to the action context. Similarly, case pairs of the form, A → C; A → D imply that action pairs (C D) may be substituted for one another as non deterministic actions—*without* regard to the situational context. This follows because the validity-insuring effect of context bears proportion to the specificity (i.e., length) of the transformation along with the specificity of the resident segment, which taken together imply domain specificity. Also, the greater the number of distinct paths by which a case situation and action pairing can be synthesized, the greater the likelihood of validity due to multiple

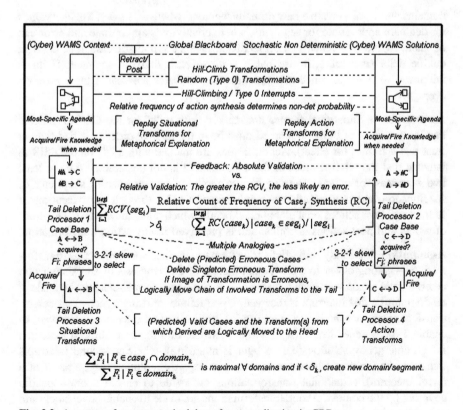

Fig. 3.2   A system of systems methodology for generalization in CBR

analogies. Thus, the need for context checking and the associated computational cycles that it steals tends to disappear with scale. Then, the fact that this is a heuristic method eliminates the need for such context checking altogether.

Assumption 2 Situations and actions are given a frame (component) format to facilitate pattern matching and segmentation as well as descriptive content to instantiate the frame (component) and are of the form, Context-Free Category (Component): context-sensitive phrase instance (parameters).

Assumption 3 Situational and action transforms may be acquired whenever a novel case is acquired. They will be acquired at the logical list head of every segment, which is within tolerance of having a maximal number of situational and action frames in common (see below). They are saved in distinct bases at the logical list head, where they are also moved to when feedback shows them to have been successfully fired. Every domain-specific case base segment is associated with a situational and action transform base segment. Each segment may be assigned one or more processors. In practice, segments are determined by domain types, which are mapped by having a maximal percentage of dynamic sets of situational and action frame (component) names in common. $\frac{\sum F_i | F_i \in case_j \cap domain_k}{\sum F_i | F_i \in domain_k}$ is maximal $\forall$ domains and if $< \delta_k$, create a new domain/segment, where $\delta_k \to 0^+$. Smaller values for delta are appropriate for segments, where creativity is appropriate and error can be tolerated and vice versa. Thus, if feedback shows a symmetric case to be in error, double delta (max of 1.0) else halve delta in each deriving segment. Delta is initialized to 0.5 (i.e., since half a segments predicates are shared by a case, on average).

Assumption 4 Situational frames are marked with a "#" on the LHS to indicate that the sequence of those so marked must be in the relative order specified (i.e., at least two # frames for effect). Action frames are marked with a "#" on the RHS to indicate that the relative order of those so marked is not constrained (i.e., at least two # frames for effect). The use of non monotonic cases enables simplicity of specification and reuse (and thus randomization [12]) through the execution of Retract and Post actions on a global blackboard, which alters the context. The order of retract and post actions is critical and is preserved in any action transformation.

Assumption 5 Initially, hill-climb transformations, i.e., non contracting transformations—getting closer to a situational match, or the synthesis of an existing non deterministic action with each transformation. Such transformations may be made at random until a first interrupt is received. Upon failure, or using a second processor, transformations are made in random order, until a second interrupt is received, to enable Type 0 generality. First interrupts translate into the time allowed to search for an emergency solution (e.g., a tiger is attacking); whereas, second interrupts translate into the time allowed to search for a cognitive solution (e.g., a proof of a novel theorem). Situational transformations are made to enable a most-specific covering by the context. Action transformations are made to enable novel stochastic non deterministic solutions. A count is maintained for the number of times a particular action is synthesized for a particular situation (i.e., case). Dividing this number by the total number of actions synthesized for a particular situation yields the dynamic stochastic probability for a non deterministic selection (i.e., presuming the involved pair of basis cases is valid). Numeric overflow is prevented by normalizing (i.e., rounding) the ratio for the number of surviving distinct non deterministic alternatives (i.e., the denominator). An alternative is deleted if this ratio is below threshold. An associated transform is likewise deleted, or transforms are moved to the logical tail (see below). These alternatives (and associated situations) may be found in distinct segments. Thus, the sum of these results for each containing segment is taken. A non deterministic alternative is created (increased in count) every time a transformation is applied. Thus, it is important to set delta high enough to eliminate spurious and intermediate results. If feedback shows a symmetric non deterministic alternative to be in error, double delta else halve delta in each deriving segment. Delta is initialized to 1.0 (i.e., since this assumes an equal distribution of symmetric non deterministic actions).

$$\sum_{l=1}^{|seg|} non \det prob(i,j) = situ_j \left| \frac{card\{action_i synthesis\}}{(\sum_{k=1}^{|action_k|} card\{action_k synthesis\})/|seg_l|} \right| > \delta_l$$

Assumption 6 Cases already in the base are assumed to be valid until proven otherwise. The relative (i.e., predictive) validity of a case depends on the number of distinct paths for its derivation (i.e., multiple analogies). This number is approximated (with scale) by a *relative count* of the number of times a particular case is synthesized. Thus, using the 3-2-1 skew, cases synthesized from more recently acquired/fired cases are given a higher relative validity, since they are more likely to be repeatedly derived. This makes sense because these solutions are immediately needed and not just stored for possible future use, where validation is less error tolerant. Similarly, the longer the runtime, the higher the relative validity. The specificity (i.e., length) of transformation is not a factor because the system converges to eliminate erroneous transformations (i.e., through relative and absolute validity testing). A Relative Case Validity (RCV) is obtained by dividing the relative count by the sum of this number for all cases in the segment. The sum of these results for each containing segment is taken, if a case may be present in more than one segment.

Segments may be linked and moved to the head when initiated or a case within them is fired. In that manner, segment storage may be reclaimed from segments that fall to the tail over time.

$$\sum_{i=1}^{|seg|} RCV(seg_i) = \frac{Relative\ Count\ of\ Frequency\ of\ Case_j\ Synthesis\ (RC)}{(\sum_{k=1}^{|seg_i|} RC(case_k)|case_k \in seg_i)/|seg_i|} > \delta_i$$

The greater the RCV, the more likely a transformed case is to be valid (i.e., presuming the involved pair of basis cases is valid). A domain-specific threshold may be set for RCV validation. Thus, if feedback shows a symmetric case to be in error, double delta else halve delta in each deriving segment. Delta is initialized to 1.0 (i.e., since this assumes an equal distribution of symmetric cases). Transformed cases whose ratios are below threshold are deleted. An associated transform is likewise deleted, or transforms are moved to the logical tail (see below). Thus, it is important to set delta high enough to eliminate chance cases. Transformed cases are also deleted if not covered by the context as the expected utility of these cases is relatively low. Absolute case validity is predicated on feedback as absolute validity is not necessarily provable [17].

It is good to have cases (including non deterministic alternatives) logically acquired by more than one segment if the intersection metric, RCVi, is within a global tolerance, t, of the segment producing the maximum value, i.e.,

$$\sum_{i=1}^{|seg|} RCV(seg_i) > \delta_i - t \geq 0.$$

This is because different segments will yield different analogies. If this tolerance is set too tightly, then too few analogies will be produced. Conversely, if it is set too loosely, then the associated processors will be slowed by comparison operations, which don't yield (valid) analogies. Thus, if feedback shows a symmetric case to be in error, halve the global tolerance else double the global tolerance in each deriving segment. The global tolerance is initialized to 1.0 (i.e., again since this assumes an equal distribution of symmetric cases).

Case actions can modify the context through the use of a global blackboard mechanism. Case segments may also communicate in this manner. Applicable segments, containing cases covered by the context, are automatically pattern-matched through parallel search without the need to specify any mnemonic segment identifiers. RCVs are multiplicative when non monotonic cases are fired.

Assumption 7 Validation is preferably absolute, but may also be relative using a domain-specific threshold—depending on the specifics of the domain. Erroneous cases and single transforms from which they were derived are deleted. If a longer transformational chain was used to create the erroneous case, the involved trans-forms are logically moved-to-the-tail of their base in their order of application. Thus, these transforms will be among the first to fall off of their tails (i.e., tail deletion) when space is reclaimed as other transforms are logically moved-to-the-head of their base when acquired or fired and validated.

Assumption 8 Validated cases and the transform(s) from which they were derived are logically moved-to-the-head of their base segment in their order of application. Thus, these cases and transform(s) will be more likely to be candidates in the solution of forthcoming problems because of temporal locality [23]. Convergence upon correct cases, and thus correct transforms, and so on follows with scale.

Assumption 9 A 3-2-1 skew may be used to select the order for the application of candidate situational as well as action transforms. Given the logical movement of acquired/fired transforms, a 3-2-1 skew insures that the order of candidate transformations in hill climbing, as well as in Type 0 searches, bears dynamic stochastic proportion to their currency of use. Thus, knowledge tends to be discovered where it is needed and in the time available—just in time discovery. Furthermore, cases and the applied sequence(s) of transformations provide metaphorical explanations for the derivation of any symmetric knowledge.

Assumption 10 This methodology is to be proven by supplying (cyber) WAMS context, teaching it (cyber) WAMS cases, supplying contexts for similar (cyber) WAMS problems—correct actions for which have never been supplied, and seeing if the system converges on finding novel (cyber) WAMS solutions. Such a result not only solves the long-standing generalization problem in CBR [9], but the context-based knowledge extension problem therein, previously described in [9, 10, 24, 25].

3.5   Discussion and Concluding Remarks

The concepts introduced in this chapter are novel, heuristic, and convergent. In particular, the acquisition of heuristic transformation cases determines new target cases, which iteratively determine new transformations. The greater the multiplicity of cases (including their non deterministic alternatives) derived through transformation, the greater their individual reliabilities will be and vice versa (multiple analogies). The definition of a dynamic case transformative system provides for the inference of symmetric cases and transforms. Subsets, supersets, and Cauchy subsequences are used to effect a dynamic representation. Case segments may communicate, using non monotonic cases, through a global blackboard mechanism. Such segments are automatically searched in parallel by context without the use of named segments. Although the methodology is self-referential, it is not subject to the limitations imposed by the Incompleteness Theorem [17]. This is because it is inherently heuristic—not logical in nature. Also, cases are used in preference to rules so that their validity can be independently ascertained and generalized. This is more of a practical than theoretical concern. The methodology may be adapted for use in heuristic state-space search. It is also particularly well-suited to applications in natural language transformation (i.e., semantic understanding), where state-space search is limited due to its inability to frame the problem.

In theory, the only competing way to realize the results promised by this chapter is to apply knowledge to the inference of knowledge. A few approaches here have met with limited success [26]. The problem is that the knowledge, which is self-referential, needs proper context for applicability. This context cannot be generated by the formal system itself due to limitations imposed by the Incompleteness Theorem [17]. Rather, it must be externally supplied, which by definition necessarily makes it an incomplete set, or it must be heuristic in nature (e.g., multiple analogies) to avoid applicability of the Incompleteness Theorem. A multiple-analogies approach underpins this chapter. A theoretical consequence of this heuristic approach is that all non-trivial learning systems must embody an allowance for inherent error in that which may be learned. Thus, despite the seeming appeal of valid deduction systems (e.g., the predicate calculus and Japan's Fifth Generation project [1]), they are inherently not scalable. The Navy requires scalable learning systems (e.g., to support strategic and tactical warfare) as a prerequisite to any financial investment in them.

The proposed technology will be realized in (cyber) WAMS for the generation of novel knowledge for wartime engagement (countering cyber warfare). The problem here pertains to the acquisition of cases for naval battle (cyber) management along with a methodology, which can induce transforms for mapping the situational knowledge onto a supplied context as well as the associated action knowledge onto non deterministic solutions (contextually equivalent components having at least distinct syntax). This methodology will be realized in newLISP (in view of its superior list processing capabilities) as a System of Systems. It will be tested against a sequence of progressively more complex problems for which no solution has been pre-programmed.

The performance of the system will be rated as a function of scale. It will be shown that the inferential error rate is inversely proportional to scale. That is, the larger the domain-specific case bases and the more processing power/time allocated, the lower the inherent inferential error rate. Contemporary cases for WAMS will be obtained through collaboration with the Naval Postgraduate School and/or the Naval War College among other sources, such as [2, 21]. Cases for cyber WAMS will be taken from the newLISP functions used for the realization of the methodology and modified, at a minimum, for syntactic differentiation, while retaining semantic equivalence.

The value of a successful experiment is that as a result, scalable intelligent systems will be able to learn outside the bounds of deductive logics—in the same manner as we humans learn. Moreover, the inference engine allows for *modus ponens*, which is something neural nets can never do. Furthermore, learning by neural nets, genetic algorithms, and scaled-up logics is *NP-hard*. The heuristic learning enabled herein is not only polynomial time, but is not subject to incompleteness [17].

On the pragmatic side, a result of a successful development effort is that creative very large scale decision support systems, for supporting the joint services in (cyber) warfare, become a boldly attainable goal. Strategic as well as tactical solutions to battle (cyber) management problems that are overlooked or bypassed today will be made available and supported by dynamic estimates of inherent error (i.e., the dynamic delta values) and metaphorical reasoning.

This chapter holds the promise of developing a scalable synthetic and creative intelligence. Performer capabilities may be found in Appendix II.

Appendix I: The 3-2-1 Skew

The 3-2-1 skew is a simple (fast) methodology for assigning knowledge relative weights on the basis of Denning's principle of temporal locality [23]. More recent knowledge tends to be proportionately more valuable. This skew is used to increase the likelihood of solving a problem by taking full advantage of the current operational domain profile (Fig. 3.3).

Knowledge is acquired at the logical head and moved there when fired. It is also expunged from the logical tail when necessary to release space. The selection of a

Fig. 3.3 The 3-2-1 skew for dynamic selection in a case-base segment

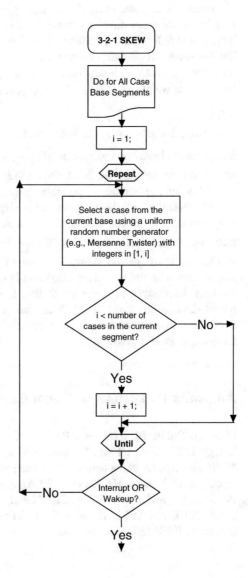

particular skew is domain specific. For example, the rate of radioactive decay is known to be proportional to how much radioactive material is left (excluding the presence of certain metals). The nuclear decay equation may be used as a skew for various radioactive materials and is given by $A(t) = A_0 e^{-\lambda t}$. Here, A(t) is the quantity of radioactive material at time t, and $A_0 = A(0)$ is the initial quantity. λ (lambda) is a positive number (i.e., the decay constant) defining the rate of decay for the particular radioactive material. A countably infinite number of other skews may be applicable.

In the following assignment of skew-weights, the skew vector, **S**, favors the logical head of the case base in keeping with temporal locality. Cases, which were most-recently acquired or fired, and thus appear at or nearer to the logical head of a case-base, are proportionately more heavily weighted under the 3-2-1 skew. Of course, this differs from a uniform skew. The closer a case is to the top of its linked list, the greater its weight or importance. A heuristic scheme (i.e., the 3-2-1 skew) for achieving this with a dependency category consisting of d rules is to assign the head case a weight of $\frac{2d}{d(d+1)}$. The map just below the head map has a weight of $\frac{2(d-1)}{d(d+1)}$.

Finally, the tail map of the segmented case base has a weight of $\frac{2}{d(d+1)}$. The ith map from the head has a weight of $\frac{2(d-i+1)}{d(d+1)}$, for $i = 1, 2, \ldots, d$. For example, using a vector of four weights, the 3-2-1 skew (**S**) is $\mathbf{S} = (0.4, 0.3, 0.2, 0.1)^T$. There are a countably infinite number of possible skews, such that $\sum s_k = 1.0$.

The evaluation of the members of a dependency category is the contiguous weighted sum of its constituent elements. A 3-2-1 skew is defined where the ith map from the head has a weight of $\frac{2(d-i+1)}{d(d+1)}$, for $i = 1, 2, \ldots, d$; where, d specifies the number of terms in the skew. The use of the 3-2-1 skew is optional (i.e., in comparison with uniform weighting) and is useful for domains where the value of the data deteriorates in linear proportion to its time of collection—valuing more recent data, more highly [27]. The use of additional time-dependent weights, depending on whether there is an additional time dependency of the value of the knowledge, is also possible.

Appendix II: Performer Capabilities

Stuart H. Rubin, PI, received a Ph.D. in Computer and Information Science from Lehigh University in 1988. He was previously an ONT Post-Doctoral Fellow, at NOSC, for 3 years and a tenured associate professor of computer science at Central Michigan University. He has over 30 Assigned Navy Patents, over 287 Refereed Publications, and received SSC-PAC's Publication of the Year Awards in 2007, 2009, 2010, and 2011. He is a SIRI Fellow and serves in leadership roles in numerous IEEE technical societies.

Thouraya Bouabana Tebibel received a Ph.D. in Computer Science from USTHB University (Algeria) in collaboration with Pierre & Marie Curie University (France) in 2007. She has been an engineer/researcher for eight years and is now a full professor of computer science at Ecole Nationale Supérieure d'Informatique in Algeria. She has over 80 refereed publications and a book edited by Editions Universitaires Européennes. She has successfully conducted numerous research projects and supervised 30 BS, 17 MS, and 4 Ph.D. theses.

References

1. Feigenbaum, E.A., McCorduck, P.: The Fifth Generation. Addison-Wesley Publishing Co., Reading, MA (1983)
2. Stilwell, A.: SAS and Elite Forces Guide Special Forces in Action. Lyons Press, Guilford, CN (2012)
3. Shi, L-X., Peng, W-L., Zhang, W-N.: Research on reusable design information of shoes based on CBR. In: 2012 International Conference on Solid State Devices and Materials Science. Physics Procedia—Elsevier, vol. 25 pp. 2089–2095 (2012)
4. Fan, Z-P., Li, Y-H., Wang, X., Liu, Y.: Hybrid similarity measure for case retrieval in CBR and its application to emergency response towards gas explosion. Expert Syst. Appl. **41**, 2526–2534 (2014) (Elsevier)
5. Khanuma, A., Muftib, M., Javeda, M.Y., Shafiqa, M.Z.: Fuzzy case-based reasoning for facial expression recognition. Fuzzy Sets Syst. **160**, 231–250 (2009)
6. Sampaio, L.N., Tedesco, P.C.A.R., Monteiro, J.A.S., Cunha, P.R.F.: A knowledge and collaboration-based CBR process to improve network performance-related support activities. Expert Syst. Appl. **41**, 5466–5482 (2014)
7. Lao, S.I., Choy, K.L., Ho, G.T.S., Yam, R.C.M., Tsim, Y.C., Poon, T.C.: Achieving quality assurance functionality in the food industry using a hybrid case-based reasoning and fuzzy logic approach. Expert Syst. Appl. **39**, 5251–5261 (2012)
8. Kumar, Y.J., Salim, N., Abuobieda, A., Albaham, A.T.: Multi document summarization based on news components using fuzzy cross-document relations. Appl. Soft Comput. **21**, 265–279 (2014)
9. Rubin, S.H.: Case-based generalization (CBG) for increasing the applicability and ease of access to case-based knowledge for predicting COAs. NC No. 101366 (2011)
10. Rubin, S.H.: Multilevel constraint-based randomization adapting case-based learning to fuse sensor data for autonomous predictive analysis. NC 101614 (2012)
11. Chaitin, G.J.: Randomness and mathematical proof. Sci. Am. **232**(5), 47–52 (1975)
12. Rubin, S.H.: On randomization and discovery. Inform. Sci. **177**(1), 170–191 (2007)
13. Solomonoff, R.: A new method for discovering the grammars of phrase structure languages. In: Proceedings of International Conference Information Processing, pp. 285–290. UNESCO Publishing House, Paris, France (1959)
14. Solomonoff, R.: A formal theory of inductive inference. Inform. Contr. **7**, 1–22 and 224–254 (1964)
15. Honavar, V., Slutzki, G. (eds.): Grammatical inference. In: Proceedings of Lecture Notes in Artificial Intelligence, vol. 1433. Springer, Berlin, Germany (1998)
16. Fu, K.S.: Syntactic Pattern Recognition and Applications. Prentice-Hall Advances in Computing Science and Technology Series, Englewood Cliffs, NJ (1982)
17. Uspenskii, V.A.: Gödel's Incompleteness Theorem. Ves Mir Publishers, Russian, Moscow (1987)

18. Rubin, S.H.: Computing with words. IEEE Trans. Syst. Man Cybern. Part B **29**(4), 518–524 (1999)
19. Bouabana-Tebibel, T., Rubin, S.H. (eds.): Integration of Reusable Systems. Springer International Publishing, Switzerland (2014)
20. Nilsson, N.J.: Principles of Artificial Intelligence. Tioga Pub. Co., Palo Alto, CA (1980)
21. The U.S. Army Field Manual No. 3–24: Marine Corps Counterinsurgency Field Manual No. 3–33.5. The University of Chicago Press, Chicago (2007)
22. Bilinski, M.: Compiler Techniques as a Defense Against Cyber Attacks. Machine Learning Series, SSC-PAC (2014)
23. Deitel, H.M.: An Introduction to Operating Systems. Prentice Hall, Inc., Upper Saddle River, NJ (1984)
24. Rubin, S.H., Lee, G., Chen, S.C.: A Case-Based Reasoning Decision Support System for Fleet Maintenance. Naval Engineers J., NEJ-2009-05-STP-0239. R1, 1–10 (2009)
25. Rubin, S.H.: Multi-Level Segmented Case-Based Learning Systems for Multi-Sensor Fusion. NC No. 101451 (2011)
26. Minton, S.: Learning Search Control Knowledge: An Explanation Based Approach, vol. 61. Kluwer International Series in Engineering and Computer Science, New York (1988)
27. Rubin, S.H.: Is the kolmogorov complexity of computational intelligence bounded above? In: Proceedings of the 2011 IEEE International Conference on Information Reuse and Integration (IRI), pp. 455–461. Las Vegas, NV (2011)

Chapter 4
Big Data Approach in an ICT Agriculture Application

R. Dennis A. Ludena and Alireza Ahrary

Abstract The advent of Big Data analytics is changing some of the current knowledge paradigms in science as well in industry. Even though, the term and some of the core methodologies are not new and have been around for many years, the continuous price reduction of hardware and related services (e.g. cloud computing) are making more affordable the application of such methodologies to almost any research area in academic institutions or company research centers. It is the aim of this chapter to address these concerns because big data methodologies will be extensively used in the new ICT Agriculture project, in order to know how to handle them, and how they could impact normal operations among the project members, or the information flow between the system parts. The new paradigm of Big Data and its multiple benefits have being used in the novel nutrition-based vegetable production and distribution system in order to generate a healthy food recommendation to the end user and to provide different analytics to improve the system efficiency. Also, different version of the user interface (PC and Smartphone) was designed keeping in mind features like: easy navigation, usability, etc.

4.1 Introduction

Information and Communication Technologies (ICT) in Agriculture is a relative new term. The first time the term was formally used was in The "Tunis Agenda for the Information Society," published on 18 November 2005, and recognized in the World Summit on the Information Society in the same year.

R.D.A. Ludena
Department of Computer and Information Sciences, Sojo University, Kumamoto, Japan
e-mail: dennis@cis.sojo-u.ac.jp

A. Ahrary (✉)
Faculty of Computer and Information Sciences, Sojo University, Kumamoto, Japan
e-mail: ahrary@ieee.org

© Springer International Publishing Switzerland 2016   109
K. Nakamatsu and R. Kountchev (eds.), *New Approaches in Intelligent Control*,
Intelligent Systems Reference Library 107, DOI 10.1007/978-3-319-32168-4_4

We need to begin understanding what is ICT in Agriculture therefore a definition is needed. There is a history of the term, but not yet a formal definition. Therefore, we propose a definition:

ICT in Agriculture is the use but not limited to of the information exchange technologies and frameworks provided by the ICT in the specific field of Agriculture in order to enhance its output by improving their related processes, reducing costs and supporting new businesses opportunities.

These concept needs to be improve but we think it could represent a good beginning. In fact, ICT Agriculture was always a part of Agriculture since ancient times, where farmers used to exchange their produce information in order to generate new business opportunities, exchange knowledge with their pairs, discuss about new farming techniques, find the best market for their produce, and a very long etcetera of activities.

With the coming of the information era and all its related services, agriculture is taking advantage of them in order to procure a better information exchange platform, which increase the number of activities that can be done using this new platform. Among the different benefits farmers can get from the use of ICT we could mention:

1. Improve the relationship with governmental programs

From ancient times farmers, as well as society, work under rules dictated by the regulation authorities. Having a better access in a short period of time to specific relevant information from the government source is a benefit not only from the business point of view but also to continuing following new standards and regulations.

2. Increase business possibilities

In ancient times, the farmer was confined to the local market surrounding its geographical area. With the use of the ICT, farmers can increase its business possibilities. Market produce requirements share their information in common affordable business platforms, into which farmers can access these requirements.

3. Distance disappearance

Having a wide produce market, thanks to the information exchange, farmers can locate their produce in non-local markets. By means of the use of ICT the access to other geographical markets without affecting so much their final price.

4. Supply chain management improvement

Through the use of ICT in Agriculture, the Supply Chain Management currently in operation can be improved drastically in all its processes by the use of the different available tools.

Although, the adoption was driven by different business-related factors, on 2010 the World Bank indicated in their Food Price Watch of 2010 the need of agriculture to introduce more innovative initiatives in order to fight the rising food prices that pushed over 40 million people into poverty. It is important to mention that the adoption of ICT in Agriculture began in developed countries, due that in their early stages the implementation cost of these initiatives was only affordable by these countries.

Currently, due to the decrease of hardware prices and the advent of mobile technology, wireless devices, Internet affordable/high speed connections and the development of the Internet of Things (IoT) have step up the penetration of ICT in Agriculture.

These processes improvements and the benefits that ICT brings to Agriculture will refresh the current market and increase the investments towards better products and conditions to the farmers.

4.1.1   World Penetration of ICT in Agriculture

In the case of Agriculture, what kind of technologies can be applied in agriculture? Specifically, ICT technologies can be represented by any device, tool or application that allows the end user to exchange, share or collect any desire information. The term ICT encloses every technology from satellite connections to mobile devices or sensor networks. New affordable devices enable developing countries to access this technology, allowing farmers to take advantage of the different characteristics previously mentioned.

Fast Internet connections, wireless networks and sensor networks are available in developed countries, but in developing countries Internet access is still limited and not widely available for all the population specially for rural areas as the ones farmers live and work. Mobile phones in the other side have a high penetration in developing countries, Fig. 4.1 show us the penetration mobile phones worldwide, amazingly only 10 % by 2009 will not be connected by a mobile phone.

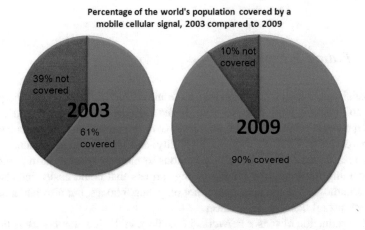

Fig. 4.1 Worldwide mobile phone penetration percentage (ITU)

The main difference is that this mobile phone penetration is much higher in developing countries, although speed and reliability are not as high as the ones available in developed countries, the signal quality is enough to carry basic services.

From the technological point of view services available for Agriculture are but not limited to:

- Media recording. Text, audio/video, graphics, process descriptions, etc.
- Databases dedicated to store/update/reference transactions of the produce amount every farmer has
- Information exchange between the different agriculture process stockholders
- Algorithm development for information extraction at affordable cost and high speed
- Improve interactivity among the different members involved in the agriculture process
- Real time weather forecast and conditions.

Some of the technological applications in use in Agriculture are:

- Wireless technology
- Office automation
- Global Positioning Systems (GPS)
- Geographical Information Systems (GIS)
- Remote sensing
- Remote automation
- Knowledge management systems
- Distance/E-Learning
- E-commerce
- Computer-aided design (CAD)
- Computer-aided manufacturing
- Agriculture resources and service management.

4.1.2 Future Prospects

Geospatial technologies are becoming more affordable in time and are the perfect match to improve land management. Instead of the only use of landscape, a farmscape could be implemented as a layer over the digital map. Some advents of this approach are: food security, sustainability, and land use management.

The use of smart phone based applications to monitor specific farm parameters are becoming more popular among young farmers that could easily interact with these technologies. In the near future, not only smartphones, but also tablet devices will fetch useful data for the farmers.

The introduction of sensor networks in the farm in the IoT framework is the next step in ICT in Agriculture. Sensor networks will provide real time information about the crop conditions, not only using text but images as well. The interaction these data will have with the weather information will allow the farmer to proactively execute different tasks in order to protect the crop quality.

4.2 ICT in Agriculture in Japan

In the past, only industrialized countries were able to afford the high cost required to implement ICT solutions in agriculture. In that context Japan was one of the first countries to adopt and research about the introduction and use of ICT in Agriculture.

Japan represents a unique set of characteristics in the case of farming, among several we can extract the most important:

1. Difficult Geography

Japanese geography is one of the most challenging in the world due to the mountainous characteristic of the land. Japanese farmers had to create innovative methods in order to take the most of the restricted area useful to grow crops.

2. Farmers' aging society

One of the biggest problems for farmers and for the Japan in general is the fast peace growing age society. Specifically, farmers are used to pass their knowledge from generations, but the current scenario is leaving farmers with few or no generations, since their kids decide to migrate to the city on order to find a working position in manufacturing companies. This phenomenon is reducing at a alarming peace the number of young farmers putting in risk the continuity of the business.

3. Globalization

Japanese agriculture has key products in the year, their high quality means high prices in most of them and together with a strict quality control of the produce by the governmental regulator, creates a challenging scenario for farmers all around Japan. This scenario selects the best products at a high price but the ones that slightly do not accomplish its high standards are not selected and farmers have to find the way to commercialize these products.

The former scenario is becoming even more challenging since the Japanese government decided to sign different Free Trade Commerce (FTC) Agreements with different countries, specially in the Pacific Rim area, to provide Japan with different types of produce at lower prices than the local farmers. This increasing competition is making even more challenging the scenario for Japanese farmers.

Later in this chapter we will discuss how Japanese Academia and companies are creating solutions based on ICT in order to improve the condition of Japanese farmers.

4.2.1 Japanese Agriculture in Numbers

According to the Ministry of Agriculture, Forestry and Fishing, the number of Japanese employed in farming dropped to 2.09 million in 2015 from 3.35 million in 2005. About 1.32 million farmers were 65 years or older in 2015.

By comparison, only 318,000 were 39 years old or younger and that number was down by 141,000 from ten years earlier (Fig. 4.2).

Prior to the analysis of the application of ICT in Agriculture it is important to understand the agriculture market in Japan by its numbers.

As we can observe in Fig. 4.3, the agriculture production is driven mainly by rice and vegetables, which they represent around 50 % of the total production output.

The high standard of the produce selection process drove companies and academic research institutions to developed different solutions that allow farmers not only to keep their high selection standard but to allow them to developed new business opportunities for their remain produce.

Because these two main trends in Japan, we will analyze their benefits to ICT in Agriculture.

4.2.2 The Academic Point of View

Japanese Academic institutions decided to use cutting edge technologies in order to provide the better solutions in the near future. Among those technologies are:

1. Sensor networks

The use of different specific-application designed sensor in a network array connected to a high-speed networks, allow information exchange among different business partners. This sensor network is the core of multiple application layers that will be built over the information provided.

Fig. 4.2 Agriculture employed population in Japan

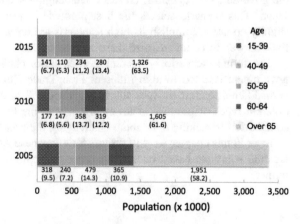

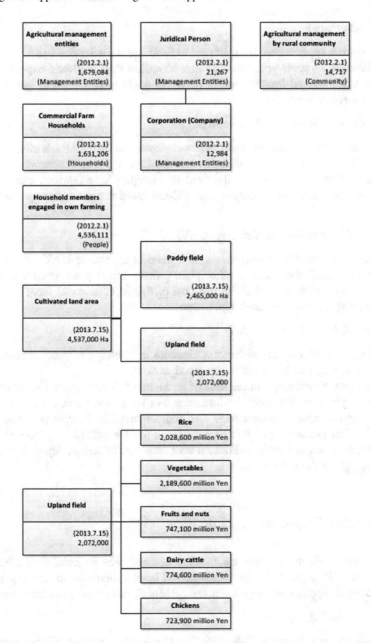

Fig. 4.3 Summary of agriculture production [28]

2. Cloud computing

Many current applications, from storage to Software As a Service (ASA) are using this common affordable platform in order to reduce their hardware expenses. The information retrieved from the sensor networks could be processed or stored in the cloud for future purposes.

3. Augmented Reality (AR) Solutions

Using AR solutions, farmers could retrieve information on their lands effortless and in real-time. The crop information is retrieved from the sensor networks. A good example of this is the use on the field of Google Glasses, which can use the retrieved information previously stored in cloud-based databases, or to process it as well in the cloud.

4. Use of Unmanned Air Vehicles (UAV)

For supervisory/monitory purposes a new trend is the use of UAVs. Among the supervisory applications are: monitor plant growth, insect pests, plant and animal diseases, natural disasters, etc. the low-cost of this solution could serve as a good replacement for satellite-based applications.

5. Control Area Network (CAN)

The data generated by the different agriculture machines, e.g. yield of pesticides, production, etc., can be stored in the cloud as well.

The main disadvantage in the case of Academic ICT for Agriculture projects is the development cost and cost performance. For small size lands, the development cost is high compared to their large size counterparts and the cost performance is low. Another problem academic projects faced are the lack of long term research funding. Due that a highly politicized area like Agriculture in Japan is research funding suffers the same issue.

4.2.3　The Corporate Point of View

Meanwhile academic projects are focused in small size projects, their corporate counterparts focus more on the industrialized level agriculture. In this application area many companies developed different products, among the most important are:

1. Fujitsu Akisai

Fujitsu's Akisai system is a SaaS (Software as a Service) product. It is not exclusive for agriculture purposes but for cattle as well. Akisai is a management system that integrates data generated in the field through sensor network and mobile devices into a much enterprise-like managerial structure. This structure consolidates functions like: business analysis, business administration, production management (agricultural production management, beef production management, cattle husbandry, soil analysis—fertilizer planning and green house environmental control), and Sales (sales management and sales delivery).

This system was launched in 2012 and it is designed to integrate a big number of farmers and its main objective is to locate their produce in a fair business environment. The system has two options depending of the management level farmers want to allow the system to do.

The biggest disadvantage from the point of view of the farmer is the cost. Knowing that farmers are financially struggling due that because they are not able to commercialize the 100 % of their produce at a single price since, paying a substantial amount of money for this service could represent more a problem than a solution if they system can not ensure to commercialize the 100 % of their produce.

Another disadvantage is that farmers have to improve their knowledge regarding some IT tools the system use. In that case farmers can find troublesome the system operation.

2. Toshiba FoodCaster Ex

FoodCaster is basically a sales management service. A key factor for the farmer is to successfully commercialize if not possible close to 100 % of their produce. In this regard, FoodCaster is based in 4 main cycle areas: Plan, Do, Check and Action. The general idea of the system is to seamlessly integrate farmer's produce and potential markets. The system can create easy to understand charts based on the sales and different sales management related parameters. The end user needs to input the required information for the charts to be made together with some automated information the system will retrieve from different authorized third party sources.

As Fujitsu's Akisai, the main disadvantage is the price farmers have to pay to take advantage of the different system capabilities.

3. Agriculture ICT Solution NEC "IT-aided Mikan Production"

"IT-aided Mikan Production" is a project accepted to promote practical research by the Minister of Agriculture, Forestry and Fisheries (MAFF) of Japan. The project is being done by different organizations among them are: Graduate School of Mie University, Mie Prefecture Agricultural Research Institute, Kumano Agriculture, Forestry, Commerce, Industry and Environment Office, etc. The system's target is to create a platform where experienced farmers can share their expertise and knowledge with novice or young farmers. The reason this platform was created is to solve the farmer's aging problem in Japan, where aging farmers cannot find young farmers to share/pass their knowledge to ensure business continuity.

This system is still in the research phase but it is mentioned their intentions to launch as a commercial application and increase the number of crops they system will work on. Figure 4.4 shows the system outline.

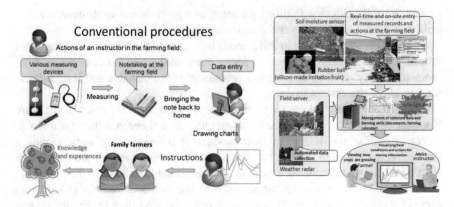

Fig. 4.4 NEC system outline

4.2.4   The Future of ICT in Agriculture in Japan

The use of highly specialized technology platforms is the path Japanese farmers are taking in order to solve some problems that not only affect farmers but the Japanese society in general.

Some Japanese University based laboratories are developing ICT Agriculture applications using remote sensing methodologies; mesh networks, monitoring and visualization of agriculture parameters, systematization of agriculture management, etc.

Japanese technological companies, alone or in tandem with Universities' laboratories, are developing and commercializing ICT Agriculture products, mainly administration, sharing tools based on software/cloud platforms.

4.2.5   Business Applications of ICT in Agriculture in Japan

There are many existing ICT Agriculture applications already working in Japan, most of them run by Companies. One common characteristic of all of them is their approach based on the information exchange to locate produce in different markets.

In order to provide to the farmer the best capabilities of both worlds, the academic side and the company side, we propose a new project called: "Novel nutrition-based vegetable production and distribution system".

In this new project the main characteristics from the academic side that are:

1. Based on research
   The project itself is a consortium between Professor Ahrary's laboratory at Sojo University in Japan and companies in the Kyushu area. The integration of the system is in charge of Sojo University.

2. Open criteria to test new ideas
   As in any academic research, different approaches can be presented and tested in order to find the most suitable for the project and its purposes.
3. High integration capability
   In this project a particular characteristic of academic work is used, integration, which is applied between specific tasks done by local companies and Sojo University.

From the enterprise side the main used characteristics are:

1. Specific knowledge
   In order to successfully complete the project, specialized companies are needed.
2. Proved experience
   This characteristic is a key factor for the project purposes. That experience will serve as a good background for future improvements.

Farmers are facing specific challenges due to different factors (globalization, population decreasing, etc.). Local Kyushu farmers' situation is not different.

The presented project tackles the following issues:

1. Lack of a market where farmers can commercialized their produce
   The system will open new markets for the farmer, not only the ones provided by normal channels, but new ones based on the use of technology
2. New market member's Information integration
   The creation of new databases of the new market members will allow future markets integration, in a regional, national or international structure.
3. Integration through technology
   Using cutting edge technology, as well as traditional methodologies, the project brings an effective support to farmers and other project members
4. Social integration and collaboration
   The project involves social groups that only met in specific opportunities into a much more interactive and social environment supported by technology.

4.3 Our Solution's Perspective

The main objective of the "Novel nutrition-based vegetable production and distribution system" is the creation of a new commercialization platform supported by technology to integrate farmers and end users in a much more social environment where information exchange is the key factor.

In order to create this platform different steps in different stages must be taken, in this section we proceed to explain the steps taken so far.

4.3.1 Questionnaire Based Analysis

In order to understand better the potential market for the proposed system, a series of questionnaires were purposed. In this section we proceed to present some of the most important ones (Table 4.1).

Figure 4.5 shows the general use of Internet regarding the search of agriculture related information, which is expected in a highly interconnected society as Japan.

Figure 4.6 show the focused information related to the Kumamoto user, which its 90 % of usage means a highly available market.

Table 4.1 Group universe

Farmers	30 people
Restaurants	30
End users	162 people

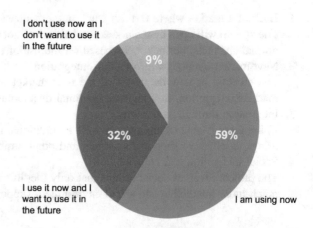

Fig. 4.5 Information gathering

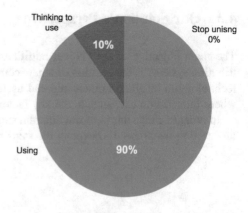

Fig. 4.6 Use of the Internet regarding vegetables in Kumamoto

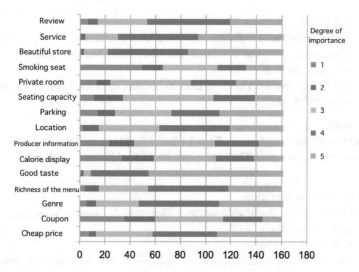

Fig. 4.7 Important points to select a shop or restaurant

Figure 4.7 show a general profile of the Kumamoto user regarding its preferences about a shop or restaurant. This statistic is important to understand in order to know how to position the product in the future.

4.3.2 Project Member Definition

As a first step, the members their functions and interconnections must be defined in order to create the project's model. In our case are:

1. Farmers
   Farmers represent no only the main users of the project's benefits but the first and most important information provider. Based on the information provided the system could begin to make the matching among members.
2. End users
   End users are, as in any application/project, the most important members and the one that the application is made for. In this project the end user represent one of two users of which the application is made for.
3. Restaurants
   Restaurants represent the "ready meal" option for the end user in which the ingredients are from farmer's produce.
4. Knowledge based database creators
   During the project development 2 knowledge based databases were created.

4.1. Nutritional information
   User's nutritional information, like: user's physical information, status
   information, physical condition, nutritional requirement, etc.
4.2. Food information
   Food/vegetable information, like: nutritional calories, traceability, seasonal
   information, etc.

5. Integrator
   In this case the integrator is Sojo University, which will receive all the information previously presented and generate two additional databases:

   5.1. Platform information
      Details about the user's device, like: mobile, tablet, Web, downloaded
      application, etc.
   5.2. Attribute information
      Information the user input in the registration step, like: gender, age, family
      structure, etc.
   5.3. Recommendation algorithm generation
      Using the information from the different databases (knowledge based,
      platform information and attribute information), a recommendation algo-
      rithm will be designed in order to provide the end user with most accurate
      and fastest answer.

   A project perspective is shown in Fig. 4.8.

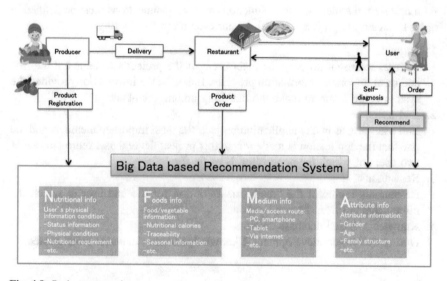

Fig. 4.8 Project perspective

4.3.3   Functions Among Project Members

Information wise, Fig. 4.9 shows the information flow among all project members. Double line indicate the information the system will receive:

1. From the vegetable producer, the system will receive periodic product availability updates
2. From the user, the system will receive, through a web app, the different symptoms the user has
3. From the restaurants, the system will receive the available daily menu based on the vegetables produced locally

Single line indicate the information the system will provide to the different members:

1. To the vegetable producer, the system will send an order (if the user decides so) to deliver the vegetables to the user
2. To the restaurants, the system will send an order to deliver the meal based on the vegetables produced locally
3. To the user, the system will send first, the recommendation for the healthy food based on the symptoms previously introduced, (vegetables or meal) and then the order confirmation

Dashed line represents the transportation of the product to the user, whichever is the option requested.

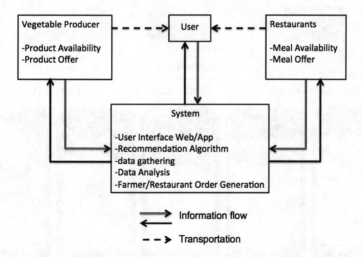

Fig. 4.9 Information flow among members

System functions are:

1. To gathered the following information: vegetable producer availability, restaurant menu availability, interact with and store all the information (symptoms) the user will provide, generate an acute food/meal recommendation.
2. To stored all the transaction information
3. Perform specific Data Analysis based on the data-at-rest and data-in-movement
4. Through the results interpretation, improve the accuracy and effectiveness of the system [1, 2].

Fig. 4.10 User interface designs

4.3.4 Interface Design

Because of the limited screen size on the Smartphone edition of the user interface, priority was given to the user's information input.

A high visibility and easy to navigate interface was designed with all navigation features like: search from symptoms list and nutrients. For the PC version, it was designed in order to allow making an easy icon selection with a reduced directory structure. For display purposes, e.g.: food, nutrients, symptomology, etc. that contains large contents a matching program was built in order to provide specific information to the user's needs (Fig. 4.10).

4.4 Big Data Approach

The main target to the project is to deliver a precise and fast meal recommendation to the end user. Therefore the system must fulfill the following parameters:

1. Quick interaction among heterogeneous databases to retrieve specific data
2. Allow the implementation of different algorithms
3. Elasticity
4. Growth capability
5. Heterogeneous device integration

Due to the project needs, a Big Data approach was selected because it could accomplish all the required parameters and also it could serve as a good test platform for other similar projects.

Prior to the development we proceed to make an analysis of Big Data and some of its issues. The analysis' outcome will be useful for this project targets and for future projects where Big Data could improve the project outcomes.

4.4.1 Data Availability

Currently, the broad use of social networking is related to almost all daily activities in any given society, e.g. Facebook, Twitter, Flickr are daily use applications that their use and the result of their use could go beyond the social environment. The continuous use of these social application generate large amount of information, including: personal information (depending on the Security privacy measures configured in every user account), activities, location, posts, etc.

In order to retrieve this information, the use of Application Program Interfaces (APIs) is a must. In the particular case of social sites, every platform releases their APIs to retrieve information from their databases following the platform guidelines.

It is well known that some of these APIs retrieve specific keywords from the databases meanwhile other have the necessary attributes to retrieve any information.

The difference among these APIs makes the quality of the retrieve information not as desired, e.g. only 1 % of public twits are available through the official API [3].

Above of all, these sample data is only available to some companies and startups, but only few researchers have access to this information. The same behavior is observed in the rest of the so-called "big" companies in the Internet, making the data not representative of the Universe, which it belongs.

4.4.2 Security

Security is a general underestimated area in almost all Information Technology (IT) areas, reactive could be the closest definition to describe the kind of behavior companies have towards security threats all around the world.

In the particular case of Big Data, Security is strongly related to the protection of user's privacy. Most of users they have different accounts in different social platforms. Using APIs and other tools is possible to establish a pattern among these different accounts. Now, how detailed could be the result of this information retrieval will be directly related to how many details could be retrieved through these APIs. These resulting patterns or behaviors could be use for individuals or groups to define specific patterns for different purposes, e.g. marketing, social behavior analysis, prediction of future behaviors/patterns, etc. [4].

In the recent decade the number of attacks related to steal user's information increased dramatically and it will continue growing since the social engineering methodologies are becoming more aggressive and companies do not change their behavior towards information security.

4.4.3 Ethics

While Big Data is an emerging trend, there is a little understanding about the ethical implications on the research being done. A very good example is the case of a research done in Harvard in 2006, in which a collection of colleagues' Facebook profiles were used to make a study of how was their friendships and interests over time [5]. The collected data was publicly released to other researches to conduct experiments on that, but they realized it was possible to de-anonymize part of the data set.

It is important that scholars reflect on the importance of accountability in order to act in ethical manner. Regarding Big Data, accountability is strongly related to their research field as well as to their research subjects. Being accountable for a specific action will entitle to specific set of rules that must be kept in mind in any step done towards the realization of the research specific goals and objectives [6].

Those characteristics make accountability a broader concept than privacy [7].

4.4.4   Inaccurate Interpretation of Big Data Results

As a large mass of information, Big Data is not self-explanatory. And yet the specific methodologies for interpreting the data are open to sort of technological and philosophical debate. Can the data represent an "objective truth" or does some subjective filter necessarily bias any interpretation or the way the data is "cleaned" [8].

Every researcher represents a data interpreter [9]. In the corporate side, data interpretation and representation depends on the non-standardize way the company decided to use. From the technological point of view, Big Data performs its analysis using tools that are freely available in the internet such us: R, Hadoop, Pig, etc., together with some statistical analysis that could shape the date the ways the researcher want it. Big Data's result interpretation must be done with the less possible bias in order to keep an ethical behavior towards them.

4.4.5   Data Set Equivalence

There is a general assumption that small sets' analytical methodologies could be applied in large data sets, which in some particular scenarios is possible, but in the majority of cases those analysis can not be applied. This issue could be describe as follows:

a. Cleaning process of collected data

   During the data collection process, so called "no interesting" data is also gathered. These data must be cleaned in order to have a well-defined data set for further analysis. This cleaning process must be done preserving the integrity of the useful data [10].

b. Assuming that the former point was successfully accomplished, the next step is the metadata generation, which should describe in the most faithful way the data set. The most important points metadata must show are how it is recorded and measured. Prior to the metadata generation, or if possible prior to the data collection, the data format to be used must be decided or, if required, its development and implementation. The reason for this early adoption is the different particularities every research topic has. Another issues that make more challenging this decision process is: the wide research topics available to choose from and its relation with the industry. An important issue here is data provenance. It is necessary to know where the data was generated in order to trace forward and backwards its accountability [11]. Having different metadata from different sources and trying to find its correspondence with another data set represent a big challenge due to the lack of standardization showed before, and trying to find a correlation between two different data set formats could affect the granularity needed to get the best results of the applied analysis.

4.4.6 Big Data Applications

Currently, big Data methodologies are being applied in many different science and industry applications, from Computer Science to marketing research and Industrial management. Elsevier's "Research Trends" 2012 publication shows different statistics based on publications found under the "Big Data" keyword search, (Fig. 4.11). From the figure we can observe that the research areas where Big Data is applied are from Computer Science to Art and Humanities [12].

About industry, the most applications areas of Big Data are: Supply Chain Management, Environmental Applications, Finance, Smart Grid, etc. Because of the wide range of applications where Big Data could be useful, Big Data is becoming a recurrent name in business applications from large size industries to small size ones. A consequence of that recent popularity is the newborn term "Data Scientist" [13].

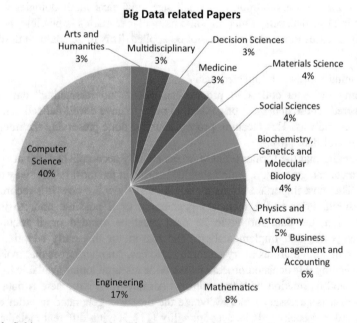

Fig. 4.11 Subject areas researching Big Data, Elservier's Research Trends

4.4.7 Project Security

Since the project is based on the integration of different technologies, one useful addition to the project is the uses of Sensor networks to be used in the different crops farmers have in their lands. The integration of Sensor networks into the present outline of the project represents a challenge but at the same time its application could bring a much richer data to the project in order to bring a better service to the end user and to provide much meaning statistics to the farmer about its produce performance.

Sensor networks belong to the relative young research area of Internet of Things (IoT), but the projects that are using or including this new paradigm are increasing exponentially. Therefore, IoT is in its infancy, the same happens with the issue of securing the exchanged information in the wireless networks that represent the core concept of IoT.

Because of the heterogeneous device characteristic of IoT networks, standardization represents one of the main issues to be solved before implementing this kind of infrastructure.

Table 4.2 shows the different efforts made by private organizations towards the creation of an IoT standard.

Because of the wireless nature, IoT networks are vulnerable to different kinds of attacks e.g. eavesdropping in addition to physical attacks, since the devices included in these networks will be unattended for long periods of time after installation.

For the project purposes, and also for any IoT related project is necessary to secure the exchanged information. The IoT automatic update environment must be secure in order to create confidence among users and above of all to create a reliable environment for the farmer to share its produce and other relevant information.

Table 4.2 Main standardization efforts

Standard	Objective	Comm. range (m)	Data rate (Kbps)
ZigBee	Based on the IEEE 802.15 Standard, designed to deliver a low cost, low data rate, long battery life communications under a reliable network	10–100	$\sim 10^2$
GRIFS	Support action project founded by the EU to maximize the interoperability of RFID	~ 1	$\sim 10^2$
EPCglobal	RFID technology integration into the electronic product code (EPC) framework	~ 1	$\sim 10^2$
6LoWPAN	Low power IEEE 802.15.4 devices integration into the IPv6 networks	10–100	$\sim 10^2$
Wireless Hart	Developed as a multi-vendor, interoperable wireless standard, WirelessHART was defined for the requirements of process field device networks	10–100	$\sim 10^2$

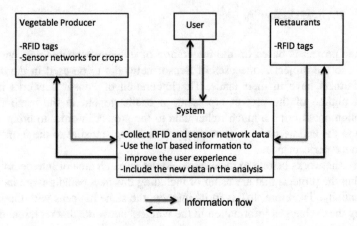

Fig. 4.12 Information exchanged by the project actors from the IoT point of view

Figure 4.12 shows the information flow of the project from the IoT point of view. It is necessary to mention that this is a first approach towards a much more geographically wide future implementation.

4.4.8 Project Security Definitions

For the specific purpose of our project, a general security IoT approach is analyzed in order to provide the necessary guidelines to establish where are the main security points and how to address their security requirements. Definitions are shown in Table 4.3.

As we could observe most of the system needs to be authenticated in order to provide a reliable information exchange. Using current technologies these requirements could be achievable but further analysis and research is required in the area of IoT security in order to provide a solid approach towards a secure information exchange [14, 15].

4.4.9 Possible Threats Against the Project Infrastructure

For the specific requirements of the project security analysis will be focused on two main aspects: authentication and data integrity.

Authentication and data integrity are two of the major problems related to IoT-based projects. Authentication is difficult due to its infrastructure requirements and servers that perform the authentication process through the information exchange with other nodes. In an IoT environment this could result not practical

Table 4.3 System security definitions

From	To	Security requirement
Vegetable producer	System	Remote access
		User identification and authentication
		Device identification and authentication
System	Vegetable producer	Remote access
		Device identification and authentication
Restaurant	System	Remote access
		User identification and authentication
		Device identification and authentication
System	Restaurant	Remote access
		Device identification and authentication
User	System	Remote access
		User identification and authentication
System	User	Communication integrity

due to that RFID tags cannot exchange too many messages with the authentication servers because their issues related to energy management and messages standardization, this issue applies as well to sensor networks. Energy issues are one of the most difficult to overcome in RFID networks as well as in sensor networks, due that there are scattered over a wide area and sometimes unmonitored, energy management is a key factor in order to ensure a long device life as well as usability. In the same manner, some authentication protocols could not be used due to their lack of standardization [16].

In this context, several approaches were developed for sensor networks. In these cases gateways that are part of the sensor networks are required to provide connection to the Internet. In the IoT scenario sensor nodes must be seen as nodes in the Internet, therefore authentication is required in order to differentiate them from sensors in the same area but not belonging to the same network. In the case of RFID several approaches were presented, but most of them have serious issues, some of them mentioned in [17–20].

The "Man in the middle" attack is considered one of the biggest threats against wireless networks as well to IoT networks. Data integrity solutions should guarantee that the data in transaction cannot be modified and the system must be able to detect this situation. Data integrity as a issue has been extensively studied for standard network applications and communication systems and early results are related to sensor networks [21, 22]. But, when a RFID networks with their own unique characteristics are included in the current Internet paradigm, different problems arise as well as unforeseen problems related to their use. Several approaches are developed or under research to solve the different new RFID related issues i.e. EPCGlobal Class-1 Generastion-2 and ISO/IEC 18000-3, both of them working in different process to protect the device memory. These approaches also consume large amount of the resources in encryption processes needed. The main used resources are: energy and bandwidth, both of them in the destination. Therefore, even using these approaches specific related problems with RFID still remain [23–27].

4.5 Conclusions

1. Japan was one of the pioneers using ICT in Agriculture. The reasons for this early adoption of ICT in Agriculture are: to help farmers to commercialize their produce, to share their knowledge among young/novice farmers, and to establish a much direct link between some potential buyers and farmers.
2. Japanese companies developing commercial ICT in Agriculture applications, are mainly focus in the development of a state of the art platform into which different kind of data will be shared, e.g. produce statistics, sensor-based information of the crops, weather information, market information, etc. the main advantage of this platform is the creation of a marketplace for agriculture purposes. Its main disadvantage is the cost involve for the farmer to use this system and the need of the farmer to use time to input specific information into the system.
3. Research centers in Japan are also using state of the art technology to generate a different approach for the farmers' produce; remote monitoring and the use of predictive algorithms are two of the main tools used in their development. Japanese academic institutions see ICT in Agriculture as a potential development area; since Agriculture includes several of the many challenges present in Japanese society. Finding a solution to solve some farmers' problems could also used to solve some of society issues.
4. The "Novel nutrition-based vegetable production and distribution system" project under development in Sojo University, represents a bold approach to a new business platform, where the output's value benefits not only the farmers' side but to the user which takes advantage of the produce benefits in order to solve simple symptoms.
5. The use of techniques such as Big Data in the project development allow us to have a elastic platform where growing will not represent a problem and future additions to the basic platform will be easy to implement.
6. Based on the diverse Big Data analysis available, we can find out hidden symptoms distribution patterns that could be useful for further analysis by the medical sector.
7. The use of sensor networks as an improvement of the project represents a much wider approach into the Internet of Things (IoT), where sensor networks installed and later unattended interact in real time with devices present in their environment as well as remote data transfer and diagnostics.
8. The IoT paradigm is still in its infancy, as well as most of the technology that surround and defines it. Therefore, standardization being one of its most important issues needs a solution, making this area rich for development.

9. Security, as well as standardization, represents a challenging task in IoT. Because of its heterogeneous device characteristic and the use of many proprietary communication protocols among those devices, creating or implementing a common secure platform with well-defined administration rules and metrics is not only a challenging task but an important to provide confidence on the system to the end user and farmers.

10. As a general rule, not only for our project purposes, it is necessary to define the security needs of every information exchange point priory to the implementation. The former becomes even more important in a challenging environment like IoT, which our project is going towards to.

11. Making a proper security definitions will allows to correctly focus efforts and technology into the most critical communication links in our project infrastructure. These definitions will also allow us to define specific metrics to be observing as a definitive definition of our system behavior.

12. As a difference of traditional wired mesh networks, IoT represents a new challenge from the Security point of view, but at the same time the threats related to IoT could be shortlisted.

References

1. Ludeña, R.D.A., Ahrary, A.: Big Data approach in an ICT agriculture project. In: Proceedings for the 5th IEEE International Conference on Awareness Science and Technology (iCAST 2013), Aizu-Wakamatsu, Japan, pp. 261–265 (2013)
2. Ludeña, R.D.A., Ahrary, A.: A Big Data approach for a new ICT agriculture application development. In: Proceedings for the 2013 International Conference on Cyber-Enabled Distributed Computing and Knowledge Discovery (CyberC 2013), Beijing, China, pp. 140–143 (2013)
3. Meeder, B., Tam, J., Kelley, P.G., Cranor, L.F.: RT @IWantPrivacy: widespread violation of privacy settings in the twitter social network. In: Proceedings of Web 2.0 Security and Privacy (W2SP 2011), Oakland, CA, USA (2011)
4. Manovich, L.: Trending: the promises and the challenges of big social data. In: Gold, M.K. (ed.) Debates in the Digital Humanities. The University of Minnesota Press, Minneapolis (2011)
5. Zimmer, M.: More on the "Anonymity" of the Facebook dataset—It's Harvard College. MichaelZimmer.org Blog (2011)
6. Beyer, M.A., Laney, D.: The Importance of "Big Data": A Definition. Gartner, Stamford (2012)
7. Jacobs, A.: The pathologies of Big Data. Commun. ACM **52**(8), 36–44 (2009)
8. Bollierm, D.: The Promise and Peril of Big Data. The Aspen Institute, Communications and Society Program, Washington (2010)
9. Latour, B.: Tarde's idea of quantification. In: Candea, M. (ed.) The Social After Gabriel Trade: Debates and Assessments, pp. 145–162. Routledge, London (2011)
10. Agrawal, D., Bernstein, P., et al.: Challenges and Opportunities with Big Data. Community white paper, Purdue University, West Lafayette, Indiana (2011)
11. Gitelman, L.: Raw Data Is An Oxymoron. Massachusetts Institute of Technology Press, Cambridge (2013)
12. Research Trends: Special Issue on Big Data, Elsevier, Issue 30 (2012)

13. deRoos, D., Eaton, C., Lapis, G.: Paul Zikopoulos, Tom Deutsch, "Understanding Big Data: Analytics for Enterprise Class Hadoop and Streaming Data. McGraw-Hill, New York (2012)
14. Ludeña, R.D.A., Ahrary, A.: Big data application to the vegetable production and distribution system. In: Proceedings for the 2014 IEEE 10th International Colloquium on Signal Processing & Its Applications (CSPA 2014), pp. 20–24, Kuala Lumpur, Malaysia (2014)
15. Ahrary, A., Ludeña, R.D.A.: Big Data approach to a novel nutrition-based vegetable production and distribution system. In: Proceedings of The International Conference on Computational Intelligence and Cybernetics (CyberneticsCom 2013), pp. 131–135, Yogyakarta, Indonesia (2013)
16. Savola, R.M., Abie, H.: Metrics-driven security objective decomposition for an e-health application with adaptive security management. In: Proceedings of the International Workshop on Adaptive Security (ASPI 2013), Article No. 6, New York, NY, USA (2013)
17. Cirani, S., Ferrari, G., Veltri, L.: Enforcing security mechanisms in the IP-based internet of things: an algorithmic overview. Algorithms 6(2), 197–226 (2013)
18. Ning, H., Liu, H.: Cyber-physical-social based security architecture for future internet of things. Adv. Internet Things 2(1), 1–7 (2012)
19. Sivabalan, A., Rajan, M.A., Balamuralidhar, P.: Towards a light weight internet of things platform architecture. J. ICT Stand. 1, 241–252 (2013)
20. Weber, R.H.: Internet of things-new security and privacy challenges. Comput. Law Secur. Rev. 26(1):23–30 (2010) (Elsevier)
21. Acharya, R., Asha, K.: Data integrity and intrusion detection in wireless sensor networks. In: Proceedings of IEEE ICON 2008, New Delhi, India (2008)
22. Juels, A.: RFID security and privacy: a research survey. IEEE J. Sel. Areas Commun. 24(2), 381–394 (2006)
23. Floerkemeier, C., Bhattacharyya, R., Sarma, S.: Beyond RFID. In: Proceedings of TIWDC 2009, Pula, Italy (2009)
24. Sung, J., Sanchez Lopez, T., Kim, D.: The EPC sensor network for RFID and WSN integration infrastructure. In: Proceedings of the Fifth IEEE International Conference on Pervasive Computing and Communications Workshops, pp. 618–621 (2007)
25. Commission of the European Communities, Early Challenges Regarding the "Internet of Things" (2008)
26. Kushalnagar, N., Montenegro, G., Schumacher, C.: IPv6 over low-power wireless personal area networks (6LoWPANs): overview, assumptions, problem statement, and goals. In: IETF RFC 4919 (2007)
27. Weiser, M.: The computer for the 21st century. ACM SIGMOBILE Mob. Comput. Commun. Rev. 3(3), 3–11 (1999)
28. Monthly Statistics of Agriculture: Forestry and Fisheries, Japanese Ministry of Agriculture Forestry and Fisheries (2014)

Chapter 5
Intelligent Control Systems and Applications on Smart Grids

Hossam A. Gabbar, Adel Sharaf, Ahmed M. Othman,
Ahmed S. Eldessouky and Abdelazeem A. Abdelsalam

Abstract This chapter discusses advances in intelligent control systems and their applications in Micro Energy Grids (MEG). The first section introduces a PID fuzzy model reference learning controller (FMRLC) implemented in the control loop of Static VAR Compensator (SVC) to stabilize the voltage levels in the islanded mode of the microgrid. FMRLC performance is compared to the conventional PI controller. The introduced results show that SVC with FMRLC has better capabilities to compensate for microgrid nonlinearity and continuous adaptation for dynamic change of the microgrid's connected loads and sources. The second section discusses performance optimization of micro energy grid. A recent heuristic optimization technique, called Backtracking Search Optimization Algorithm (BSA), is proposed for performance optimization of micro energy grids with AC/DC circuits, where it is used for the selection of the scheme parameters of PWM pulsing stage

H.A. Gabbar (✉)
Energy Safety & Control Lab, University of Ontario Institute of Technology,
2000 Simcoe Street North, Oshawa, ON L1H7K4, Canada
e-mail: hossam.gabbar@uoit.ca

A. Sharaf
Sharaf Energy Systems Inc., 147 Berkley Drive, Fredericton, NB E3C 2P2, Canada
e-mail: profdramsharaf@gmail.com

A.M. Othman
University of Ontario Institute of Technology, 2000 Simcoe Street North, Oshawa,
ON L1H7K4, Canada
e-mail: ahmed.abdelmaksoud@uoit.ca

A.S. Eldessouky
Canadian International College, Electrical Engineering Department,
New Cairo Campus: Land 6, Center Services Area, Fifth Settlement, New Cairo, Egypt
e-mail: ahmed_eldessouky@cic-cairo.com

A.A. Abdelsalam
Department of Electrical Engineering, Faculty of Engineering, Suez Canal University,
El Salam District, Ismailia, Egypt
e-mail: aaabdelsalam@eng.suez.edu.eg

A.M. Othman
Zagazig University, Zagazig, Egypt

© Springer International Publishing Switzerland 2016   135
K. Nakamatsu and R. Kountchev (eds.), *New Approaches in Intelligent Control*,
Intelligent Systems Reference Library 107, DOI 10.1007/978-3-319-32168-4_5

used with a novel Distributed Flexible AC Transmission System (D-FACTS), type called Green Plug-Energy Economizer (GP-EE). The following section includes simulation models and results to illustrate the merits of the proposed intelligent control designs and their use to achieve high performance micro energy grids.

Keywords Static VAR compensator · D-FACTS · Micro energy grids · BSA · Fuzzy logic control · Model reference control

5.1   Introduction

Artificial Intelligence (AI) is the area of computer science focusing on creating hardware/software that can mimic human behaviors of intelligence by memorizing, acquiring, modifying and extending knowledge, and applying reasoning, in order to solve problems and adapt based on new situations. The ability to create intelligent system started after 50 years of research into AI programming techniques. Advances in processors (speed and power), memory devices, and programming (architecture and techniques) brought the dream of smart system to reality. AI systems are algorithms that use both hardware technology and software techniques to mimic human thought and way of thinking. Computers by their own cannot reason and have no logic thinking. Studying and developing AI algorithms have different objectives. Scientists' main concern is to better understand how human solve problem. Engineers' objective is implement AI algorithms to solve real world problems by simulating human solving problems capabilities.

AI algorithms family includes Expert Systems (ES), Fuzzy Logic (FL), Artificial Neural Network (ANN), Genetic Algorithms (GA) and hybrid algorithms that implement a combination of two or more of those algorithms as shown in Fig. 5.1. Each of those algorithms has its own feathers that fit different engineer applications of different problem domain and different objectives.

Expert systems are intelligent systems that depend on logic thinking to emulate the decision making of human expert. They are based on classical set theory with sets of sharp boundaries and rule-base formed by if-then statements. The main advantage of ES is the ability to encode experience of multiple expertise in one knowledge domain into its knowledgebase. Accordingly, desired improvement of

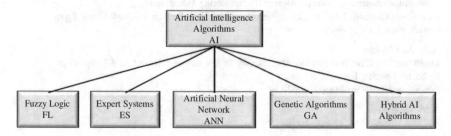

Fig. 5.1 Artificial intelligence family

Fig. 5.2 Structure of artificial
neural network

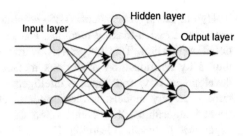

decision making based on wide knowledgebase, which can never be achieved by single expert, can be achieved by ES. The feature that can distinguish ES among other AI algorithms is the ability to comprehend its knowledge structure and have explanation facilities for its conclusions and decisions. That is because its knowledgebase is encoded linguistically. More features have been added to ES to dynamically improve its knowledgebase by rule induction and generalization tools.

Fuzzy logic is a tool to handle imprecise or ambiguous knowledge. It is similar to ES in its structure and inference mechanism. However, in FL, classical sets are replaced by fuzzy sets that have gradual boundaries allowing gradual transition between sets. Consequently, FL is less sensitive to uncertainty and noisy inputs offering better presentation of systems with those characters. Similar to ES, FL knowledgebase is presented by linguistic variable and if-then rules at the rule-base structure that preserve the ability to inject knowledge directly from expertise and comprehend conclusions and decisions.

Artificial Neural Network has analogy structure to the human brain. It consists of processing units (neurons) with nonlinear activation function interconnected by weighting links and arranged in multiple layers as shown in Fig. 5.2. This parallel processing structure of ANN allows high-speed processing that distinguishes it among all other AI algorithms. Each neuron represent infinitesimal portion of the nonlinear surface of the system to be modeled while connected weighting links represent the contribution of this portion to the overall surface. The learning mechanism of ANN adjusts the weights of interconnected links to match nonlinearity of the system to be modeled providing machine learning capabilities. ANN is a black-box modeling tool that can complement and support the understanding of human experiences.

Genetic Algorithms are based on the principle of genetics and natural selection "survival for the fittest". They adapt time searching techniques to select the most suitable solution that has the highest fitness with considering certain constraints. GA can be considered main entrance for heuristics techniques. Contrary to conventional search techniques that stuck in local minima, genetic algorithms are search tools that grantee global minimum in highly nonlinear search surface. The mutation process involves forcing the search process to jump to unexplored search area and avoid trapping the search in local minima. Despite of their advantages, Genetic Algorithms cannot be used in on-line processing because of their sequential processing nature. They are best working in off-line processing in higher level of optimization and control hierarchy.

Hybrid Intelligent Systems (HIS) combine different knowledge implementation schemes, decision-making modules, and learning strategies to handle a computational application. This integration overcomes the constraints of individual techniques by hybridization of various methods. These concepts are related to the development of various types of intelligent system architectures. Many HIS contain three necessary models: artificial neural networks, fuzzy inference systems, and genetic algorithms. The combination of FL and ANN permits adding human experience to machine learning. Hence, knowledgebase of neuro-fuzzy systems is maximized. It is common practice to implement GA in off-line high level control while implementing neuro-fuzzy systems in direct control loop.

5.1.1 How Important Is AI to Engineering Applications?

Why we need AI? Does AI have a better performance than mathematical or analytic solutions? AI is not always a better alternative to well-developed mathematical solutions. That is because well-developed mathematical models are comprehensible and easy to integrate to other developed model. However, mathematical models are developed to describe physical systems in defined operating range with some approximation. Within certain limits, approximation is acceptable yet, switching to wide operating range could result in detuned performance. AI has capabilities to describe complicated system in wide operating range. The following reasons are some of situations at which AI performs better than other analytic methods.

(a) The presence of human knowledge that cannot be described by analytic methods.
(b) No clear knowledge (analytic or linguistic) about the problem domain with presence of collected data (weather forecasting for example).
(c) Complicated problem domain with large number of inputs and outputs that makes analytic solution inaccurate and infeasible.

5.1.2 Artificial Intelligence History

The beginning of AI was noted back before the beginning of the electronics technologies that were used as the foundation of AI Logic. The invention of computers in 1943 helped in the establishment of AI. The technology was finally available, or so it seemed, to simulate intelligent behaviors.

The theory and insights brought by AI research will set the trend in the future of computing. The advancements in the artificial intelligence techniques have, and will continue to affect all our work and lives trends (Fig. 5.3).

There are many important considerations to mimic human experience and emulation the behavior of the human brain. One of the best ways to evaluate the

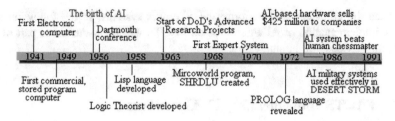

Fig. 5.3 History of AI

Fig. 5.4 Alan Turing

intelligence of a machine is the Alan Turing's test developed by British computer scientist Alan Turing. He stated that a computer would deserve to be called intelligent if it could deceive a human into believing that it was human (Fig. 5.4).

Artificial Intelligence has vast applications. It can be found in Computer science, Finance (stock market forecasting), Power Systems (smart microgrid), Medical systems (Disease Diagnostic System), Heavy industry, internet (Google search algorithm), Transportation, Telecommunications, Toys and games, Music, Aviation, News and publishing.

The focus of this chapter is on the implementation of AI in power systems, and in particular, on smart grid applications. Power systems are complex networks with too many connected components. The term smart grid refers to intelligent and cost-effective power grids with higher reliability, resiliency, and performance. In order to incorporate renewable energy technologies in buildings, and small communities, the term microgrid was introduced. The recent development of microgrids as connected to utility grids with different types of power circuits (AC-DC), requires adding power electronics to match power characteristics of the utility grid. Adding power electronics reduces power quality that may harm connected loads and generators. Flexible AC Transmission Systems (FACTS) [1] are devices that enhance the efficiency and performance of power systems. In the following sections, implementation of AI algorithms on FACTS will be discussed as part of the implementations within smart grids.

5.2 FACTS Implementing AI Algorithms

Microgrids (MGs) are an efficient solutions for power system to manage, control and integrate renewables (clean energy with no emission that provide less dependency on limited fossil fuel for power generation) as Distributed Sources (DS) within utility grid [2, 3]. They have two modes of operation, connected and islanded [4]. However, the presence of renewable technologies within MGs causes problems of stability due to their high degree of uncertainty, resulting in voltage fluctuations and low power quality [5]. The phenomenon of voltage fluctuations is evident for islanded MGs where power flow is limited. Uncontrolled power flow could negatively affect all MG components (loads, generators and distribution lines) [6].

One of the devices that are commonly used to improve MG performance is static VAR compensator (SVC) [7]. The SVCs are widely used as reactive power compensators. They have the ability to respond quickly to the fast dynamics of the power system to reduce the effect of the transient variation and maintain voltage steady state as required. They are implemented to enhance stability and reduce oscillation of power systems [8, 9]. Active researches discussed SVCs performance dependency on the implemented control algorithms that control the flow of reactive power [10–14].

Fuzzy Logic controllers [15] have great advantage over conventional control techniques due to three reasons: (1) They can present nonlinear mapping surface by membership functions and inference mechanism inherited in their structure; (2) They are independent on mathematical modeling and require only behavioral performance to set out linguistic variables and rule basis; (3) They are better to present uncertainties about system noisy measurements.

The efforts of the implementation of fuzzy controllers with SVC have been recorded in many publications. In [16], a fuzzy controller is implemented to control, activate and deactivate the number of thyristor switched capacitors (TSCs) to maintain operational voltage as required. In addition, Genetic Algorithm is used to find optimal firing angle of SVC. The system has no ability to adjust the load time variation, where fuzzy controller has no adaptation and genetic optimization cannot be implemented online. Moreover, the simulation conducted was based on a connected mode MG. Hence, the effect of control algorithms with SVC is minimum due to the presence of utility grid that has the ability to compensate for load variation. Similar work can be found in [17]. Another effort of implementing FC is found in [18, 19] where the authors in [18] designed two parallel controllers, instantaneous reactive power compensator and fuzzy controller. The fuzzy controller is equipped with rule adaptation mechanism; however, it has the same inputs of the fuzzy controller. Hence, both FC and its adaptation mechanism can be combined in one controller. The main contribution in this work can be considered as a linear controller presented by instantaneous reactive power compensator supported with parallel FC to compensate for system nonlinearity. Yet, the control algorithm has no adaptation mechanism that can compensate for load variation.

The first example that will be introduced in the next sections, shows the implementation of a Fuzzy Model Reference Learning Control algorithm (FMRLC) in the control loop of the SVC. Fuzzy algorithm is able to compensate for the nonlinearity of complicated nonlinear power systems. Hence, the system's performance can be kept as required during wide ranges of operating conditions. Moreover, the implementation of the learning mechanism keeps updating the controller parameters to compensate for load variation within the MG. The update process considers loads dynamic change of MG parameters. Adaptation mechanism measures system performance and compares it with desired performance presented by a reference model. Accordingly, adequate adaptation action is formed to update controller parameter. A simulation of MG in islanded mode is used to verify the control algorithm.

Another FACTS device that effectively enhances power system performance by implement AI algorithms is Green Plug-Modulated Filter Compensator (GP-FC). It is used as green energy efficient plug compensation scheme. A Simple modulated filter/Capacitor compensation scheme is controlled by an integrated genetics algorithm with Adaptive Neuro-Fuzzy Inference System (ANFIS) controller which is a member of a family of Energy efficient; Soft Starting Switched/Modulated FACTS based Compensation Devices. Those devices are used for single phase and three phase motorized, inrush and nonlinear loads. Active and Reactive powers have direct impact on the energy efficiency, efficient utilization, power factor, power quality and voltage profile of the system. There are many techniques to supply and compensate for load reactive power requirements of nonlinear/inrush/motorized type loads. So fast control action is needed. New power technologies, such, offer a number of advantages in control of power systems, including speed and accuracy as FACTS, LC Switched Compensators, high-power CSI and VSI-electronic converters and drives acy of the controlled response. Advanced control and improved semiconductor switching of these devices have provided distinguished solutions for power quality enhancement [20–24].

The functions of the Energy Efficient, soft starting and reactive compensation green plug scheme acts as power factor correction, power quality enhancement, efficient utilization, dynamic voltage control, inrush current reduction and dynamic speed reference tracking. GP-FC may be used in AC power system for various applications, from controlling system reactive power to improving voltage regulation and power factor by reducing transient/inrush content in voltage and current supplied to the dynamic nonlinear/inrush type motorized load. Beside the power demand requirements, some contingencies and negative sequence/ripple content are among other factors that are crucial when it happens to power quality issue [6, 7]. The most important point is to find the optimal dynamic self- regulating switching patterns for the GP-FC devices that can be adapted by AI techniques. In recent years, AI theory applications have received increasing attention in various areas of power systems such as operation, planning, and control. The effect of different controllers as conventional and adaptive AI controllers can be compared to conclude the most effective controller.

Genetic algorithms guarantee global optimization for any problem space. The most effective component of GA is fitness function that forms the search hyper surface required for optimization process.

Adaptive Neuro-Fuzzy Inference (ANFIS) is a merging system between neural networks and fuzzy logic. Therefore, it combines benefits of both systems. Fuzzy logic is a very efficient tool in control systems where it is less sensitive to uncertainty and noisy environment. It is powerful tool to encode fuzzy knowledge of expertise using linguistic variables and inference system. On the other hand, Neural Networks (NN) is powerful in pattern classification and pattern recognition. They can be merged with the fuzzy system to adapt fuzzy parameters to reach the best combination of those parameters leading to the required control procedure. Therefore, positive performance from both AI algorithms can achieve global benefits.

The system electromechanical modes can be a good indicator for its dynamics and its associated transient performance; the system electromechanical modes, such as the speed deviation response ($\Delta\omega$) and the mechanical rotor angle response ($\Delta\delta$), can be good indicators for system dynamics and its associated transient performance.

The significance of setting the GP-FC parameters for related operating case and its effect on transient response is evident. Without proper adaptation of GP-FC parameters, the benefits achieved by installing the GP-FC may be lost and impair system performance may be occurred [8–13].

The second example that will be introduced in the next sections, shows the implementation of ANFIS with GA optimization to Green Plug-Filter Compensator (GP-FC) scheme using a dynamic Multi-Loop Error Driven regulator. The objective of the control algorithm is to improve power quality and utilization in Distribution Systems, especially for single phase induction motor. The pulsing sequence of GP-FC utilizes tri-loop dynamic error-driven weighted modified PID controller. GP-FC scheme proved its effectiveness in improving power quality, enhancing power factor, reducing transmission losses and limit transient over voltage and inrush current conditions on the AC interconnected system.

We will start first with description of AI algorithms implemented in the two cases then description of the cases will be introduced.

5.2.1 FMRLC Algorithm

Fuzzy model reference learning control (FMRLC) [25, 26] uses reference model to describe the required performance to the adaptation mechanism. The term learning reflects memorizing capability of the fuzzy controller where enhancement in system

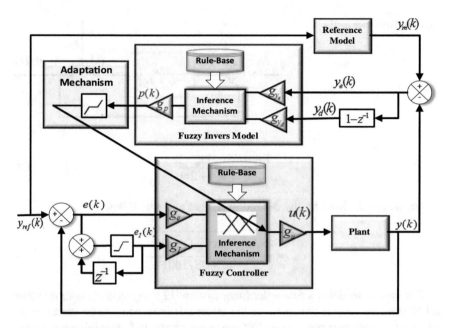

Fig. 5.5 The FMRLC algorithm used for the control loop of the SVC voltage regulator

performance can be experienced by frequent exposure of the system to the same dynamic range of state variables. The algorithm was successfully implemented to control the speed of induction motor drives [27, 20].

Figure 5.5 shows the structure of the control algorithm implemented in this work. The algorithm consists of 1—reference model, 2—inverse model, 3—adaptation mechanism, and 4—fuzzy controller. Reference model is the model that presents required performance achievable by the system. Hence, power system time constant (including the SVC) is the parameter representing achievable performance of the system. For simplicity, MG can be assumed as aggregated resistive and reactive components. Accordingly, MG time constant can be calculated based on assumed power flow in the grid in terms of rated active power of the grid and rated reactive power of the SVC as follows:

$$\tau = \frac{P_{rated}}{2\pi \times f \times Q_{SVC}} \tag{5.1}$$

where P_{rated} is the maximum active power flow within the system, Q_{SVC} is the maximum reactive power the SVC can inject to or absorb from power system and f is the grid frequency. Equation (5.1) indicates a guide line to select reference model time constant however, a larger time constant is recommended if performance of the reference model is not achievable by MG. The reference model is a disturbance rejection 2nd order system as shown in Fig. 5.6. Its forward transfer function is of type one to insure disturbance rejection with zero steady state error and given by:

Fig. 5.6 Reference model

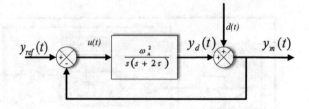

$$\frac{Y_d(s)}{Y_m(s)} = \frac{\omega_n^2}{s(s+2\tau)} \tag{5.2}$$

where ω_n is calculated by the desired overshoot and τ as follow:

$$\omega_n = \frac{1}{\tau} \sqrt{\frac{\pi^2 + \ln^2(\%OS/100)}{-\ln(\%OS/100)}} \tag{5.3}$$

The inverse model is a first order fuzzy system (PI fuzzy system) with rule-base set to describe inverse dynamics of the system. It is responsible to calculate the adequate amount of control action that forces the system to follow reference model. Based on knowledge of system dynamics, fuzzy inverse dynamics can be encoded in inverse model rule base. Table 5.1 shows designed rule-base of the inverse model that is implemented for the SVC.

The inputs to inverse model are given by the equations:

$$y_e(k) = y_m(k) - y(k) \tag{5.4}$$

Table 5.1 Rule-base for the fuzzy inverse model

y_I ＼ y_e	NB	NE	ZE	PO	PB
NB	NB	NB	NB	NE	ZE
NE	NB	NB	NE	ZE	PO
ZE	NB	NE	ZE	PO	PB
PO	NE	ZE	PO	PB	PB
PB	ZE	PO	PB	PB	PB

NB: Negative big
NE: Negative
ZE: Zero
PO: Positive
PB: Positive big

Fig. 5.7 Fuzzy controller
membership functions

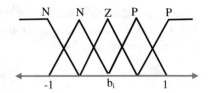

$$y_d(k) = \left(\frac{y_e(k) - y_e(k-1)}{T}\right) \tag{5.5}$$

where y, y_m, y_e and y_d are the output of the process, the output of reference model, the error between the output of reference model and the process, and the derivative of that error respectively. The third part of control algorithm is learning and adaptation mechanism that is responsible for encoding the output of inverse model into the rules at rule-base of the fuzzy controller. The fuzzy controller rules describe nonlinear control surface that compensates and linearizes the overall system to match the reference model. The adaptation mechanism is also responsible for adapting the control surface to compensate for time varying parameters of the loads connected to the power system. Triangular membership functions are used for fuzzy controller as shown in Fig. 5.7 where its centers b_i subjected to adaptation by adaptation mechanism. The update membership centers b_i is as follows:

$$b_i(k) = b_i(k-1) + \eta g_p p(k) \tag{5.6}$$

where, η is the learning factor, g_p is the adaptation gain and $p(k)$ is the output of inverse model. Both η and g_p can be combined as a one gain (g_p). To avoid generating control actions exceeding the limits of the process input, the following two equations are added:

$$b_i(k) = b_{max}, \quad b_i(k) \geq b_{max} \tag{5.7}$$

$$b_i(k) = b_{min}, \quad b_i(k) \leq b_{min} \tag{5.8}$$

where b_{min} and b_{max} are the minimum and maximum control action values respectively. Equation (5.6) shows that the FMRLC algorithm provides both adaptation and learning capabilities. This is due to the fact that adaptation process is independent of the inputs of main fuzzy controller. Also, adaptation process is seeking a certain performance defined by the reference model regardless of any change of process parameters. To avoid learning oscillation that could cause oscillation and instability of the system, a learning dead band was added to adaptation mechanism according to the following equation:

$$b_i(k) = \begin{cases} b_i(k-1) + g_p p(k) & -p_{DB} \geq g_p p(k) \geq p_{DB} \\ b_i(k-1) & \text{otherwise} \end{cases} \tag{5.9}$$

where p_{DB} is the dead band limit of learning process. Note that ηg_p is replaced by g_p in Eq. (5.9). The main fuzzy controller is a PI fuzzy controller with inputs described by the following equations:

$$e(k) = r(k) - y(k) \tag{5.10}$$

$$e_I(k) = T\left(e(k) + \sum_{i=1}^{k-1} e(i)\right) \tag{5.11}$$

where e, and e_I are the error between reference input r and measured output y and its integral respectively in current sample k. The dynamic range that is covered by membership functions is set to interval $[-1, 1]$. The adjustment of the signals dynamic range to dynamic range covered by the membership functions is carried out by input-output scaling factors (g_x where x represents the label of the signals at which gain is placed on.

In order to avoid accumulated large values during transient period from integral part [accumulation of error from first sample to the current sample k presented by Eq. (5.11)], the digital integrator is followed by a limiter. Such large values (over the dynamic range of the proper signal) could cause slowdown of the controller performance. The limiter maximum and minimum values are set in order of magnitude of output signal dynamic range.

Minimum is used for premises activation and supremum for implication. Center Of Area (COA) is used for the defuzzification process.

5.2.2 Tri-loop Dynamic Error-Driven ANFIS-PID with GA Optimization

The dynamic controller based on Tri-loop driven error is used to modulate the switched filter compensator. The error signal between a reference load voltage (V_{L-ref}) and measured one (V_L) is fed to control unit of PWM generator. The PWM signal is used for switching MOSFET gate to control modified VSC controller as shown in Figs. 5.8 and 5.9.

The global error Composed of four individual multi error in four control loops including voltage stabilization, inrush current limiting and synthesize dynamic power. Each multi dynamic leading loop is used to minimize the global error based on a tri-loop functional error signal in addition to other supplementary motor current limiting and/or feeder currents for loss reduction.

The four loops can be briefly explained as:

1. V_L—Load Bus Voltage Stabilization Loop by tracking the error of load voltage and regulating it to unity.
2. I_L—Dynamic rms-Current Minimization Loop to compensate any sudden current change that may be caused by inrush current or induction motor starting current.

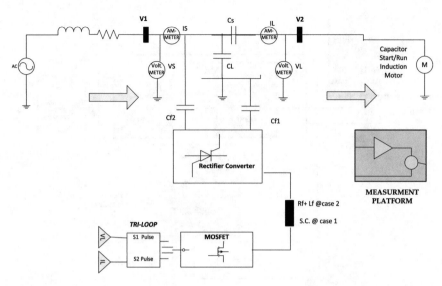

Fig. 5.8 Dynamic controller based on tri-loop driven error

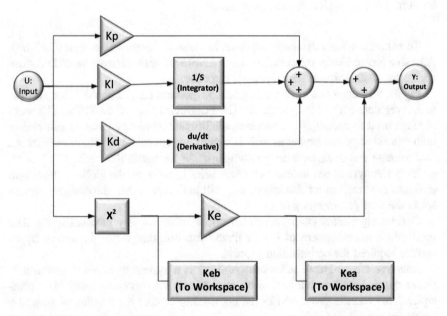

Fig. 5.9 Internal structure of PID

3. P_L—Excursion Damping Loop.
4. I_L—Motor Load Inrush/Ripple/Transient Current Damping Loop to reduce the harmonic ripple content in distribution system.

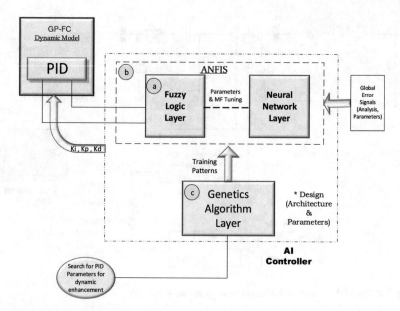

Fig. 5.10 Working layers of the adaptive controller

To enhance system dynamic response, Integrated Genetics algorithm (GA) with Adaptive Neuro-Fuzzy Inference System (ANFIS) is applied to set the PID control parameter to fine-tune system dynamic response.

Figure 5.10 depict an integrated Genetics algorithm (GA) with ANFIS controller to self-regulate PID tri-loop stage for GP-FC Device applied on SPIM. The technique is used to accomplish a better feasibility and efficiency where it can realize both criteria of power saving as well as quality improvement of the source current, load voltage and dynamic reactive compensation for the SPIM loads.

Each GA system has inherent characteristics to work on the global minima and trying to not stuck in local minima, if it will not success then the designer should select the best parameters pattern.

Genetic algorithms guarantee global optimization for any problem space. The most effective component of GA is fitness function that forms the search hyper surface required for optimization process.

Adaptive Neuro-Fuzzy Inference (ANFIS) is a system of artificial neural networks that is based on integration between neural networks and fuzzy logic principles. That attitude gives ANFIS the potential to capture the benefits of both in a single framework [14–18].

The fuzzy system is a very efficient tool in the controlling actions while the neural networks (NN) are powerful in patterns classifications and patterns recognitions. NN will be merged with the fuzzy system to adapt the parameters of the fuzzy systems to reach the best collection of fuzzy parameters lead to the required controlling procedure [19–21].

The structure of control algorithm can be as three layer controller. PID is presented at lower control layer to generate the direct control action to GP-FC. The second control layer is an ANFIS adaptation mechanism to adapt PID controller parameters. The top control layer represents off-line genetic algorithms optimization system to globally set ANFIS parameter for different operating conditions.

5.3 Application 1: MG Stability Enhancement Using SVC with FMRLC

A MG in islanded mode is used to verify the performance of proposed FMRLC algorithm. This is because MG in islanded mode has very limited power flow and hence, it is hard to achieve stability. The structure of the MG power system is shown in Fig. 5.11. The system is composed of PV bus supplied by two sources, wind turbine and asynchronous generator, and PQ bus with two loads, linear load and induction motor. The PV and PQ buses are connected via distribution line of 100 m length. Both wind turbine and asynchronous generator are 300 kVA. The induction motor is 150 kVA while the linear load consists of 100 kW active load and 140 kV capacitive load connected continuously to the bus.

A 100 kVAR Static VAR Compensator (SVC) is connected at load side. The SVC controller was modified to switch between PI conventional control algorithm and PI-FMRLC algorithm.

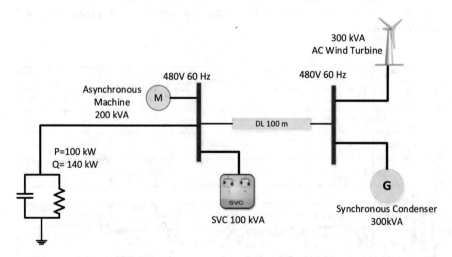

Fig. 5.11 Single line diagram of the power system used in the simulations

5.3.1 Simulation Results

The simulation is conducted using the system described in Sect. 5.3 for 8 s. Both machine load and wind speed are subjected to variation pattern to examine the ability of the controller to reject MG voltage disturbance.

The reference model parameters were calculated from Eqs. (5.1) and (5.3). The time constant τ can be calculated by $\tau = \dfrac{600\,\mathrm{kVA}}{2\pi \times 60 \times 100\,\mathrm{kVA}} = 0.016\,\mathrm{s}$, however, it was set to 0.05 s to allow more relaxed performance. The required OS% is set to 1 %, hence $\omega_n = 24\,\mathrm{rad/s}$. A PI-FMRLC was used in control loop of the SVC. Five membership functions are used for fuzzy controller and three for inverse model. Figure 5.12 shows simulation result. The top curves show the bus voltage (pu) for FMRLC and PI controller. It can be noted that FMRLC algorithm was able to maintain the line voltage to required level with minimal disturbance compared to conventional PI controller. The second curve shows the error between reference model output and measured bus voltage for FMRLC algorithm. The learning algorithm was able to improve controller performance with time. On the other hand, Conventional PI controller shows steady performance that does not improve with time as expected.

Figure 5.13 shows bus voltage and voltage error in early period and later period of simulation as shown. It can be noted how the performance improved during later period of control. The measured overshoot was 0.5 % for the PI-FMRLC algorithm while it was 1 % using the PI controller. voltage settle within 5 % of its maximum deviation from desired bus level after 0.25 s (four to five times the time constant) for the PI-FMRLC (measured during last training period) while it takes 0.8 s with the PID controller.

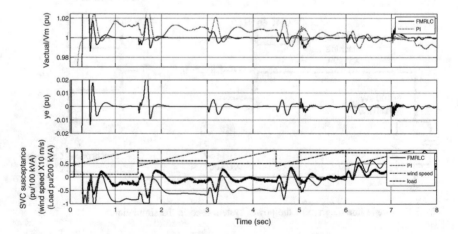

Fig. 5.12 Measured performance of the SVC voltage regulator for both FMRLC algorithm and PI controller. From *up to down* 1—the bus voltage, 2—voltage error between bus voltage and reference model, and 3—the SVC susceptance, wind speed and load

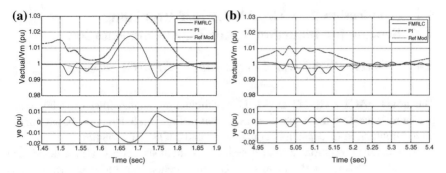

Fig. 5.13 Bus voltage and voltage error in intervals from **a** 1.45 to 1.9 s and **b** 4.95 to 5.4 s

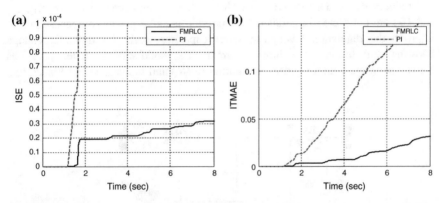

Fig. 5.14 **a** Integral-of-time multiplied absolute error (ITMAE) and **b** integral square error (ISE) for both algorithms

Figure 5.14 shows Integral-of-Time Multiplied Absolute Error (ITMAE) and integral square error (ISE) for the three cases for the 8 s simulation. The measurement of both error parameters (ITMAE and ISE) was performed after the 1st 1.2 s of simulation to avoid building high values during MG transient operation that would hide details of system performance during the rest of simulation period. While the ISE measures the system oscillation or steady state error, ITMAE measures improvement of system performance with time by giving error at later control stage more significance than error at early control stage. The curves show that PI-FMRLC is outperforming PI controller. The PI controller has larger slope for ISE and exponential rising curve for ITMAE (no improvement of performance with time indicating no learning capabilities).

5.4 Application 2: A GA-ANFIS Self Regulating Scheme for Induction Motor Filter Compensation

5.4.1 Green Plug-Filter Compensator (GP-FC) Scheme

The novel FACTS GP-FC device is a switched/modulated filter-capacitor compensator, it has two main schemes, as shown in Figs. 5.15 and 5.16, based mainly on a combination of capacitors connections and MOSFET switches.

In general description, a hybrid series and shunt capacitors with a tuned arm power filter constitutes the main GP-FC components. The series capacitor C_s is connected in series with the transmission line to offset dynamically part of the line inductance. Such reduction improves the inrush condition and inherent voltage drop and reduces the feeder reactive power loss. Three shunt capacitor banks (C_{f1}, C_{f2} and C_L) are connected in parallel; they provide reactive power compensation and improve the regulation level. The series capacitor works for dynamic voltage boosting and limiting the inrush current. The solid-state switches path has converter-bridge with six pluses in addition to (R_f) and inductance (L_f) branch that

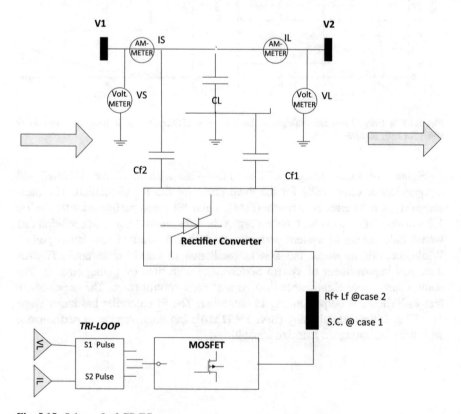

Fig. 5.15 Scheme I of GP-FC

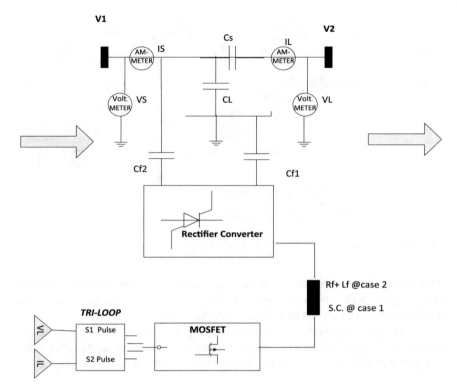

Fig. 5.16 Scheme II of GP-FC

structures a tuned arm filter of the DC side. The device is mounted between the AC and DC sides of the converter bridge.

The two HEXFET/MOSFET switches (S_1 and S_2) are controlled by two complementary switching pulses (P_1 and P_2) that are supplied by the dynamic tri-loop error driven modified VSC controller, which will be described later. The first pulse P_1 directs S_1, while the second pulse P_2 directs S_2. The procedure of complementary PWM pulses can be explained as follow:

Case 1: If P_1 is high and P_2 is low, the resistor and inductor will be fully shorted and the device will provide the required compensation to the load
Case 2: If P_1 is low and then P_2 is high, the resistor and inductor will be connected into the circuit as a tuned arm filter.

5.4.2   Case System Description

The digital simulations using Matlab/Simulink/Sim-Power Software Environment is applied to a simple AC study system, which has AC source with 240 V and

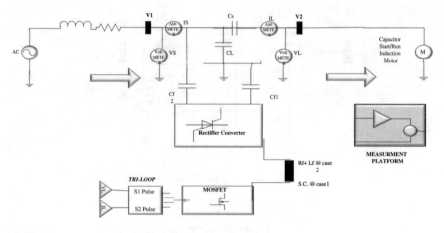

Fig. 5.17 Concerned system with single phase IM loads

transmission line represented by R_s and L_s single phase capacitor-run induction motor. Figure 5.17 depicts a single line diagram of the studied AC system. The detail parameters of the system are given in Appendix.

The sample study system is controlled by applications of both scheme I and scheme II of switched smart filter compensated device using Green Plug Filter Compensator GP-FC devices to the Single Phase Induction Motor (SPIM) Load.

5.4.3 Simulation Results

The MATLAB/SIMULINK/SimPower Platform is used as environment for the proposed GP-FC for the two different schemes. The digital simulation is carried out with and without the controlled GP-FC in order to show its performance in voltage stabilization, harmonic reduction and reactive power compensation at normal operating condition. The dynamic responses of voltage, current, active power, reactive power, apparent power, power factor, frequency spectrum (for voltage and current), (THD)v and (THD)i at source bus and load with Comparison of harmonics at each bus is made for two different cases, with and without GP-FC. Voltage and current harmonic analysis in term of total harmonic distortion (THD) is given. It is obvious that voltage harmonics are significantly reduced, also the THD of current waveform at each bus is decreased. Platform of measurement system is shown in Fig. 5.18 (Table 5.2).

Figures 5.19 and 5.20 show the ability of proposed technique to damp the oscillations in load voltage profile. Also, related to Fig. 5.22, Installed GP-FC with proposed integrated controller improved the system power factor, while the system without GP-FC has poor distributed power factor (Fig. 5.21).

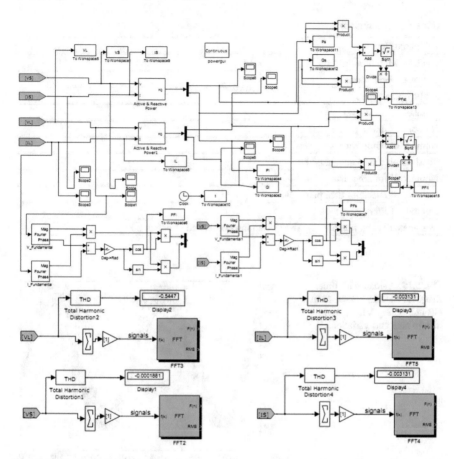

Fig. 5.18 System measurement platform

Table 5.2 GA and ANFIS system parameters

GA parameters	
Input variables	x(1) = KP, x(2) = KI, x(3) = Kd, x(4) = Ke
Variables lower bound	LB = [0 0 0 0];
Variables upper bound	UB = [100 30 15 10];
Options. PopulationType	Double vector
Options. PopulationSize	30
Options. EliteCount	Adapted in simulations
Options. CrossoverFraction	Adapted in simulations
Options. MigrationDirection	Forward
Options. MigrationInterval	25

(continued)

Table 5.2 (continued)

GA parameters	
Options. MigrationFraction	0.1
Options. Generations	200
ANFIS parameters	
Number of nodes: 40	
Number of linear parameters: 60	
Number of nonlinear parameters: 20	
Total number of parameters: 80	
Number of training data pairs: 34	
Number of checking data pairs: 24	
Number of fuzzy rules: 14	
Designated epoch number –> ANFIS training completed at epoch 300	

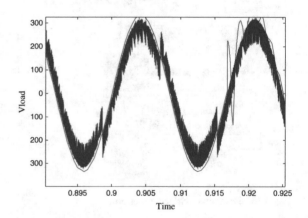

Fig. 5.19 V-load with time. *Red* with GP-FC, *blue* without GP-FC, noting VL disturbance and variations

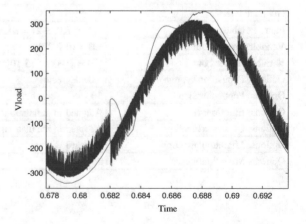

Fig. 5.20 Vload with time, more zooming. *Red* with GP-FC, *blue* without GP-FC, noting VL disturbance and variations

Fig. 5.21 Iload with time.
Red with GP-FC, *blue* without
GP-FC

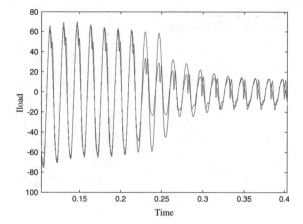

Fig. 5.22 Source PF with
time. *Red* with GP-FC, *blue*
without GP-FC

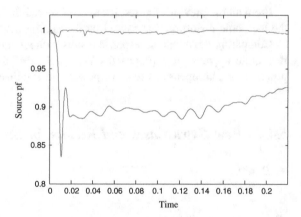

Figures 5.23 and 5.24 indicate that the magnitude of oscillations for the source
current and reactive power reduced by installing GP-FC, as shown in red and blue
response.

Fig. 5.23 Qsource with time.
Red with GP-FC, *blue* without
GP-FC

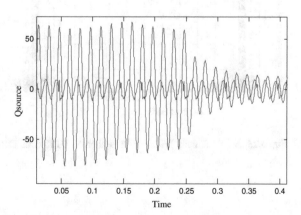

Fig. 5.24 Isource with time. *Red* with GP-FC, *blue* without GP-FC

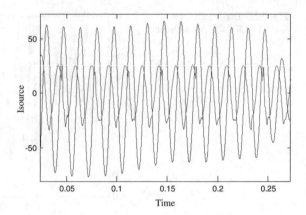

From all the previous figures, it can be observed that controlled GP-FC mitigates the harmonic distortion that caused by the nonlinearity of IM loading condition.

Comparing the dynamic response results without and with the proposed GP-FC, it is quite apparent that proposed GP-FC enhances the power quality, improves power factor, compensates reactive power and stabilizes the buses voltage.

5.4.4 Few Extra Cases and Results for Simulated System

- *Base Case*

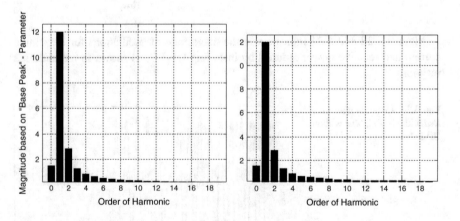

Fig. 5.25 Vs FTT, with THD = 0.68 % then Vl FTT, with THD = 15.41 %

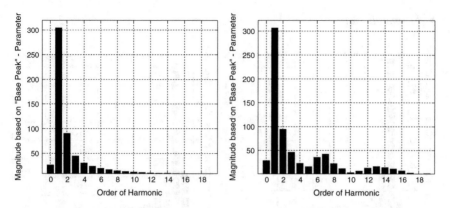

Fig. 5.26 Is FTT, with THD = 9.48 % then Il FTT, with THD = 9.48 %

- *Scheme I run*

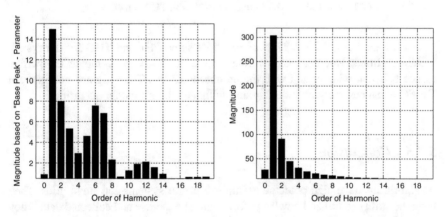

Fig. 5.27 Vs FTT, with THD = 0.53 % then Vl FTT, with THD = 7.44 %

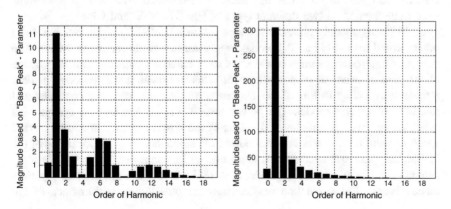

Fig. 5.28 Il FTT, with THD = 7.2 % then Is FTT, with THD = 1.5 %

• *Scheme II run*

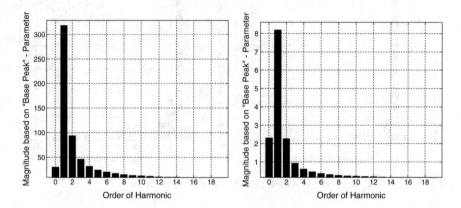

Fig. 5.29 Vs FTT, with THD = 0.68 % then Vl FTT, with THD = 6.46 %

From Figs. 5.25, 5.26, 5.27, 5.28 and 5.29 show FFT and THD for the system variables (Iload, Isource…etc.) without and with GP-FC to compare the performance between without and with the proposed GP-FC. Those figures confirm that GP-FC enhances the power quality and mitigates the harmonic distortion.

5.4.5 Conclusion

AI algorithms have high potential to improve power systems and smart grids. They have the ability to utilize knowledge form human expertise and collected data. They can perform better than conventional analytic methods for highly nonlinear systems with uncertainty and noisy environment. This chapter provided two AI algorithms, FMRLC and ANFIS with GA optimization mechanism. The two AI algorithms were implemented on two different types of FACTS, SVC and GP-FC.

A PI fuzzy model reference learning control (FMRLC) algorithm has been implemented to control the SVC voltage regulator in a nonlinear power system. The design of controller parameters was conducted based on human expert knowledge of the control behavior of power system. This information was encoded in both input/output scaling factors of the adaptation mechanism and fuzzy controller and rule-base of inverse model. The power system was subject to load change of a synchronous machine (nonlinear dynamic of the power system). In addition, wind speed of wind turbine was changed during the simulation. The performance of the PI-FMRLC was examined against conventional PI controller. To accurately measure control algorithms performance, both IMTAE and ISE are used as a performance index. The simulation results show an outstanding performance of the

FMRLC. It is not only able to fast compensate to the load variation but also improve its performance during learning process. The results show that learning index decays with time, indicating high level of learning progress the algorithm achieves during simulations.

The second algorithm was implemented to FACTS based Device GP-FC for Soft Starting, Energy Efficient Utilization, Loss Reduction/Power Factor Correction, power quality improvement and power factor correction.

Integrated Genetics algorithm (GA) with Adaptive Neuro-Fuzzy Inference System (ANFIS) dynamically regulates the gains of PID tri loop stage for Green Plug-Filter Compensator (GP-FC) applied to single phase Induction motor. A Tri Loop dynamic error controlled technique is applied to minimize overloading and high inrush motor currents, in addition to its regulation of motor dynamic speed ensuring efficient utilization and system stability. The technique is used to adjust the feeding of PWM switching of GP-FC by finding the optimal control gain settings that dynamically minimize the absolute value of the global dynamic error. Digital simulations are provided to validate the effectiveness of the proposed GP-FC with control algorithm in improving the power quality and system stability. Inappropriate parameters of the PID can affect the dynamic performances of the system, so it is important to adapt them in real time by ANFIS. The integrated GA with ANFIS technique is used to adjust controllers' gains to minimize total absolute error.

The same device is being extended to three phase utilization motorized/nonlinear/inrush type/switched type loads as well as other topologies for three phase Feeder Loss Reduction, Voltage Regulation, Power Quality and Power Factor Correction for Distributed Renewable Energy-Smart Grid Applications.

Appendix

- *Appendix for Simulated System Parameters*

SPIM (capacitor run)	Vs = 208/220/240 V fs = 60 Hz Speed = 1740 rpm Ws = 377 rad/s P = 3–5 KVA@ 0.75–0.88 pf
Feeder	Rs = 0.1–0.15 Ω Ls = 2–3 mH
GP-FC	Cf1 = Cf2 = 75–225 µf CL = 22–66 µf Rf = 0.15–0.85 Ω Lf = 3–5 mH

(continued)

(continued)

SPWM	Sa = Sc not Sb Fs/w = 1750 Hz
PID	Kp = 0–100, Ki = 0–30 Kd = 0–15, Ke = 0–10
Tri-Loop	To = 20–40 ms T1 = 5–10 ms Delay = 20–40 ms KVL = 1 KIL = 0.5–0.75 KPL = 0.25–0.5 Kh = 0.25

References

1. Rogers, K.M.: Power system control with distributed flexible AC transmission system devices. Master thesis, University of Illinois at Urbana-Champaign (2009)
2. Katiraei, F., Iravani, R., Hatziargyriou, N., Dimeas, A.: Microgrids management. IEEE Power Energy Mag. **6**, 54–65 (2008)
3. Kroposki, B., Lasseter, R., Ise, T., Morozumi, S., Papathanassiou, S., Hatziargyriou, N.: Making microgrids work. IEEE Power Energy Mag. **6**, 40–53 (2008)
4. Hatziargyriou, N.D., Meliopoulos, A.P.S.: Distributed energy sources: technical challenges. In: IEEE Power Engineering Society Winter Meeting, New York (2002)
5. Xiongfei, W., Guerrero, J., Zhe, C.: Control of grid interactive AC microgrids. In: IEEE International Symposium on IEEE International Industrial Electronics (ISIE) (2010)
6. Khan, U.N., Yan, L.: Power swing phenomena and its detection and prevention. In: 7th EEEIC International Workshop on Environment and Electrical Engineering, 5–11 May 2008
7. Hill, D.J., Hiskens, I.A.: Incorporation of SVCS into energy function methods. IEEE Trans. Power Syst. **7**(1), 133–140 (1992)
8. Genc, I., Usta, Ö.: Impacts of distributed generators on the oscillatory stability of interconnected power systems. Turk. J. Electr. Eng. Comput. Sci. **13**(1), 149–161 (2005)
9. Biswas, M.M., Das, K.K.: Voltage level improving by using static VAR compensator (SVC). Glob. J. Res. Eng.: J. Gen. Eng. **11**(5), 12–18 (2011)
10. Noroozian, M., Andersson, G.: A robust control strategy for shunt and series reactive compensators to damp electromechanical oscillations. IEEE Trans. Power Delivery **16**(4), 812–817 (2001)
11. Hsu, Y.Y., Cheng, C.H.: Design of a static VAR compensator using model reference adaptive control. Electr. Power Syst. Res. **13**(2), 129–138 (1987)
12. Mishra, Y., Mishra, S., Dong, Z.: Rough fuzzy control of SVC for power system stability enhancement. J. Electr. Eng. Technol. **3**(3), 337–345 (2008)
13. Kumar, R., Choubey, A.: Voltage stability improvement by using SVC with fuzzy logic controller in multi-machine power system. Int. J. Electr. Electron. Res. **2**(2), 61–66 (2014)
14. Ali, S., Qamar, S., Khan, L.: Hybrid adaptive recurrent neurofuzzy based SVC control for damping inter-area oscillations. Middle-East J. Sci. Res. **16**(4), 536–547 (2013)
15. Yen, J., Langari, R.: Fuzzy Logic: Intelligence, Control, and Information. Prentice Hall (1998)
16. Calderon, J., Chamorro, H., Ramos, G.: Advanced SVC intelligent control to improve power quality in microgrids. In: 2012 IEEE International Symposium on Alternative Energies and Energy Quality (SIFAE), 25–26 Oct 2012

17. Lo, K., Sadegh, M.: Systematic method for the design of a full-scale fuzzy PID controller for SVC to control power system stability. IEE Proc. Gener. Transm. Distrib. **150**(3), 297–304 (2003)
18. Wang, J., Fu, C., Zhang, Y.: SVC control system based on instantaneous reactive power theory and fuzzy PID. IEEE Trans. Industr. Electron. **55**(4), 1658–1665 (2008)
19. Fang, D.Z., Xiaodong, Y., Chung, T.S., Wong, K.P.: Adaptive fuzzy-logic SVC damping controller using strategy of oscillation energy descent. IEEE Trans. Power Syst. **19**(3), 1414–1421 (2004)
20. Hingorani, N.G., Gyugyi, L.: Understanding FACTS: concepts and technology of flexible AC transmission system. Institute of Electrical and Electronics Engineers (2000)
21. Sharaf, A., Hassan, A., Biletskiy, Y.: Energy efficient enhancement in AC utilization systems. In: IEEE Canadian Conference on Electrical and Computer Engineering CCECE 07, Vancouver, Canada (2007)
22. Sharaf, A., Huang, H., Chang, L.: Power quality and nonlinear load voltage stabilization using error-driven switched passive power filter. In: Proceedings of the IEEE International Symposium on Industrial Electronics (1995)
23. Anillaga, J., Bradley, D.A., Bodge, P.S.: Power System Harmonics. Wiley (1985)
24. Sabin, D.D., Sundaram, A.: Quality enhances. IEEE Spectr. **2**, 34–38 (1996)
25. Layne, J., Passino, K.: Fuzzy model reference learning control for cargo ship steering. IEEE Control Syst. Mag. **13**(6), 23–34 (1993)
26. Layne, J., Passino, K.: Fuzzy model reference learning control. In: IEEE Conference on Control Applications, Dayton, OH (1992)
27. El-dessouky, A., Tarbouchi, M.: Optimized fuzzy model reference learning control for induction motor using genetic algorithms. In: The 27th Annual Conference of the IEEE, 2001, IECON '01, Denver, CO. Industrial Electronics Society (2001)

Chapter 6
Control Through Genetic Algorithms

Nicolae Paraschiv, Marius Olteanu and Elena Simona Nicoara

Abstract Many real world applications require automatic control. This chapter addresses genetic algorithms to achieve the control, based on their numerous advantages for the difficult problems. First of all a unitary approach of the control through the perspective of the systems theory is presented. There are described examples of control in biology, economy and technical areas in order to highlight the general system behaviors: preventive control, reactive control or combined control. In the next section, fundamentals of genetic algorithms theory are featured: genetic representation, genetic operators, how it works and why it works. Further, two process control systems based on genetic algorithms are described: a chemical process control involving mass transfer, where the genetic algorithms are used in the system identification for a NARMAX model, an important issue with respect to model based control and a job shop scheduling process in manufacturing area where the genetic algorithm is the tool to model the optimization process control.

Keywords Genetic algorithm · Automatic control · Job shop scheduling problems

6.1 Unitary Approach to Automatic Control

Starting from the generality of the "system" concept, in the following, aspects that highlight the unity regarding automatic control approaches are covered. There are characterized two essential control strategies, differentiated by the measures considered in the control law, namely the controlled variable and the measured disturbances. The advantages and the disadvantages of the strategies are emphasized.

N. Paraschiv (✉) · M. Olteanu · E.S. Nicoara
Petroleum-Gas University of Ploiesti, Ploiesti, Romania
e-mail: nparaschiv@upg-ploiesti.ro

M. Olteanu
e-mail: molteanu@upg-ploiesti.ro

E.S. Nicoara
e-mail: snicoara@upg-ploiesti.ro

© Springer International Publishing Switzerland 2016   165
K. Nakamatsu and R. Kountchev (eds.), *New Approaches in Intelligent Control*,
Intelligent Systems Reference Library 107, DOI 10.1007/978-3-319-32168-4_6

In addition, the combined control strategy which associates the advantages of the two essential strategies is described. There are provided some examples from the technical, economical and biological fields in order to accentuate the unitary approach to automatic control.

6.1.1 Preliminary of Automatic Control Systems

A system is defined as an ensemble of interconnected elements that interact with each other and with the environment in order to reach a specific goal [1]. This ensemble (the system) has additional properties that are not possessed by the components. Such a property is represented by the dynamic of the system or its evolution with respect to time, evolution imposed by internal or external factors. Physical systems obey the principle of causality according to which causes precede effects.

Currently a system can be analyzed using three perspectives: abstract, structural and topological, the present chapter following the abstract approach. Given a system, the abstract or informational approach implies the existence of two categories of variables for the system: inputs (u_i) and outputs (y_i), as illustrated in Fig. 6.1. Input variables are also called causes or independent variables and the outputs are called dependent variables or effects. For a real system at least one input has to exist; the systems that do not fulfill this condition are called trivial and are not physically feasible.

An important category of systems is represented by the automated systems, characterized by an autonomous evolution in time, without direct human intervention. Automated systems are classified, in accordance to their functionality, in the following categories: monitoring systems, control systems, safety systems and optimization systems. The object of the following discussions is represented by the automatic control system (ACS). An ACS is composed by two subsystems: the controlled subsystem and the command subsystem.

To get an automatic control of a system means to guide it towards a reference state and to maintain that state without human intervention. From the control perspective, the inputs of a system can be classified as commands and disturbances. The commands are driven by the command subsystem and the disturbances, having random fluctuations, are inaccessible to it. Given a system, the number of controlled variables matches the number of commands and they are found among the outputs. Defining command—controlled variable pairs represents an important phase in designing an ACS.

Fig. 6.1 Abstract representation of a system S

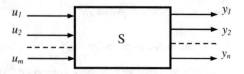

6.1.2   The Feedback Control Strategy

The concept of *feedback* was introduced in the domain of process control by Norbert Wiener in his work [2]. In the following, for the "controlled subsystem" it will be used the term "controlled plant" or "controlled process" and for the "command subsystem" it will be used the term "feedback controller" or "feedback compensator", connected, as illustrated in Fig. 6.2.

The feedback control strategy (FBCS) obeys a principle that imposes adjusting the command by the feedback controller when error occurs. The error can be produced by the changing of the reference state (the set point) or by random fluctuations of the disturbances. The controller keeps adjusting its command until the error is removed.

A system using FBCS has a reactive-corrective action, which means it reacts when an error occurs and its target is error removing. The main advantage of a system using FBCS is represented by the fact that its actions are independent of the errors cause. As a consequence, the universal control laws (universal controllers) can be used to compute the command. The most well known is PID controller, that uses the following relation to compute the command $u(t)$:

$$u(t) = u_0 + K_p e(t) + K_i \int_0^t e(\tau)d\tau + K_d \frac{de(t)}{dt}, \tag{6.1}$$

where u_0 represents the command value when there is no error ($e = 0$), e represents the error, K_p, K_i and K_d are tuning parameters.

The effect of the set point (r) changing upon the output is much faster than the effect of disturbance changing, and therefore it is preferred a version of PID algorithm which uses only the derivative of the controlled variable, as shown in the following relation:

$$u(t) = u_0 + K_p e(t) + K_i \int_0^t e(\tau)d\tau + K_d \frac{dy(t)}{dt}, \tag{6.2}$$

for which it can be used the computing scheme shown in Fig. 6.3.

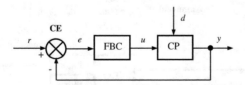

Fig. 6.2 Block diagram for a feedback control system: *CE* Comparator Element; *FBC* Feedback Controller; *CP* Controlled Process; *r* reference (set point); *e* error; *u* command; *y* controlled variable; *d* disturbances

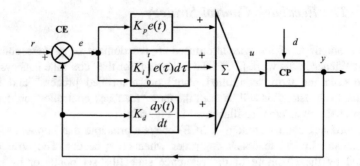

Fig. 6.3 Block diagram for the PID algorithm with controlled variable derivation

The feedback control has also an important drawback: the existence of a time delay to remove the error, delay depending on the controlled process inertia. During this time delay the controlled variable value is different than the set point value and the control system passes through a transient state.

The fundamental principle of FBCS is universal, in the sense that it can be applied to any controllable, observable and accessible system [3]. In the following some examples of real systems from biology, economics and technology will be presented, examples meant to constitute arguments for the unitary approach to FBCS.

A first example, depicted in Fig. 6.4, refers to the feedback control of the human body temperature [4].

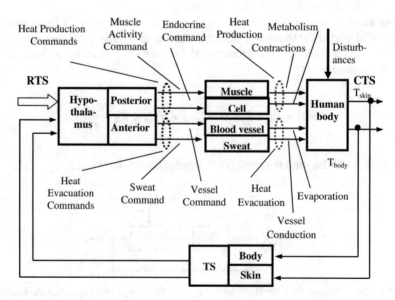

Fig. 6.4 Human body temperature feedback control: *RTS* Reference Thermal State; *CTS* Current Thermal State; *TS* Temperature Sensors

The normal temperature of a healthy human body lies around the value 36.5 °C, that in our approach is the Reference Thermal State (RTS). The function that maintains constant this value is a vegetative function that falls under the hypothalamus area of responsibility. At this level, the current temperature mirrored in the Current Thermal State (CTS) is compared with the Reference Thermal State.

The commands transmitted by the hypothalamus (through its anterior and posterior sections) regard heat producing, respectively heat evacuation. Heat producing will increase the temperature of the body, heat evacuation will decrease the temperature. The support elements of these two commands, namely the muscles activity and the endocrine glands activity for heating, respectively the sweating and the distension of blood vessels for cooling, are depicted in Fig. 6.4. As an example, contraction of muscles produces heat, while evaporation of water at skin surface induces cooling.

Regarding the transmission of temperatures signals, the biological temperature sensors TS are engaged. TS deal with thermo-sensitive fibers: A delta for cold, C for heat. These sensors convert the temperature variations sensed in the electrical signals transmitted to the hypothalamus through nerves.

An argument based on the cause-effect relation proves the functionality of the simplified human body temperature feedback control structure in Fig. 6.4. For example, if the ambient temperature decreases (disturbance occurs) then the skin temperature decreases (cooling sensation occurs). Information regarding this variation of temperature will be caught by the sensors and will be transmitted to the hypothalamus, which will activate the command section related to muscles activity. As an effect, contractions (tremble) will occur, and consequently the temperature increases to the value of RTS.

The control system in this example acts as *regulator operation*, because the reference is practically constant. The reasoning described leads to a relevant characteristic of the feedback control: the effect is removed while the disturbance is maintained.

The second example [4], illustrated in Fig. 6.5, belongs to the economical field and presents a feedback control system for a trading company. The company management acts such that the objectives in the Reference Economical State (RES) are achieved. Among these objectives, the most important are those related to efficiency and implicitly to profitability. To achieve these objectives, the company management has at its disposal many types of resources: raw material, energy, financial, informational and human resources. They are used in the production structures in order to obtain market capitalized goods or services by the sales office. Data regarding production and sales are processed by the functional departments to obtain information that is transmitted to the trading company management as economical results.

If by operation of some disturbances (such as the market state) a difference between the objectives and the economical results occurs, some commands are issued in order to remove this difference. It is worthy to mention that bringing the results to the objectives is accomplished in a time interval while the trading company may experience loss. The effects (namely the differences between objectives

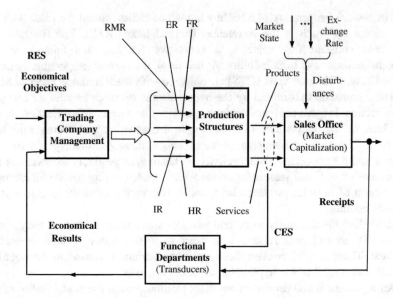

Fig. 6.5 Feedback control of a trading company: *RMR* Raw Material Resources; *ER* Energy Resources; *IR* Informational Resources; *FR* Financial Resources; *HR* Human Resources; *CES* Current Economic State; *RES* Reference Economic State

and the actual results) are also removed when any disturbance occurs, such as the exchange rate.

The third example belongs to the technical field; the aim of the system is to maintain to a reference value the temperature of a fluid that is heated in a heat exchanger (Fig. 6.5). The necessary heat results from the steam condensation (Hot Fluid In) (Fig. 6.6).

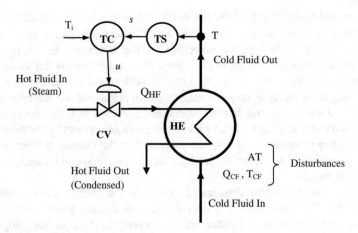

Fig. 6.6 Feedback control of temperature: *HE* Heat Exchanger; *TC* Temperature Feedback Controller; *TS* Temperature Sensor (Transducer); *CV* Control Valve (Actuator); T_i Temperature Reference (Set Point); *T* Controlled Temperature; *AT* Ambient Temperature

Among the many input variables that influence the temperature T (QCF—Cold Fluid Flow Rate, TCF—Cold Fluid Temperature, QHF—Hot Fluid Flow Rate, AT—Ambient Temperature etc.). Hot Fluid Flow Rate QCF was chosen as command; the rest of measures are considered disturbances.

If, for example, the ambient temperature decreases, a difference between the temperature T and the reference T_i occurs. To remove this difference, the controller TC will command the opening of control valve CV; the effect will be that the Hot Fluid Flow Rate and implicitly the temperature T increase. In the feedback control structure, alongside of the process, besides the temperature sensor ST and the temperature controller TC, there is the control valve CV, which is a remote controlled valve.

Out of the detailed examples, the unitary character of the feedback control strategy emerges, all cases having the structure highlighted in Fig. 6.2. Additionally, the presence of two adjacent components comes into sight: the sensor and the actuator. These components allow the access to the process to obtain information form it (the sensor) and to implement the command to it (the actuator). Structurally, these can be integrated either alongside the controller in the automation device or alongside the process in the extended process (the fixed part).

6.1.3 The Feedforward Control Strategy

In the following, for the "controlled subsystem", referred in Sect. 6.1.1, it will be used the term "controlled plant" or "controlled process" and for the "command subsystem" it will be used the term "feedforward controller" or "feedforward compensator", connected as illustrated in Fig. 6.7.

The feedforward control strategy (FFCS) obeys a principle that imposes adjusting the command by the feedforward controller when variation of one or many measured disturbances (MD) occurs. The aim of the controller is to maintain the controlled variable to the reference value.

A system using FFCS has a reactive-preventive action, which means it reacts when MD change and its target is to compensate their influence before a deviation of the controlled variable occurs. The main advantage of a system using FFCS is

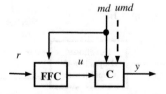

Fig. 6.7 Block diagram for a feedforward control system: *FFC* Feedforward Controller; *CP* Controlled Process; *r* reference (set point); *u* command; *y* controlled variable; *md* measured disturbances; *umd* unmeasured disturbances

that when MD change, does not exist a transient state for the controlled variable; in other words, this variable does not differ from the set point value. Consequently, to determine the command, some algorithms required by the controlled process behavior are needed.

The feedforward control has also an important drawback: when the unmeasured disturbances change, errors that can not be removed occur.

The principle that FFCS is based on has an universal character, in the sense that it can be applied to any controllable, observable and accessible system [3].

In the following, significant examples of real systems from economics and technical areas will be presented, examples meant to constitute arguments for the unitary approach to FFCS.

The example in the economical field, illustrated in Fig. 6.8, deals with the same trading company referred in the Feedback Control Strategy section. As Fig. 6.8 depicts, the company management receives through the marketing department information regarding market state and trends (therefore regarding a disturbance). In this context, one can say that commands are computed based on the economical objectives and this information.

When a disadvantageous market position occurs (adverse to the supply of the company such as: demand reduction, new competitor(s) emergence etc.), the management will adopt those decisions and will generate those commands adequate

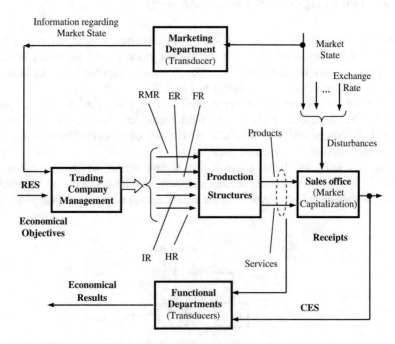

Fig. 6.8 Feedforward control of a trading company: *RMR* Raw Material Resources; *ER* Energy Resources; *IR* Information Resources; *FR* Financial Resources; *HR* Human Resources; *CES* Current Economic State; *RES* Reference Economic State

to maintain the results at the economical objectives level. The appropriateness (or efficaciousness) of the commands is conditioned by the management ability to quantify the impact of the new market state on the economical results of the company.

If other disturbances manifests, such as difficulties in raw materials supply or variations in exchange rate, outside the forecasted boundaries, the system wastes its functionality. Because the company management is not informed about these disturbances, corresponding measures will not be applied and as a consequence this condition will have an effect upon the economical results. As Fig. 6.8 shows, the management lack of reaction is explained by the absence of information regarding the current economical state of the company.

The next example belongs to the technical area and handles the feedforward control for the temperature of the fluid that is heated in a heat exchanger. As Fig. 6.9 depicts, the steam flow rate as manipulated variable, is determined by processing in real time the data provided by the sensors associated to the measured disturbances: Q_{CF} (Cold Fluid Flow Rate) and T_{CF} (Cold Fluid Temperature).

Processing the data is made in a model with two sections:

- a static section based on the thermal balance;
- a dynamic section associated to the inertia of the disturbances-controlled variable channel.

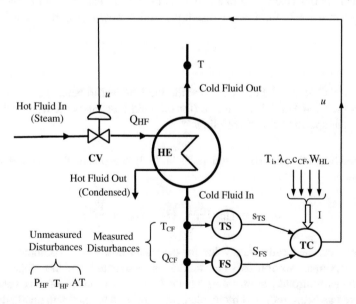

Fig. 6.9 Feedforward control of temperature: *HE* Heat Exchanger; *TC* Temperature Feedforward Controller; *TS* Temperature Sensor (Transducer); *FS* Flowrate Sensor (Transducer); s_{TS}, s_{FS} Signals from sensors; *CV* Control Valve (Actuator); *I* Inputs; T_i Temperature Reference (Set Point); *T* Controlled Temperature; *AT* Ambient Temperature, T_{HF}, P_{HF} Temperature, Pressure of Hot Fluid (Steam)

Fig. 6.10 Thermal balance illustration: W_{HP} Heat Produced Flow Rate; W_{HU} Heat Used Flow Rate; W_{HP} Heat Produced Flow Rate

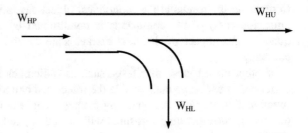

In the thermal balance are present the caloric flow rates associated to the following types of heat (Fig. 6.10): the heat produced by steam condensation, the heat used to warm the cold fluid, the heat lost in the environment.

According to representation in Fig. 6.10, the thermal balance equation has the form:

$$W_{HP} = W_{HU} + W_{HL}. \tag{6.3}$$

Heat is produced by steam condensation, and the caloric flow rate is as follows:

$$W_{HP} = Q_{HF} \cdot \lambda_C, \tag{6.4}$$

where Q_{HF} is the steam flow rate and λ_C is the latent heat of steam condensation.

To warm the cold fluid heat is used; the caloric flow rate in this case is:

$$W_{HU} = Q_{CF} \cdot c_{CF} \cdot (T_i - T_{CF}), \tag{6.5}$$

where
Q_{CF} is the Cold Fluid Flow Rate (for the fluid subject to heating);
T_{CF} is the Cold Fluid Temperature (for the fluid that enters the heat exchanger);
c_{CF} is the specific heat of cold flow

Embedding relations (6.4) and (6.5) in (6.3) we obtain the thermal balance equation:

$$Q_{HF\,st} = Q_{CF} \cdot \frac{c_{CF}}{\lambda_C} \cdot (T_i - T_{CF}) + \frac{W_{HL}}{\lambda_C}. \tag{6.6}$$

and from this, the steady state value for the steam flow rate is computed.

The dynamic section of the model is associated to the delay in the disturbances-controlled temperature channel. If the dynamics of this channel is considered as being reflected by an element of order one, the mathematic dynamic model has the form:

$$a_T \frac{dQ_{HF}}{dt} + Q_{HF} = Q_{HF\,st}, \tag{6.7}$$

respectively

$$a_T \frac{dQ_{HF}}{dt} + Q_{HF} = Q_{CF} \cdot \frac{c_{CF}}{\lambda_C} \cdot (T_i - T_{CF}) + \frac{W_{HL}}{\lambda_C}, \tag{6.8}$$

where a_T represents the time constant associated to the delay in the disturbances-controlled temperature channel.

The relation (6.8) indicates as inputs for the controller TC, besides the measured disturbances the reference T_i, the latent heat of steam condensation λ_C, the specific heat of cold flow c_{CF} and heat loss flow rate W_{HL}. These four last measures are grouped in the inputs vector I, presented in Fig. 6.9.

The system is built such that any change of a measured disturbance (Q_{CF} or T_{CF}) does not influence the controlled parameter T to deviate from the reference value T_i. For example, if the flow rate Q_{CF} increases, the controller TC will determine such an evolution for the steam flow rate Q_{HF} [accordingly to relation (6.8)] that maintains the temperature at the value T_i. If an unmeasured disturbance is changed or the reference is changed a non-removable error will occur.

The presented examples prove the unitary character of the feedforward control strategy, all cases having the structure highlighted in Fig. 6.7. The presence of two adjacent components comes into sight: the sensor and the actuator. These components allow the access to the process to obtain information form it (the sensor) and to implement the command to it (the actuator). Also in the case of FFCS, the sensor(s) and the actuator can be integrated alongside the controller in the automation device or alongside the process in the extended process (the fixed part).

6.1.4   The Combined Control Strategy

In the previous sections the advantages and the disadvantages of the feedback and the feedforward control strategies were highlighted. In order to keep the advantages and to limit the disadvantages, a combined control strategy (CCS) is the solution. In the following, for the "controlled subsystem", referred in the Sect. 6.1.1, it will be used the term "controlled plant" or "controlled process" and for the "command subsystem" it will be used the term "combined controller" (that includes the feedback controller and the feedforward controller). All these three entities are interconnected as Fig. 6.11 presents.

A system using CCS has:

- a reactive-preventive action when a measured disturbance is changed (the feedforward controller is active) and
- a reactive-corrective action when the reference and/or any unmeasured disturbance is changed (the feedback controller is active).

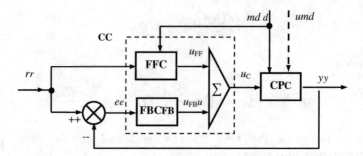

Fig. 6.11 Block diagram for a combined control system: *CE* Comparator Element; *CC* Combined Controller; *FBC* Feedback Controller; *FFC* Feedforward Controller; *CP* Controlled Process; $\sum$ Summation Element; *r* reference (set point); *e* error; u_{FB} feedback command; u_{FF} feedforward command; u_C combined command; *y* controlled variable; *md* measured disturbances; *umd* unmeasured disturbances

The command u_C prompted to the controlled process is obtained by summing the commands u_{FB} and u_{FF} generated by the two controllers when they are active. The CCS, in contrast with FFCS, comes up with an advantage: the controlled variable does not pass through a transient state when changes of the measured disturbances occur. In addition, the existing error when the reference and/or an unmeasured disturbance occur is removed. The disadvantages of the two strategies (the transient state for the controlled variable and the inability to remove some deviations of the controlled variable against the set point value) are partially rejected in an adequate manner. In the following the examples in the economics and technics fields are resumed in the CCS context.

The example in the economic field handles the same trading company referred in the Feedback Control Strategy presentation. As Fig. 6.12 describes, the company management receives information regarding the market state and trends (therefore regarding a disturbance) from the marketing department, and information regarding the economical results (practically the receipts obtained by selling the products and/or services) form the functional departments. In this context, one can say that the commands are elaborated taking into account the economical objectives, the market state and the economical results.

When other disturbances occur (such as: demand reduction, new competitor(s) emergence, exchange rate modification etc.) or RES changes, a deviation of CES against RES is present. The trading company management will adopt therefore such decisions and will generate such commands to converge the results to the economical objectives. The former reasoning proves that the disadvantages of FBCS and FFCS are reduced, while their advantages are maintained.

The second example covers the technical area and handles the combined control for the temperature of the fluid that is heated in a heat exchanger. As Fig. 6.13 depicts, the command u_C applied to the Control Valve has the components u_1 and u_2, corresponding to the feedback, respectively feedforward sections. These two components can be computed adequately applying relations (6.2) to u_1, and (6.8) to u_2.

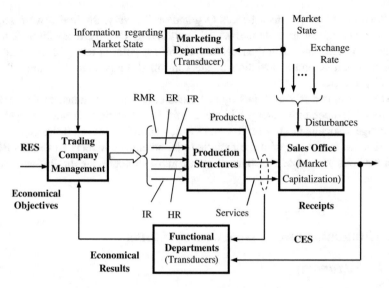

Fig. 6.12 Combined control of a trading company: *RMR* Raw Material Resources; *ER* Energy Resources; *IR* Information Resources; *FR* Financial Resources; *HR* Human Resources; *CES* Current Economic State; *RES* Reference Economic State

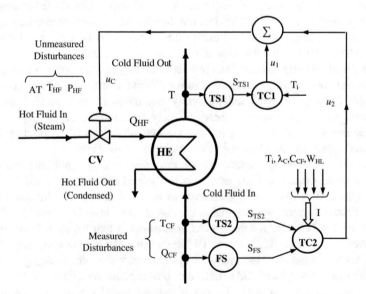

Fig. 6.13 Combined control of temperature: *HE* Heat Exchanger; *TC1* Temperature Feedback Controller; *TC2* Temperature Feedforward Controller; *TS1, TS2* Temperature Sensors (Transducers); *FS* Flowrate Sensor (Transducer); S_{TS1}, S_{TS2}, S_{FS} signals from sensors; u_1, u_2, u_C command signals; Σ Summation Element; *CV* Control Valve (Actuator); *I* Inputs Vector; T_i Temperature Reference (Set Point); *T* Controlled Temperature; *AT* Ambient Temperature; T_{HF}, P_{HF} Temperature, Pressure of Hot Fluid (Steam)

When the measured disturbances Q_{CF} and/or T_{CF} vary, the feedforward section is active, and the steam flow rate is such modified to keep the temperature T to the value T_i. When other disturbances vary or the set point is changed, the feedback section is active, and a deviation of the controlled temperature T against the set point T_i occurs.

This example also shows that by combined controlling no transient state for the controlled temperature emerges if the considered disturbances are changed (the advantage associated to the FFCS), and the deviation is removed if the unmeasured disturbances or the set point are modified (the advantage associated to the FBCS). Simultaneously, the disadvantages of both control strategies are diminished or even removed.

6.2 Fundamentals of Genetic Algorithms

6.2.1 Taxonomy

Being available a vast academic (and empirical) knowledge of algorithms, real-world problems and research areas, a comprehensive understanding of Genetic Algorithms (GA) requires first of all an adequate framing into the research scope.

GAs ⊂ Evolutionary Computation ⊂ Computational Intelligence ⊂ Artificial Intelligence. Computational Intelligence (CI) is a sub-symbolic paradigm of Artificial Intelligence that addresses difficult problems by imitating (often natural) systems which solve this kind of problems in a practical way, even if a formal mathematical description of the problem is lacking. Its methodology is a non-standard one: intelligence occurs from collective interactions of a large number of simple entities that work independently, continuously, in a parallel and interconnected way. Knowledge is here diffusely represented, in an implicit way. Evolutionary Computation (EC) follows the evolution "ability" to produce highly optimized structures, as observed in the biological evolution.

GAs do not learn by accumulating knowledge, but by modifying the global structure of the system; in fact, they are rather trained than programmed. According to CI approach, a dynamic system evolves and reaches good solutions with no visible central control, while remains consistent. An "invisible hand" produces a self-organizing behavior for the system; interesting is that the causal principles in the lower-level are not known. The CI algorithms only imitate good behaviors in nature, such as mind processes, immune system protection, ants foraging and so on.

A similar idea places GAs into the bio-inspired paradigm of Distributed Intelligence, in contrast with the organizational paradigm and the ontological paradigm [5]:

- the entities (the candidate-solutions) have individual goals, not shared goals;
- the system (the population) function upon the stigmergy principle: the entities are not aware on others, they rather have the ability to apprehend relevant local information;
- the actions of an entity do not promote attaining the goals of the other entities.

David E. Goldberg, one of the prominent researchers in Evolutionary Computation, evaluates GAs as dynamic complex systems which behave differently in different points of the search space [6]. Even the simplest GAs are nonlinear, discrete, stochastic, big algorithms that operate on problems of near infinite diversity.

GAs ⊂ Evolutionary Algorithms ⊂ Evolutionary Search Algorithms. Evolutionary Search Algorithms (ESAs) are best suited to problems with huge search space where very little information on the searching success is available. More precisely, ESAs are search algorithms adequate to problems where [7]:

1. the solutions can be evaluated by a function with values in an ordered set;
2. there are many possible solutions besides those who perfectly solve the problem;
3. the objective is to find the best solution in a given computational time.

Such problems are named fitness-based search problems.

Synthetically, ESAs are prescriptions to "guess" good solutions, based on knowledge of the problem kept only in a population of previously evaluated solutions [7]. Among ESAs many methods were developed; the "guessing" method makes the difference, and the performance of an ESA is determined based on the number of efficient "guessing" trials in a given time.

Evolutionary Algorithms (EAs) manage the population of candidate-solutions, named individuals, using two (generally stochastic) functions: generate new individuals and reject individuals. The "evolutionary" label has two reasons: the manner of storing knowledge about individuals and its usage to generate new individuals. Beginning regularly with very restricted knowledge of the environment, an EA learns to solve the problem with a reasonable level of competency even in dynamic conditions [8]. The evolutionary principles used by the EAs may be applied in various ways; therefore, many different approaches were designed: hill-climbing search, simulated annealing, genetic algorithms, genetic programming, evolution strategies, evolutionary programming, classifier systems, differential evolution. The diversity occurs in relation to types of individuals alteration, to data structures used for solutions representation and/or to methods of parents selection [8]. GAs reach the solution(s) focusing on the genetic operators as found in the biological evolution: natural selection, genetic inheritance and random mutations.

GAs ⊂ Metaheuristics. The stochastic character of GAs and the best-suited-for problems (fitness-based search problems) place the GAs in the class of metaheuristics. A GA is a population-based metaheuristic that uses learning strategies to optimize an objective function. Glover and Kochenberger in [9] define methaeuristic as a problem solver that manages an interaction between local improvement methods and higher level strategies to search a solution in a robust way, while escaping from local optima regions of the search space.

The main ideas used to build genetic algorithms—genetics and species evolution theory—are presented hereafter.

6.2.2 Inspiration and Strategy

John H. Holland, the father of genetic algorithms, which investigated during over forty years this evolutionary paradigm both theoretically and practically, stated in 1992 that [10]: *"Living organisms are consummate problem solvers. They exhibit a versatility that puts the best computer programs to shame. This observation is especially galling for computer scientists, who may spend months or years of intellectual effort on an algorithm, whereas organisms come by their abilities through the apparently undirected mechanism of evolution and natural selection. Pragmatic researchers see evolution's remarkable power as something to be emulated rather than envied"*. Related to this golden character of living organisms, which also belongs to other natural systems [11]—atomic structures, molecules, crystalline structures, water, solar systems, galaxies—there are two different perspectives: evolutionism and creationism. The evolutionism point of view pretends that species are formed by evolution, by successive transformations starting from other organisms, and not by spontaneous generation.

The *natural selection*, on one hand, was considered by Charles Darwin the central mechanism that guides the species modification during many generations. On the other hand, *heredity*, proved by the parent of genetics, Gregor Mendel, related to plants and bees, put focus on the inheritance of traits from generation to generation. Once combined the Darwin's natural selection theory with the theory of genetics issued by Mendel and Weismann's selectionism, the neo-darwinian paradigm occurred. Hence, a link between the evolutionary units (the genes) and the mechanism for the evolution (the natural selection) was made [12].

In biology, evolution and heredity are highly correlated with genes and chromosomes. A *gene*, simply speaking, is the basic physical and functional unit of heredity, made up of DNA,[1] which determines in the organism a specific feature. The alternative forms (or values) of a gene, given by the small differences in their sequence of DNA bases are called *alleles*. Different alleles of a gene produce different expressions for the corresponding feature. In the nucleus of each cell, the DNA molecule is packaged into thread-like structures called *chromosomes*,[2] only visible during cell division [13]. The organism's complete set of DNA, including all

---

[1]DNA is the hereditary material in all living organisms; nearly every cell of the body has the same DNA. The information in DNA is stored as a code formed of four chemical bases [13].

[2]Valid definition for the large majority of living organisms, whose cells are structured in nucleus and cytoplasm.

of its genes, forms the *genome* of that organism, found in almost every cell of it. Each genome comprises all the genetic information needed to build and maintain the organism [13]. A very interesting aspect is the special conformation that the genome adopts: consistent with a fractal globule and knot-free, which enables the cell to include DNA with high density,[3] simultaneously remaining perfectly functional by its ability to easily fold and unfold any segment in the genome [14]. Two organisms whose alleles differ in at least one gene have different *genotypes*. The *phenotype* of an organism designates, in contrast, the ensemble of visible physical features of the individual; it is determined by the hereditary base (its genotype) and the environment conditions. In other words, the phenotype is the behavioral expression of the genotype in a specific environment [15].

Cell division stands for the process where the cell divides in two usually identical cells and involves DNA replication: creation of a copy of every gene in the genome. Most of the processes in the organism are based on cell division: growth, development and reproduction. Inheritance of traits from parents is particularly explained by the fact that copies of alleles in parents' genes occur in the cells of offspring.

In the theory of evolution, the ability of an individual to survive and to reproduce, called *fitness*, measures its performance to fit into the environment. Note that the fitness exhibits the phenotype. In time, after many generations, the individuals with better fitness became more common; this process is the so-called *natural selection*.

Occasionally, a permanent alteration in the DNA sequence that forms a gene occurs. This deviation, called *mutation*, may be a faulty DNA replication or an incomplete repairing of the DNA deterioration. It is either beneficial, harmful or neutral to the health of the organism, and its propagation to the next generations leads to visible changes in the species. During the meiosis type cell division, a genetic recombination emerges. This process, named *crossover*, consists in DNA segments interchanging between homologous chromosomes.

According to the neo-darwinian paradigm, the species *evolution* occurs strictly dependent on the changes from generation to generation of the inherited features of the organisms, and the causes of these changes are: variation, reproduction and natural selection. When the hereditary differences between organisms become more and more common or more and more rare in a population, we say that the population evolved.

The first works reporting simulation of the genetic evolution belong to Nils Aall Barricelli (1954) and Alex Fraser (1957). Here, the focus was on simulation and not on the usage of the evolution procedures. The origin of evolutionary algorithms is considered to be in the '60, when John H. Holland, professor of psychology, electrical engineering and computer science designed a programming technique called *genetic algorithm*, as a result of his investigations on mathematical analysis of the adaptation process in nature. Holland concluded that in nature combining

---

[3]About three trillion times bigger than in a microchip.

Table 6.1 Comparative taxonomy between species evolution and optimization problems solving

Species evolution	Optimization problems solving
Environment	Problem
Individual	Candidate-solution (or individual)
Chromosome	Abstract representation of the candidate-solution (or chromosome)
Gene (elementary segment in the chromosome)	Gene (elementary position in the chromosome)
Allele (possible value of a gene)	Allele (possible value of a gene)
Genotype	Search space (chromosome space)
Phenotype	Objective space (space of the objective values associated to the candidate-solutions; often, fitness space)
Performance	Performance (or fitness, quality)

groups of genes by crossover represents an essential part of the evolution, and his algorithm imitated the natural evolution regarding both crossover and mutation. In the book "*Adaptation in natural and artificial systems*" [16], published in 1975 and appreciated as the cornerstone in the evolutionary algorithms, Holland describes how evolutionary processes in nature can be applied for optimizing artificial systems. Here, the first GA was depicted. The theory of GAs extrapolates the concepts in biology in order to "evolve" a population of initial candidate-solutions of a problem. In Table 6.1 a comparative taxonomy between species evolution and optimization problem solving is presented, upon GA theory is based on.

The individual comprises information about the location of the corresponding solution in the decision space and its optimality.

The framework of a simple GA is depicted as follows:

```
1. t <- 0 (first generation)
2. pseudo-random initialization of the population P_t
3. evaluate(P_t)
4. while evolution is not ended
   4.1. t <- t + 1
   4.2. P'_t <- Ø
   4.3. selection in P_t
   4.4. crossover for the selected parents
   4.5. insert the descendents in the new population P'_t
   4.6. mutation for P'_t
   4.7. evaluate(P'_t)
   4.8. P_t <- P_t ∪ P'_t
5. return the best solutions in P_t
```

Initially, a population of individuals is (pseudo) randomly generated and each individual is evaluated based on the fitness value. While evolution is not ended, in every generation the genetic operators are applied: selection of parents to create new individuals, crossover of parents and mutation of offspring.

Crossover and mutation implement the *generate_new_individuals* stochastic function in the evolutionary algorithm scheme. By these two operators, the algorithm explores new regions in the search space and therefore tries to avoid a premature convergence to a local optimum solution. Over the generations, the quality of the candidate-solutions in the current population P_t increases based on this specific performance-guided process where the random factor plays an important role.

The evolution of the population is ended when a good enough solution is found or the a priori settled runtime was reached or any other criterion or a combination of criteria is satisfied. The "evolution" stands for the multifold moving in the search space during many generations, towards better and better areas. The solution of the GA is the best solution in the population on the last generation.

Generally, using a GA to solve an optimization problem implies certain elements:

- A proper genetic representation for the candidate-solutions;
- A procedure to generate an initial population of candidate-solution;
- An adequate fitness function;
- Specific genetic operators.

Related to that, any solution must have a corresponding representation, representations and solutions must mutually correspond each other, any representation produced by the genetic operators must have a corresponding solution, the offspring must inherit the useful features of the parents.

6.2.3 Theoretic Foundation of GA

We can state that the origins of Genetic Algorithms are represented by the fusion of *combinatorial techniques*, *evolutionary principles* and *genetic combinatorial techniques*, that took place in the 1960s, although the ideas behind all these techniques and principles had appeared and evolved in time beginning with the 13th century (by the works of Ramon Llull) [17].

The success of genetic algorithms in solving extremely difficult and complex problems was not readily accepted and understood by the researchers in applied mathematics. The direction of theoretical investigations aimed at explaining their complex behavior which is in contrast to the basic operators and the simplicity of the simple genetic algorithm [18].

The Schemes Theorem. The main hypothesis on which current research in the domain of GA is based is called *schemes theorem*. A *scheme* is a template that makes possible to encode the similarities between chromosomes, as such we can assert that "a single scheme encodes a set of chromosomes that have identical genes" [19].

Using binary encoding, a scheme is composed by an array of 0, 1 and * symbols, the symbol * means it can be replaced by any of the 0 and 1 symbols. As an example, the following schema S = (01**1) encodes four chromosomes: (01001), (01011), (01101), (01111). In terms of set theory a scheme represents a hyperplane in an n dimensional space, where n is equal to the number of genes in a chromosome, in our example, a five dimensional space $\{0, 1\}^5$. An important property of schemes is the number of 'wild card' symbols *, r. A scheme that has r symbols can be considered an element of the following set $\{0, 1, *\}^r$. Another important property is called *scheme order* $o(S)$ defined as the number of 0 and 1 symbols in a scheme. The defining length $\delta(S)$ represents the maximum distance between the first and the last fixed position in a scheme. This property is useful for capturing how compact the scheme is.

How adequate a scheme is to a specific problem can be computed by taking into account the fitness or adequacy of all the chromosomes that correspond to a certain scheme. Using the notation $\{v_{i1}, v_{i2},...,v_{ip}\}$ for chromosomes, the fitness is:

$$eval(S, k) = \sum_{j=1}^{p} \frac{eval(v_{ij})}{p} \tag{6.9}$$

where
k current generation,
p the number of individuals corresponding to scheme S

In the following the process of selection is considered, with the hypothesis of a new population having the same size and the possibility that a chromosome can be selected more than once. Having asserted these ideas, the probability to select a chromosome is:

$$p_i = \frac{eval(v_i)}{F(k)}, \tag{6.10}$$

where $F(k)$ represents the total fitness of the population for step k.

Following the selection process, the number of chromosomes corresponding to scheme S, $\xi(S, k + 1)$ can be computed using the following formula [20]:

$$\xi(S, k+1) = \xi(S, k) \cdot \frac{eval(S, k)}{\overline{F(k)}}, \tag{6.11}$$

where $\overline{F(k)} = F(k)/N_{pop}$ is the mean fitness of the population and N_{pop} is the number of individuals in the population, equal to the number of individual selections.

Equation (6.11) clearly shows that the number of chromosomes matching a specific scheme increases proportional to the fitness of the scheme. In [20] it is

Fig. 6.14 Crossover for two different schemes

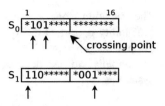

demonstrated that the fitness of a scheme grows exponentially with respect to $\xi(S, k + 1)$.

The processes of crossover and mutation are responsible for introducing new members in the population. If we consider that the crossover operator has one crossover point, than the surviving probability for a scheme S, p_s is:

$$p_s(S) = 1 - p_c \cdot \frac{\delta(S)}{m - 1}, \qquad (6.12)$$

where p_c is the probability of crossover, $\delta(S)$ is the defining length for that scheme and m is the total number of possible crossing points, an example being illustrated in Fig. 6.14.

The schemes S_0 and S_1 presented in Fig. 6.14 have defining lengths of $\delta(S_0) = 2$ and $\delta(S_1) = 11$ respectively; as a consequence, they have different chances of surviving, the scheme with a smaller defining length, S_0 having bigger chances.

The effect of mutation can be quantized with the aid of scheme order o, previously defined; a scheme with a bigger number of fixed symbols having a bigger chance to be modified by the mutation process. The resulted surviving probability is:

$$p_s(S) \approx 1 - o(S) \cdot p_m, \qquad (6.13)$$

where p_m is the mutation probability.

The three processes of selection, mutation and crossover with their specific operators have a cumulated effect on a scheme S for the k generation given by:

$$\xi(S, k + 1) \geq \xi(S, k) \cdot \frac{eval(S, k)}{\overline{F(k)}} \cdot \left[1 - p_c \cdot \frac{\delta(S)}{m - 1} - o(S) \cdot p_m \right], \qquad (6.14)$$

also called the *scheme theorem* [19].

It can be concluded that the schemes with above average fitness, having small defining length and small order are the fundament on which the genetic algorithm builds better solutions from one generation to another. These schemes have been called *building blocks* and the theory which uses them as starting point *Building Block Hypothesis*.

Markov Chains as Models for Genetic Algorithm. A *Markov Chain* is a theoretical model of a system which can be in a finite number of countable states.

The transition from one state to another at one moment k can be described as the probability $p_{ij}(k)$ that the system being in ith state initially (at $k - 1$ moment) will jump to the jth state [21].

Formally we can describe the probabilities for all the possible states s for a system as:

$$\Pr[0|k], \Pr[1|k], \ldots, \Pr[s-1|k], \qquad (6.15)$$

where $Pr[i|k]$ means the probability that the ith state occurs, given the kth state. An important concept for Markov chains is the *transition matrix* $Q_{ij} = Pr[i|j]$ which has an $s \times s$ size. If the process started with an initial distribution $p(0)$, then the *Chapman-Kolmogorov* equation [22] states that:

$$p(t) = Q^t p(0). \qquad (6.16)$$

For the Simple GA, if the population contains N individuals and the search space is n results a number of states given by the number of combinations:

$$\binom{n+N-1}{n}. \qquad (6.17)$$

The search space Ω will be $\{0, 1, 2, \ldots, n - 1\}$ and the population can be represented by $v = (v_0, v_1, \ldots, v_{n-1})$, where v_k is the number of copies of the individual k in the population, and we can write:

$$\sum_{k=0}^{n-1} v_k = N. \qquad (6.18)$$

With these notations, the fitness function is a function $f: \Omega \to \mathfrak{R}^+$ [22].

Author of work [22] identifies the effects of selection (fitness proportional selection), crossover ($r(i, j, k)$—the probability to generate k if i and j cross) and mutation ($U_{i,j}$—$n \times n$ matrix that gives the probability that j mutates to i) over the population u as presented below (Table 6.2):

An *absorbing state* of the Markov chain is a state with a probability of transition equal to 1. A genetic algorithm using only selection and mutation has the property of generating every possible population with different probabilities of occurring.

Table 6.2 Effects of selection, crossover and mutation

Process	Probability of i being selected	Transition matrix
Selection	$\Pr[i|v] = \dfrac{v_i f(i)}{\sum_{j \in \Omega} v_j f(j)}$	$\Pr[u|v] = N! \prod_{i \in \Omega} \dfrac{\Pr[i|v]^{u_i}}{u_i!}$
Crossover	$\Pr[k|v] = \sum_{i,j \in \Omega} r(i,j,k)\,\Pr[i|v]\,\Pr[j|v]$	
Mutation	$\Pr[i|v] = \sum_{j \in \Omega} U_{i,j}\,\Pr[j|v]$	

Elitism for a genetic algorithm is defined as the strategy of recording and selecting always the best individual in a specific generation. It has been demonstrated that when such a strategy is used and all the operators are applied, the algorithm will converge to a population containing the optimum [22].

Research has been conducted in order to improve the convergence of the genetic algorithm. One of the research directions is based on modifying the mutation rate using the annealing strategy inspired by the physical process of cooling, an example of cooling schedule being this:

$$\mu(t) \geq \frac{1}{2} t^{-\frac{1}{Nl}}, \tag{6.19}$$

where $\mu(t)$ is the mutation rate, N is the number of individuals, t represents the time (or the generation) and l is the length of the encoding string.

One of the main problems in applying Markov Chains to practical situations is the big size of the matrices involved. For a population of 50 members, the search space is 2^{50} and the number of possible populations is immense: 10^{688}. To be able to cope with such a big number of states, one method indicated in [23] is to group together the states that seem to be similar. However, a series of questions arises regarding the similarity degree necessary to group two states and in which way the probabilities between the groups formed have to be computed.

6.2.4   Components of GA

Efficient solving of a problem using evolutionary techniques requires first of all identifying the appropriate combination of representation both for the problem and the algorithm [7].

The goal of an optimization problem is to

$$\text{minimize } f : X \to \mathbb{R} \tag{6.20}$$

while $e(x) \leq 0$ is satisfied. Here, X is the search space, $x \in X$ is a decision vector, f represents the objective function, $X_f \in X$ is the feasible set (it comprises all the decision vectors that satisfy the constraint), and $Y_f = f(X_f)$ is the objective space. The maximization problem is similarly handled.

Genetic Representation. In GAs, the genetic encoding pattern for the candidate-solutions strongly influences the horizon of the searching procedure. Moreover, for a given problem various styles for building the search space exist, and some are easier to explore than others.

To search solutions in a space Sp with a GA, first of all mapping this space to the space of chromosomes Cr by a genetic encoding function is needed. In fact, Cr is the space manipulated by the GA:

Table 6.3 The main genetic encodings

Genetic encoding	Example of chromosome	Example of individual (candidate-solution)
Binary	01010111	Five (out of eight) robot components selected to be changed, namely the 2nd, the 4th, the 6th, the 7th and the 8th
Valued	5, 102, 20, 45, 207	Number of units (of five types products) to be manufactured in a shift
Permutation	3,6,1,5,4,2 B, D, A, C, F, E	Route of a salesman in a graph with six locations
Tree		Hierarchical structure for a team of seven members

$$\rho : Sp \rightarrow Cr. \tag{6.21}$$

Ideally, the function ρ must be a bijective function such that a candidate-solution is uniquely represented by a chromosome and a chromosome is the representation of a single candidate-solution.

Generally, a chromosome in GA is a string over a finite alphabet (binary or many-valued) or a tree or any other structure adequate to the solution representation. The literature reports four fundamental genetic encodings, as Table 6.3 depicts.

A particular representation limits the forms of genetic operators to be used and has impact over the evaluation way for the candidate-solutions.

Evaluation of the Candidate-Solutions. Similarly to the choice of the genetic encoding scheme, the choice of the fitness function has a significant role over the simplicity of the search. A fitness function must reflect in a realistic manner the value of the individuals and, simultaneously, must be easy to evaluate.

The fitness function is generally identical with the objective function: the closer to the minimum value of the objective function the chromosome is, the more fit in the population the individual is. Table 6.4 covers the main differences between uniobjective and multiobjective optimization problems regarding the fitness function.

Table 6.4 Uni/multi-objective problems parallel regarding fitness

	Uniobjective optimization problems	Multiobjective optimization problems
Feasible set	Totally ordered related to the objective function	Partially ordered related to the partial objective functions
Fitness function	Identical to objective function (in general)	Aggregated value of partial objective functions, or
		Alternating objectives, or
		PARETO dominance

Genetic Operators. The evolution through GA involves, in each generation, selection of some parents, recombination of genetic material in the parents in order to obtain offspring (crossover) and mutation.

Selection. The chance for each individual to be chosen as parent depends partially on its fitness and partially is random. Selection therefore favors some individuals to reproduce more than others. There are many types of selection, such as: roulette-wheel selection, tournament selection, elitist selection.

Crossover. Offspring must inherit the useful features of the parents. This is the most important criterion to choose the crossover operator and also to design the genetic representation. Examples of crossover operators are: one-point crossover, two-point crossover, arithmetic crossover, PPX, UX, OX, MSXF.

Mutation. The role of mutation in the genetic search is to locally improve the solutions generated by crossover and to keep the diversity in the population in order to avoid the premature convergence of the GA. To do this, mutation is applied with a reduced probability, as in nature. Another reason for this is that the GA uses stochastic procedures, but it is not designed to be a random search. Bäck ş.a. [15] differentiate gene mutations (when a particular gene is altered) by chromosome mutations (when the order of genes in the chromosome or the chromosome length is altered) and by genome mutations (when the genome is modified).

GA Parameters. A GA is a stochastic-based prescription to search a solution in a difficult space and there are some minimal parameters to tune this search: the population dimension (constant or variable), crossover rate or probability, mutation rate or probability, the maximum generation or the maximum time to end the evolution. The results of GA, as stochastic algorithm, are analyzed over many runs for each of many values of the parameters.

6.3   Chemical Process Control Using Genetic Algorithms

The *distillation process* has at its core the temperature difference between two or more components of a mixture which is distilled, the bigger the difference the easier the process of separation. Assuming a binary mixture, by successive vaporizations and condensations at *equilibrium*, the lighter component concentrates in vapor phase and the heavier component in liquid phase, these concentrations increasing or decreasing for every equilibrium stage.

Quantitative evaluations of these phenomena are based on the laws of Dalton and Raoult, Gibbs phase rule, different models for vapor pressure, various state equations and the concept of fugacity in the case of big deviation of the mixture from ideality. There are many methods to design the industrial distillation column, having different complexity (inverse proportional to the number of simplifying hypothesis), methods based on the laws and equations described previously. Having such complex processes, different optimization methods are very often implied in the design.

In the following section, it is presented the process of distillation for the binary mixture of propane-propene (C_3H_8-C_3H_6) that takes place in an industrial distillation column having specific operating conditions (pressure, temperature, composition), from the point of view of process control.

The main objective of the distillation process is the separation of the mixture in its components that are more valuable than the initial mixture. A product quality is in direct relation to its purity. To measure the purity/quality of a specific component, expensive and complex systems are necessary, called chromatographs. Sometimes it is preferred to take regular samples (hourly, for example) and analyze them by specialized laboratories. A greater purity than the process specification puts a greater demand on the plant and to the overall energy consumption, resulting in a bigger operating cost, a cost that is not reflected in a bigger income. The trade-off between product price and the operating costs can be optimized by operating the column in such a way as to have a minimum energy consumption correlated with the product price on the market [24].

A complex system such a distillation column needs simultaneous accomplishing of a number of demands [25]:

- maintaining the *material balance*—mean feed flow equals the sum of final product flows resulting in a constant *holdup*, allowed variations of these flows should not exceed a certain limit;
- maintaining products purities—the process control system has to eliminate in the shortest time the effects of disturbances upon the final products quality, the most important disturbances usually are: feed flow variations, feed composition, the pressure inside the column, temperature of cooling and heating agent;
- safe operation of the column—a number of constraints regarding safety limits for the process variables has to be fulfilled.

Illustrated in Fig. 6.15 is a binary distillation column for which there are presented the possible commands (Reflux flow—L, Distillate flow—D, Bottom flow— B, Cooling agent flow—Qc and Heating agent flow—Qr) and commonly measured variables (Distillate concentration—xD, Bottoms concentration—xB, Column pressure—P, Reflux drum level—HRD and Bottom/Reboiler Level—HB). Feed flow F represents usually the most important disturbance for a distillation column.

The choice for the 'command—measured variable' pair is a very important step in designing a control system for a distillation column [26]. Controlling an industrial distillation column for the propane-propene mixture imposes a control structure that is appropriate in the case of a high pressure column (about 21 bar gauge) that results in a low α coefficient (relative volatility coefficient for the mixture, usually $\alpha = 1.12$), which means a great effort to separate the two components [27] and as a direct consequence a high reflux flow L or a big reflux ratio R (usually $R = 10$–12), defined as:

$$R = \frac{L}{D}. \tag{6.22}$$

Researches regarding the propane-propene distillation column control systems has shown that the number of available commands for the distillation column is five

Fig. 6.15 Binary distillation
column commands and
measured variables

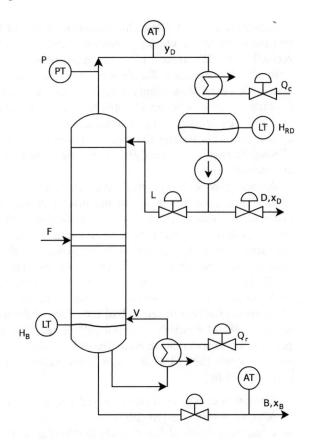

[24], a number of control structures could be chosen, among which the LB structure
has a good overall performance (LB structure implies that the L and B streams are
available for controlling the compositions of distillate and bottoms—xD and xB the
other commands being used for maintaining the column pressure and material
balance. The top (distillate) composition xD has a rigid specification (for example
xD = 0.92), imposed by economic factors as opposed to the bottom composition
which can be regarded as flexible (allowed variance interval of xB = 0.01–0.1).

6.3.1 Advanced Control Structures for Distillation Columns

The distillation process is known for its complexity, one aspect being its various
interacting loops that make the process difficult to control in order to achieve great
accuracy for the concentration set-points and also the shortest possible transient
period. The process of mass and heat transfer that takes place on every stage of the
column is characterized by very long time constants that induce difficulty in

designing the control system. One important aspect of controlling such processes is that the set-point has usually a constant value, some of the process variables are controlled by PID controllers so the main control system has to compensate the effect of a few important disturbances. A classical approach to designing a control system for a distillation column (based on linear transfer functions) is most often of little use, the process having multiple interactions and nonlinearities, in practice rule of thumbs are being used by the operators [28]. The weakest part of designing the control system is usually the lack of dynamic data for the process as a whole, although relatively accurate dynamic data can be available for the controller, valves and transducers.

Advanced control structures have been used beginning with the 1960s, the distillation process being one of the main processes considered for testing the efficiency of different types of control algorithms. Because feedback control has modest performance when the process has large time delays (for example when measuring chemical composition or molecular weight), feedforward control has been applied sometimes in combination with feedback control.

The advances in the computation capabilities of available hardware and software made possible the implementation of control algorithms that use process models and compute the control law in real time. One of the most used advanced control structure, Model Predictive Control (MPC) has been proposed in the 1960s but practical implementations in oil and chemical industries have been made beginning with the 1980s [29]. MPC is one of many predictive techniques [30] that can be characterized by:

- Utilization of a process model in order to predict the process output at future time moments (called horizon)
- Computing a control law that has to minimize a specific objective function
- A strategy to apply the control commands and to update the available information regarding the process evolution (Fig. 6.16).

Fig. 6.16 The structure of a model predictive controller

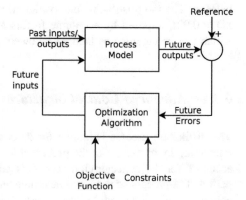

6.3.2   System Identification Using NARMAX Models and Genetic Algorithms

A model can be defined as a representation of the essential aspects of an existing system (or one that will be designed in the future) and presents the knowledge regarding the process in a form that can be further processed [31].

NARMAX models (Nonlinear AutoRegressive and Moving Average with eXogenous inputs) have been utilized since 1981, having the following mathematical expression:

$$
\begin{aligned}
y(k) = F[&y(k-1), y(k-2), \ldots, y(k-n_y), \\
&u(k-d), u(k-d-1), \ldots, u(k-d-n_u), \\
&e(k-1), e(k-2), \ldots, e(k-n_e)] + e(k)
\end{aligned}
\tag{6.23}
$$

where $y(k)$ is the system output, $u(k)$—system input, $e(k)$—noise, n_y, n_u and n_e—maximum lags for the output, input and noise sequence [32].

In practice, the function F (6.23) can be expressed in polynomial form, neural, wavelet or fuzzy models. In the following is presented an identification method based on genetic algorithms that uses a polynomial form for the F function:

$$
\begin{aligned}
y(k) =\; & a_{11}y(k-1) + a_{12}y(k-2) + \cdots + a_{1n_{y1}}y(k-n_{y1}) \\
& + a_{21}y(k-1)^2 + a_{22}y(k-2)^2 + \cdots + a_{2n_{y2}}y(k-n_{y2})^2 \\
& + a_{31}y(k-1)^3 + a_{32}y(k-2)^3 + \cdots + a_{3n_{y3}}y(k-n_{y3})^3 \\
& + b_{11}u(k-1) + b_{12}u(k-2) + \cdots + b_{1n_{u1}}u(k-n_{u1}) \\
& + b_{21}u(k-1)^2 + b_{22}u(k-2)^2 + \cdots + b_{2n_{u2}}u(k-n_{u2})^2 \\
& + b_{31}u(k-1)^3 + b_{32}u(k-2)^3 + \cdots + b_{3n_{u3}}u(k-n_{u3})^3.
\end{aligned}
\tag{6.24}
$$

The maximum number of coefficients for the past inputs and outputs has been taken to be 10 (n_{y1}, n_{y2}, n_{y3}, n_{u1}, n_{u2}, n_{u3}), so a total number of 60 polynomial terms can be included. Coding of the solutions in the genetic algorithm utilize real encoding, the coefficients (a_{11}, a_{12}, …, b_{11}, b_{12}, …) having values between $-1\ldots1$ and where grouped in C structures. The fitness function or the model error was computed with the relation:

$$
E = \frac{1}{N} \sum_{i=0}^{N-1} (y(i) - y_{id}(i))^2,
\tag{6.25}
$$

where N represents the number of samples taken from the process data, y and y_{id} are the actual output and the identified output. The actual data was generated from a special built simulator for the propane-propene distillation process, real experiments on a real distillation column being almost impossible to undertake.

Fig. 6.17 Schematic
representation of the
identification process

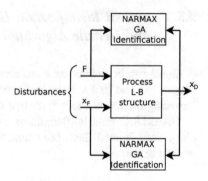

The identification of the process has been carried out on the LB process structure illustrated schematically in Fig. 6.17.

By using well known genetic operators selection—rank and tournament, continuous and discrete crossover with a specified probability p_c, uniform mutation over all the population having mutation probability p_m, different experiments have been made for various parameters of the genetic algorithm (population size, number of generations—stopping criteria, different values for mutation and crossover probability) for which step signals have been applied to the process inputs.

The following table summarizes the parameters of the genetic algorithms and the results obtained after a series of experiments where different model structures have been tried (Table 6.5).

The process model obtained is encoded for a compact notation, having the expression:

$$\begin{aligned}
y[k-1] &= 0.1578, \quad y[k-2] = 0.1573, \quad y[k-3] = 0.1565 \\
y[k-4] &= 0.1567, \quad y[k-5] = 0.1575, \quad y[k-6] = 0.172. \\
u[k-1] &= -0.003, \quad u[k-2] = 0.0153, \quad u[k-3] = 0.0151 \\
u[k-4] &= 0.0157, \quad u[k-5] = 0.0153
\end{aligned} \qquad (6.26)$$

In order to compare the output of the simulated process and the output of the identified process model, Fig. 6.18 represents both.

The results obtained in a simulated environment could be further refined with respect to real-time constraints for real systems by hardware implementations of the genetic algorithm. The implicit parallelism of the genetic algorithm represents a serious advantage when one wishes to increase its execution time, specialized

Table 6.5 Experiments using different GA parameters

Nr.crt.	Model	Select	Cross	Mut	Npop	Ngen	Fitness	Var
1	600/500	Tour	Cont/0.8	0.1	1500	1000	1×10^{-6}	0.01×10^{-9}
2	600/500	Tour	Cont/0.8	0.1	1500	400	49×10^{-6}	0.05×10^{-9}

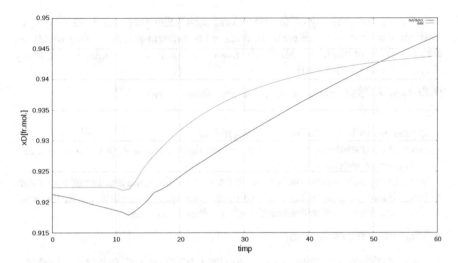

Fig. 6.18 NARMAX output and simulated process output

circuits such as FPGA's could then be used to implement the genetic algorithm population and also to speed up its fitness evaluation using different architectures and topologies for the final hardware implementation.

6.4   GA-Based Control Approach for Job Shop Scheduling

Among the multiple processes in the real world that can be efficiently computationally controlled (automated), planning and scheduling processes in manufacturing fill up a wide area; the numerous ERP tools are only few examples. It's worthy of note that many particular processes require though specific models and/or computational implementations.

6.4.1   Job Shop Scheduling Processes

Any production plant is a complex system where a management system is implemented, often hierarchical, either in an explicit way or an implicit way. Its levels are: general management, production planning and production scheduling [33]. Regarding scheduling and scheduling control, a worthwhile aspect is to mention the multiple facets such a process holds: the complexity of the process, the parameters variability and dynamics, the optimization criteria etc. [34, 35].

The most frequently addressed scheduling process is linked to JSSP (Job Shop Scheduling Problems) and the main two reasons for this are its generality and its complexity. The majority of the real world scheduling processes is JSSP-type or JSSP-similar, involving different products with different manufacturing routes on multiple different machines. More information about the computational complexity of JSSP is further presented.

Definition A JSSP process is a scheduling process where [36]:

- *n* jobs are to be scheduled on *m* machines,
- each job passes through each machine one time at the most (a job is therefore a sequence of several operations),
- the jobs are heterogeneous: the routes on the machines, the number of operations per job and the processing times of operations are not identical for all the jobs,
- the minimization of makespan for all the jobs is required.

There are three *constraints*:

- the precedence constraint: for each job the operations have to be scheduled in a given order;
- the non-preemption constraint: once started an operation, it can not be interrupted;
- the resource capacity constraint: a machine processes a single operation at a time, an operation is executed by a single machine at a time.

If many objectives are required to be simultaneously satisfied, we refer to a multi-objective JSSP. If the scheduling process is flexible (regarding the sequences of operations or the set of adequate machines for an operation), we refer to a flexible JSSP. In [37] the multi-objective flexible JSSP within the systems theory context is widely described.

A JSSP *solution* is a valid schedule for the set of jobs. It can be represented in many ways: as Gantt chart, as disjunctive graph, PERT chart, CPM diagram, permutation of the set of operations plus associated timing, or as logic-type formulation. An *optimal solution* is a solution that minimizes the objective function.

Garey et al. [38] reported in 1976 that JSSP is a NP-hard combinatorial problem. Later on, the analysis of the vast research regarding JSSP instances revealed that instances with more than two jobs and more than two machines are even strongly NP-hard [34]. Therefore, for the industrial instances the unconventional models to design the process and the unconventional algorithms to solve the problem are suitable for.

JSSP Handling. Controlling JSSPs using (only) conventional optimization methods has a reduced efficiency, because the search space is discrete, vast, often nonlinear and multimodal; such condition makes very hard to explore the space with a weak method (at least in time-computational terms).

The procedural models, regularly based on metaheuristic approaches, proved in exchange to be more adequate controlling models for the industrial JSSP instances. The central characteristic of a procedural model is to place on a secondary level the model of the process to be controlled, and to give the main role in modeling to the algorithmic procedure that controls the system. Lately, such scheduling problems are tackled using procedural models, especially because they comprise difficulties in rigorous mathematical characterization.

These unconventional techniques include, among others: genetic algorithms and evolutionary algorithms in general, multi-agent systems (such as: ACO, PSO, Wasp Behavior Model, negotiation-based techniques), and they subscribe to a so-called procedural control of the automatable processes. A JSSP-specific procedural modeling is found in [39]. When the user searches more than one solution to the scheduling problem in a reasonable time, the population-based techniques should be used; such techniques often provide multiple equal quality solutions or, if not, other near best quality solutions can be easily delivered. Such requests are fair in the scheduling area: sometimes unpredictable events force rescheduling, sometimes certain conditions regarding the sequence of scheduled jobs must be slightly modified.

A genetic algorithm (GA) comes with several advantages: it simultaneously operates with many candidate-solutions, it can easily connect to other models and implementations, it has a low development cost, it can easily use parallel processing.

The first attempt to solve JSSPs with GAs was made by Davis [40] thirty years ago, for a simple scheduling instance. Hence, the pertinence of GAs to JSSP was put to a test. Later on, much research and many practical studies led to developing more adequate genetic operators, genetic representations, fitness functions (to evaluate the schedules), particular GA mechanisms, all of that made to best "genetically handle" the JSSPs [34–37, 40–48].

6.4.2   GA-Based Approach Specific to JSSPs

Genetic Representation and Evaluation. In 1991, Nakano and Yamada [41] use a binary representation for the candidate-solutions (which are schedules in this case); to each pair of operations (o_1, o_2) which are processed on the same machine a binary variable is associated: 1 if o_1 precedes o_2 in the schedule and 0 if o_2 precedes o_1. The advantage of such representation consists in possibility of using simple genetic operators, but the relative big dimension of the chromosome and the computational effort to repair the unfeasible new individuals are serious drawbacks for this encoding.

In 1994, Bean [44] proposed a random keys representation, where a chromosome is a sequence of random numbers in [0, 1]. These values (alleles of genes) are used to sort the jobs in the schedule using a mapping function from the $[0, 1]^n$ space to the literal space. Because crossover and mutation are applied to the chromosomes, not to the sequences of jobs, the offspring are always feasible. To note though that representations and solutions do not mutually correspond to each other: a sequence of jobs corresponds to many random keys arrays.

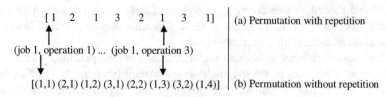

Fig. 6.19 Permutation representations for a candidate-solution in JSSP

A similar encoding proposed in [48] represents a schedule as a vector of priorities and delay times for all the operations. If q is the number of operations to be scheduled, the chromosome is an array of random keys over the alphabet $U(0, 1)$:

$$(pr_1, pr_2, \ldots, pr_q, d_1, d_2, \ldots, d_q). \tag{6.27}$$

The first q genes are operations priorities and the last q genes are used to obtain the delay times for the operations, according to the formula:

$$delay_j^l = d_j^l * 1.5 * \tau Max, \quad 1 \leq j \leq n. \tag{6.28}$$

Here, $delay_j^l$ is the delay of operation j at iteration l, d_j^l the value of gene j at iteration l, and τMax the maximum duration of all operations. This real valued representation based on random numbers allows using simple genetic operators and, additionally, ensures the feasibility of offspring.

The most appropriate genetic representation of schedules remains permutation (with or without repetitions) of the set of operations to be scheduled. As permutation with repetition, a gene of the chromosome is a job index, whose value is repeated as many times as operations per that job exists [42], as Fig. 6.19 shows.

Although the memory used is relatively low, either evaluating or modifying a chromosome or procuring information about a certain operation in the schedule need for us to decode at least partially the schedule. In order to avoid this drawback, a representation as permutation without repetitions may be employed. In this case, a gene encodes as well an operation, but in a direct manner, as a doublet (*job, operation*). For example, the allele (2, 1) marks the first operation of the second job. An extension of this representation, made in [43] for the flexible JSSP, encodes every operation as a triplet of form (*job, operation, machine*).

A correspondence between the genotype space, the space of individuals and the objective space for the permutation without repetition representation is showed in Fig. 6.20.

Whatever permutation type is used, a valid chromosome is decoded by means of two elements: the sequence of operations and the start processing times for every operation. The first element is explicit, the second is obtained by translating this sequence in active, semi-active or non-delay schedule [46, 47].

Another type of schedule representation is the sequence of rules, for example priority rules, which selects operations in the set of operations to be scheduled. In

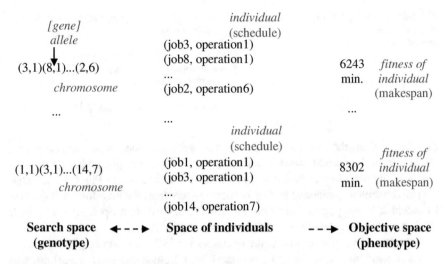

Fig. 6.20 Correspondence between genotype space, space of individuals and objective space for permutation without repetition representation (*Source* [35, p. 66])

this case the search space is the space of rules, the schedule is iteratively generated by an algorithm based on the rules, but complex crossover operators are necessary [34].

Regarding the *fitness function*:

- if the uni-objective case is considered, the schedule makespan gives the value of the fitness of individual;
- if the multi-objective case is considered, the fitness is evaluated considering the values of all the objectives, using either aggregation, or Pareto dominance relations or alternating objectives.

A more detailed analysis of representations and evaluation functions for the schedules is found in [35].

Particular Genetic Operators. Any problem tackled with GAs imposes constraints on the crossover and mutation operators by means of the chosen genetic encoding. The literature in this area reports many *crossover operators* such as: OX (Order Crossover [40]), UX (Uniform Order-Based Crossover), PPX (Precedence Preservative Crossover [45]), SXX (Subsequence Exchange Crossover), MSXF (Multi-Step Crossover Fusion), PMX (Partially Mapped Crossover), GPX (Generalized Position Crossover [42]), GOX (Generalized Order Crossover), GUX (Generalized Uniform Crossover), THX (Time Horizon Exchange), taskCrossover, CX (Cycle Crossover), LOX (Linear Order Crossover), POX (Precedence Operation Crossover) and operators for binary and real valued encodings.

For example, OX applied to two chromosomes using permutation without repetitions encoding generates one descendant as follows: first of all two positions in the chromosome are chosen, all genes between these positions are copied in the

Fig. 6.21 Applying OX on
two permutation
chromosomes

| Parent 1 | $(1,1)(2,1)\underline{(2,2)(1,2)}(2,3)(2,4)$ |
| cutting range: | $\underline{(2,2)(1,2)}$ |

Parent 2	$(2,1)(1,1)(2,2)(2,3)(2,4)(1,2)$
available genes,	$(2,4)(2,1)(1,1)(2,3)$
in proper order:	

| Descendant: | $(1,1)(2,3)\underline{(2,2)(1,2)}(2,4)(2,1)$ |

descendant from the first parent, then all non-selected genes in the second parent starting from the second position and followed by the genes placed before the first position fill up the descendant starting from the same position (as Fig. 6.21 depicts).

The descendant produced by OX is not necessary feasible; therefore, a validation to accept it in the population or an additional transformation operator is required. PPX, for example, certifies the feasibility of the result.

A wide presentation of crossover operators for JSSP is made in [49].

As regards *mutation*, the most operators are chromosome level mutations, and consist in moving pieces of genetic material from a position to another. Examples of mutations are: OBM (Order Based Mutation), SBM (Swap Based Mutation), frame-shift, translocation and inversion. None of them ensures the feasibility of the result. As an example, the translocation operator moves segments of adjacent genes from a position to another, hereby:

- if the chromosome before mutation is:

$$\sigma = (1,1) \mid (2,1)(2,2)(1,2)(2,3)\underline{(1,3)(2,4)}(2,5) \qquad (6.29)$$

the chromosome after mutation is:

$$\sigma = (1,1) \mid \underline{(1,3)(2,4)}(2,1)(2,2)(1,2)(2,3)(2,5). \qquad (6.30)$$

If the result of crossover or mutation is not valid, one of the following is performed:

- apply a legalization algorithm for the result;
- quit applying the operator at current step;
- recurrently apply the operator until a valid result is generated.

Case Study. The JSSP process taken as example covers an industrial plant with seven machines which operates weekly fifteen types of products (food supplements), such as: Multimineral (60 capsules per bottle), Calcium (100 tablets per bottle), Spirulina (90 tablets per bottle), Spirulina Powder (1 kg powder per package). The functions of the machines are as follows: M1 blends raw materials, M2 shapes the tablets/capsules, M4 dries tablets, M5 packs tablets/capsules in bottles, M3 packs powder in packages, M6 prints individual bottles/packages, M7

Fig. 6.22 The machines
routes in the shop

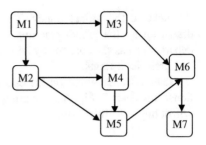

builds up the collective packs (Fig. 6.22). Each product has a certain route on the machines, together with particular processing times on the machines.

The single objective is to minimize the makespan for a weekly production plan, formed by 22 given jobs corresponding to the 15 products. Knowing that a job of a certain type demands a certain number of bottles or packages, the loading level for the shop for the week is easily determined.

Suppose that the total number of operations in the 22 jobs is 121. Using the permutation without repetition representation for the schedule, and jobs indices between 1 and 121, a chromosome is the following:

$$(4, 1)(18, 1)(4, 2)(5, 1)(18, 2) \ldots (8, 4).$$

GA-Based Control Approach. In a hierarchical approach for multiple range production systems [33], the scheduling function occurs on the third level. For the start, the link between the production planning (placed on the second level), the global planning, the operational control and scheduling is made (Fig. 6.23).

The global planning, covering one to three years, makes reference to the organizational production system definition (flow shop/open shop/job shop, unique products/multiple products, uni-objective/multi-objective etc.), to the life cycles of

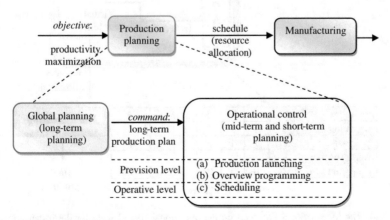

Fig. 6.23 The place of planning function in a production system (*Source* [33, p. 834])

products, to the global policy. The long-term production plan is the command drawn up by the global planning side for the operational control side. This last department has the role to rhythmically perform this plan regarding volume, sorts and prescribed terms.

Hereinafter, zooming on the operational control segment gives us the informational flowchart in Fig. 6.24, starting from the long-term production plan up to the scheduling of every operation on the corresponding machine. Here, i_1 (the

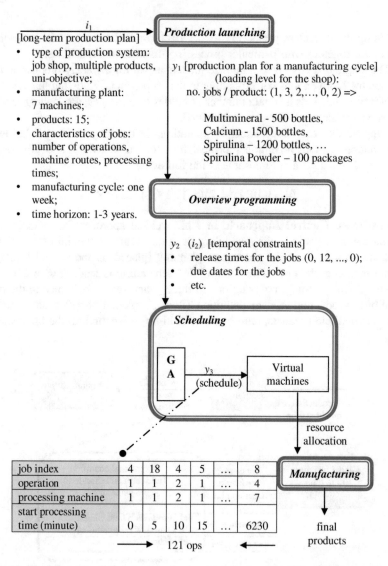

Fig. 6.24 The informational flowchart on the operational control level for the instance (*Source* [35, p. 110])

long-term production plan) is the input to the operational control segment, more specific to the production launching stage. y_1 (production plan for a manufacturing cycle, a week in our example) is the output of the production launching stage. The temporal constraints, i_2, are a direct input to the scheduling segment, alongside of y_1 and even i_1, as indirect inputs.

Based on these inputs, the scheduling stage of the control system acts accordingly with the genetic algorithm procedure and generates as output (y_3) the schedule associated to the set of jobs. By the specific nature of a population-based algorithm, we consider the virtual machines being the controlled process at this level. Once obtained or chosen the final generation best solution, this is run by means of resource allocation and therefore the physical manufacturing department obtains the final products.

6.5 Conclusions

The current chapter approaches the domain of process control through the technique of Genetic Algorithms. In the first part there are presented fundamental problems regarding automatic control emphasizing the feedback and feedforward control strategies. Advantages from the two strategies are united in a combined control strategy that offers a unitary approach to control, aspects underlined by practical examples from the areas of biology, economics and technology.

Genetic algorithms as a technique of Artificial Intelligence have at origin the darwinian principles of evolution, genetic laws and combinatorial principles. Their place in the field of Artificial Intelligence and Computational Intelligence and also their paradigm are discussed in section two, together with the main theoretical foundation on which the complex behavior of these algorithms is explained.

Two practical applications of Genetic Algorithms are presented in the sections three and four. The first one regards a process encountered very often in chemical engineering—fractionating distillation of a binary mixture. The nonlinearity and complexity of the process are important arguments for the application of advanced process control techniques in order to obtain good dynamical and steady state performances for the separation of the mixture components. Every advanced process control technique makes use of a process model, one form being the NARMAX model. The capabilities of Genetic Algorithms to explore complex and multivariable solution spaces are used in this case to optimize the NARMAX model online.

The second example is a well known NP-hard combinatorial problem, Job Shop Scheduling Process for a pharmaceutical industrial plant. Conventional optimization methods have reduced efficiency when dealing with discrete, nonlinear and multimodal search spaces. The application of a genetic algorithm in this case has clearly advantages because it can operate simultaneously with many candidate-solutions, it can connect to other models and implementations, has a low development cost and it can use parallel processing, an important aspect of the Genetic Algorithm that can be explored by designing a hardware implementation with the aid of recently developed high-performance FPGA circuits.

References

1. Cîrtoaje, V.: Automatic Systems Theory—Analysis in the Complex Domain (in Romanian). Petroleum-Gas University in Ploiesti Publishing House, Ploiesti (2013)
2. Wiener, N.: Cybernetics or Control and Communication in the Animal and the Machine. Wiley, New York (1948)
3. Penescu, C.: Systems—Concepts, Description, Linear Systems (in Romanian). Technical Publishing House, Bucharest (1975)
4. Paraschiv, N., Rădulescu, G.: Introduction in the Science of Systems and Computers (in Romanian). MatrixRom Publishing House, Bucharest (2007)
5. Parker, L.E.: Distributed intelligence: overview of the field and its application in multi-robot systems. J. Phys. Ag. **2**(1), 5–14 (2008)
6. Goldberg, D.E.: Genetic Algorithms in Search, Optimization and Machine Learning. Addison-Wesley, Reading, MA (1989)
7. Sharpe, O.J.: Towards a rational methodology for using evolutionary search algorithms. Ph.D. thesis, School of Cognitive and Computing Sciences, University of Sussex (2000)
8. Fogel, D.B.: Evolutionary Computation: Toward a New Philosophy of Machine Intelligence. IEEE Press, Piscataway, NJ (1995)
9. Glover, F., Kochenberger, G.A.: Handbook of Metaheuristics, International Series in Operations Research and Management Science. Springer, Springer Science & Business Media (2003)
10. Holland, J.H.: Genetic algorithms. Sci. Am. **267**(1), 44–50 (1992)
11. Nicoară, E.S.: Metaheuristics (in Romanian). Petroleum-Gas University in Ploieşti Publishing House, Ploieşti (2013)
12. Kutschera, U., Niklas, K.: The modern theory of biological evolution: an expanded synthesis. Naturwissenschaften **91**(6), 255–276 (2004)
13. Genetics Home Reference, Handbook. Lister Hill National Center for Biomedical Communications. US National Library of Medicine, National Institute of Health. Reprinted from http://ghr.nlm.nih.gov/handbook (2015)
14. Lieberman-Aiden, E., van Berkum, N.L., Williams, L., Imakaev, M., Ragoczy, T., Telling, A., Amit, I., Lajoie, B.R., Sabo, P.J., Dorschner, M.O., Sandstrom, R., Bernstein, B., Bender, M. A., Groudine, M., Gnirke, A., Stamatoyannopoulos, J., Mirny, L.A., Lander, E.S., Dekker, J.: Comprehensive mapping of long-range interactions reveals folding principles of the human genome. Science **326**(5950), 289–293 (2009)
15. Bäck, T., Fogel, D.B., Michalewicz, Z.: Evolutionary Computation 1, Basic Algorithms and Operators. Institute of Physics Publishing, Bristol and Philadelphia (2000)
16. Holland, J.H.: Adaptation in Natural and Artificial Systems. University of Michigan Press, Ann Harbor (1975)
17. Olteanu, M.: Genetic algorithms origins. PG Bull. Tech. S. **2**, 122–128 (2005)
18. Mitchell, M.: An Introduction to Genetic Algorithms. MIT Press, USA (1998)
19. Dumitrescu, D.: Genetic Algorithms and Evolutionary Strategies—Applications in Artificial Intelligence and Connected Domains (in Romanian). Blue Publishing House, Cluj-Napoca (2006)
20. Michalewicz, Z.: Genetic Algorithms + Data Structures = Evolution Programs, 3rd edn. Springer, Germany (1992)

21. Seneta, E.: Non-negative Matrices and Markov Chains. Springer Series in Statistics. Springer, New York (1981)
22. Reeves, C.R., Row, J.E.: Genetic Algorithms—Principles and Perspectives, A Guide to GA Theory. Kluwer Academic Publishers (2002)
23. Spears, W.M., DeJong, K.A.: Analyzing GAs using Markov models with semantically ordered and lumped states. In: Proceedings of the Foundations of Genetic Algorithms Workshop, pp. 85–100 (1996)
24. Marinoiu, V., Paraschiv, N.: Chemical Processes Automation (in Romanian), Vol. 1, Vol. 2. Technical Publishing House, Bucharest (1992)
25. Shunta, J.P., Buckley, P.S., Luyben, W.L.: Design of Distillation Column Control Systems. Instruments Society of America, USA (1985)
26. Shinskey, F.G.: Process Control Systems. McGraw-Hill, USA (1967)
27. Strătulă, C.: Distillation, Principles and Computing Methods. Technical Publishing House, Bucharest (1986)
28. Harriott, P.: Process Control. McGraw-Hill, USA (1964)
29. Morari, M., Lee, J.: Model predictive control. Past, present and future. Comput. Chem. Eng. **23**, 667–682 (1999)
30. Camacho, E.F., Bordons, C.: Model Predictive Control. Springer (1999)
31. Eykhoff, P.: System Identification. Wiley, USA (1974)
32. Billings, S.A.: Nonlinear System Identification. Wiley (2013)
33. Nicoară, E.S., Paraschiv, N., Filip, F.G.: A hierarchical model for multiple range production systems. In: 19th IFAC World Congress, pp. 833–838, CapeTown (2014)
34. Pinedo, M.L.: Scheduling. Theory, Algorithms, and Systems, 3rd edn. Springer Science-Business Media, LLC, New York (2008)
35. Nicoară, E.S.: GA-based control of multi-objective flexible job shop scheduling processes (in Romanian). Ph.D. thesis, Petroleum-Gas University in Ploieşti, Romania (2011)
36. Brucker, P., Schlie, R.: Job-shop scheduling with multi-purpose machines. Computing **45**(4), 369–375 (1990)
37. Nicoară, E.S., Filip, F.G., Paraschiv, N.: Simulation-based optimization using genetic algorithms for multi-objective flexible JSSP. Stud. Inform. Control. **20**(4), 333–344 (2011)
38. Garey, M.R., Johnson, D.S., Sethi, R.: The complexity of flowshop and jobshop scheduling. Math. Oper. Res. **1**, 117–129 (1976)
39. Nicoară, E.S.: Procedural optimization models for multiobjective flexible JSSP. Informatica Econ. **17**(1), 62–73 (2013)
40. Davis, L.: Job shop scheduling with genetic algorithms. In: Greffenstette, J.J. (ed.) Proceedings of the International Conference on Genetic Algorithms and Their Applications, pp. 136–140. Morgan Kaufmann (1985)
41. Nakano, R., Yamada, T.: Conventional genetic algorithms for job-shop problems. In: 4th International Conference on Genetic Algorithms, pp. 477–479, San Diego (1991)
42. Bierwirth, C.: A generalized permutation approach to job shop scheduling with genetic algorithms. OR Spektrum. **17**, 87–92 (1995)
43. Kacem, I.: Scheduling flexible job-shops: a worst case analysis and an evolutionary algorithm. Int. J. Comput. Intell. Appl. **3**(4), 437–452 (2003)
44. Bean, J.: Genetics and random keys for sequencing and optimization. ORSA J. Comput. **6**, 154–160 (1994)
45. Bierwirth, C., Mattfeld, D., Kopfer, H.: On permutation representations for scheduling problems. In: Voight, H.M., et al. (eds.) PPSN (Proceedings of Parallel Problem Solving from Nature), vol. IV, pp. 310–318. Springer, Berlin (1996)
46. Bierwirth, C., Mattfeld, D.C.: Production scheduling and rescheduling with genetic algorithms. Evol. Comput. **7**(1), 1–17 (1999)
47. Jensen, M.T.: Robust and flexible scheduling with evolutionary computation. Ph.D. thesis, Aarhus University, Denmark (2001)

48. Goncalves, J.F., deMagalhaes Mendes, J.J., Resende, M.G.C.: A hybrid genetic algorithm for the job shop scheduling problem. Comput. Ind. Eng. **45**(4), 597–613 (2003)
49. Aytug, H., Khouja, M., Vergara, F.E.: Use of genetic algorithms to solve production and operations management problems: a review. Int. J. Prod. Res. **41**, 3955–4009 (2003)

Chapter 7
Knowledge-Based Intelligent Process Control

**Mihaela Oprea, Sanda Florentina Mihalache
and Mădălina Cărbureanu**

Abstract In the last decades, the number of process control applications that use intelligent features has increased. This is mainly due to the complex and critical character of the process to be controlled. The intelligent process control systems works better than conventional control schemes in the domains of fault diagnosis (detection, cause analysis and repetitive problem recognition); complex control schemes; process and control performance monitoring and statistical process control; real time Quality Management; control system validation, startup and normal or emergency shutdown. Conventional control technologies use quantitative processing while knowledge-based integrates both qualitative and quantitative processing (having as target the increase of efficiency). This chapter presents an overview of intelligent process control techniques, from rule based systems, frame based systems (object oriented approach), hybrid systems (fuzzy logic and neural network). The focus is on expert systems and their extension, the knowledge based systems. Finally, an industrial case study is presented with conclusions to knowledge based systems limitations and challenges associated to real time implementation of the system.

7.1 Introduction

There is a tendency to design control systems with high autonomy and this leads to the development of automata with a higher or less intelligence degree. Autonomy means intelligence and therefore the design of autonomous control systems uses the

M. Oprea (✉) · S.F. Mihalache · M. Cărbureanu
Department of Automatic Control Engineering, Petroleum-Gas University of Ploiesti, Ploiesti, Romania
e-mail: m_oprea@yahoo.com

S.F. Mihalache
e-mail: sfrancu@upg-ploiesti.ro

M. Cărbureanu
e-mail: carbureanumada04@yahoo.com

© Springer International Publishing Switzerland 2016
K. Nakamatsu and R. Kountchev (eds.), *New Approaches in Intelligent Control*, Intelligent Systems Reference Library 107, DOI 10.1007/978-3-319-32168-4_7

implementation of control intelligence techniques. Intelligent process control is a subfield of the extensive field of intelligent machines.

A control technique is intelligent if uses methods and procedures specific to biological systems to design a controller for a specific process.

Intelligent process control systems with a high degree of autonomy works very well in the presence of significant process uncertainties and environment for extended time periods, compensating the malfunctioning without direct external intervention.

Process control can be implemented with decision support systems, and thus, can use a data-driven approach, an analytical approach or a knowledge-based approach. A proper expertise domain knowledge modelling is an important step toward the successful use of a knowledge-based approach to solve problems from that domain (see e.g. knowledge modelling in an air pollution control decision support system described in [58]).

Artificial intelligence (AI) provides a variety of techniques that can be applied in process control: artificial neural networks, fuzzy inference systems, adaptive neuro-fuzzy systems, knowledge-based systems, expert systems, case-based reasoning, machine learning, data mining, swarm intelligence, intelligent agents [65]. This chapter presents the AI techniques most used in process control, a brief review of the literature, and focuses on the knowledge-based intelligent process control, giving an industrial case study (pH control in a wastewater treatment plant).

The hybrid intelligent systems are an important research direction not only in the process control domain, but also for a better understanding of the natural concept of intelligence.

7.2 An Overview on Intelligent Process Control Techniques

The main classical artificial intelligence techniques used in process control are knowledge-based systems, rule-based systems, expert systems and case-based reasoning systems. They will be described in the next subchapter. The last three techniques are particular types of knowledge-based systems.

The basic computational intelligence techniques applied, so far, in process control either as standalone techniques or in combination with other techniques, include artificial neural networks (ANN), fuzzy inference systems (FIS), adaptive neuro-fuzzy systems (ANFIS), and genetic algorithms (GA). Other promising techniques include swarm intelligence techniques, such as Ant Colony Optimization (ACO), Particle Swarm Optimization (PSO), Artificial Bees Colony algorithm (ABC).

7.2.1 Artificial Neural Networks

Artificial neural networks are universal approximators capable to learn complex mappings [64]. An artificial neural network can be defined as a graph with artificial neurons as nodes and weighted connections between the neurons as arcs. An artificial neuron is a nonlinear processing unit that has a number of inputs, one output and an activation function. The architecture of an ANN is given by the graph structure.

There are several types of artificial neural networks: multi-layer perceptron (MLP) or feed forward artificial neural network, Radial Basis Function (RBF) neural network, Hopfield, Kohonen, Elman, Boltzmann, probabilistic ANNs etc. Each ANN type has a specific topology that is recurrent (i.e. has a feedback) or non-recurrent.

Figure 7.1 shows the architecture of a feed forward artificial neural network (FFANN) with an input layer (with n input neurons), an output layer (with m output neurons) and a number of hidden layers. The number of input neurons and the output neurons is derived from the problem that is solved with the ANN, while the number of hidden layers and the number of neurons for each hidden layer are derived by experiments. In a FFANN each layer is connected with the next layer, except the output layer. The connection between two neurons has a weight, i.e. a numerical value in the interval [0, 1], that express the degree of connection. A strong connection has a value of one, or a closer value, while a weak connection has a value closer to zero or a zero value. In this last case, the connection can be eliminated. The FFANN depicted in Fig. 7.1 is a totally connected ANN, i.e. each neuron is connected with all the neurons of the next layer, except the neurons of the output layer. If the weak connections are eliminated, the ANN is a partially

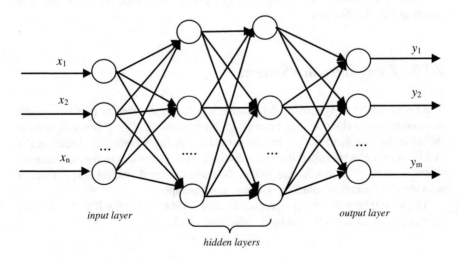

Fig. 7.1 The architecture of a feed forward artificial neural network

connected ANN. The weights are initialized randomly or heuristically (by using knowledge from the expertise domain), with values from the interval [0, 1]. The final values of the ANN weights are derived during the ANN training, by learning a set of examples (named the training set), i.e. a set of known inputs and outputs vectors: $X = (x_1, x_2, ..., x_n)$ and $Y = (y_1, y_2, ..., y_m)$.

The FFANN is trained by using a backpropagation learning algorithm. After the training phase, the FFANN is tested and validated on specific sets (testing set and validation set). The whole set of examples is usually divided in three sets: the training set, the testing set and the validation set. The training set must have enough training examples (much more than the other two sets), for a good ANN learning performance. The training algorithm can be optimized by introducing a variable learning rate and a momentum parameter.

The main advantage of an ANN is given by its learning or self-organizing capability. ANNs can be applied with success in a variety of applications: optimization problems (e.g. optimal process control), time series forecasting, object recognition, speech recognition, character recognition etc. They are proper methods for nonlinear control.

Artificial neural networks were proposed for the first time as a solution for the adaptive control of nonlinear dynamical systems in [54]. Later on, a variety of applications and improved ANN-based solutions were proposed for intelligent process control. The majority of them are based on two types of ANNs: FFANN and RBF neural networks. Examples of such solutions are given in [7, 53, 86].

A review of ANNs applications to statistical process control is presented in [88].

An extensive review of the neural networks applications in chemical process control is described in [29]. Three major control schemes are analyzed: predictive control, inverse-model-based control and adaptive control. In the majority of applications, MLP or FFANN were used for process control.

Some recent solutions of intelligent process control based on ANNs are proposed in [35, 42, 50, 83].

7.2.2 Fuzzy Inference Systems

The fuzzy inference systems (FIS) provide a solution to describe into a linguistic manner the behaviour of the process to be controlled. The process is usually described by a mathematical or linguistic model. The fuzzy controllers based on FIS architecture are more suited than conventional PID controllers in the case of nonlinear systems. The results are better than conventional control when there is additional information about the systems nonlinearities.

The architecture of a fuzzy inference system is presented in the Fig. 7.2. A fuzzy inference system is composed by 5 functional blocks:

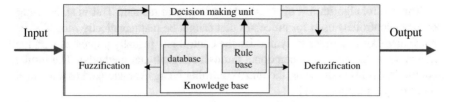

Fig. 7.2 The general architecture of a fuzzy inference system

- the fuzzification block: at this level the crisp value of the input is transformed into a set of matching degrees to membership functions describing the fuzzified input;
- the defuzzification block: the fuzzy set resulted from rule inference mechanism is transformed into a crisp value (Mamdani or Takagi-Sugeno fuzzy rules);
- the database block: the storage unit for membership functions parameters for both input and output fuzzy variables;
- the rule base block: the storage unit for the IF-THEN rules that describe system behaviour;
- the decision making block: the inference mechanism is applied to the rules from rule base block.

The mathematical modelling of the process usually starts with determining the membership functions for input and output variables. The tuning parameters for these membership functions are found using trial and error methods. This stage is followed by the construction of the fuzzy rules. The IF-THEN rules are the result of observations and experience of process operator. The last stage is applying the operations of inference to the fuzzy rules and then establishing the crisp output value and the modelling result. The standard structure of a fuzzy control system is presented in Fig. 7.3.

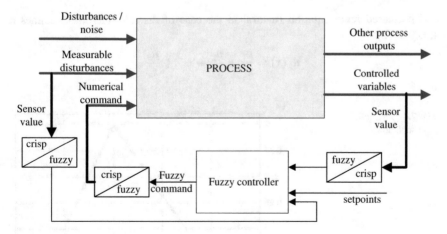

Fig. 7.3 The standard structure of a fuzzy control system

The control algorithm is synthesized in a linguistic manner. This is an advantage of fuzzy controllers used for processes that cannot be mathematically modelled in a convenient form in order to synthesize the control law (usually processes with high nonlinear behaviour, non-operational through conventional methods). The resulted control law is also nonlinear, the fuzzy controller being characterized as a nonlinear input-output mapping [31].

The standard structure of a fuzzy control system suggests that the control system can operate in feedforward and/or feedback manner. The fuzzy controller can have rules about measured disturbances if there is knowledge on the way that they affect the controlled variables (the feedforward control). The usual fuzzy controllers have two inputs: the error signal and derivative error signal (feedback control).

The input interface of the controller scales the measured process values and makes the conversion in linguistic variables through fuzzification stage. The process value is scaled and labelled according to the representation of fuzzy sets on the discourse universe. Most operations defined for fuzzy sets are for continuous universes. The operations include discrete universes as special cases. The control algorithm is computer-based and there are two possible representations of fuzzy sets: the functional representation and the paired representation.

The functional representation uses functions $\mu_A(x) = f(x)$ where A is the fuzzy set, $\mu_A(x) \in [0, 1]$ is the membership function of the element x. The fuzzy controllers use different membership functions with triangular shape, trapezoidal shape, Gauss curve like, exponential curves, bell shaped curves, sigmoid known from neural networks and singleton. For example, in the case of triangular shape from Fig. 7.4, the function is described by:

$$f(x; a, b, c) = \begin{cases} 0 & x \leq a \\ \frac{x-a}{b-a} & a \leq x \leq b \\ \frac{c-x}{c-b} & b \leq x \leq c \\ 0 & c \leq x \end{cases}$$

The paired representation (natural in the case of discrete universes) defines a fuzzy set by

$$\mu_A(x) = \frac{\mu_1}{x_1} + \frac{\mu_2}{x_2} + \cdots + \frac{\mu_n}{x_n}$$

Fig. 7.4 Triangular representation

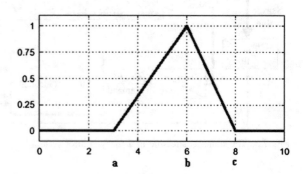

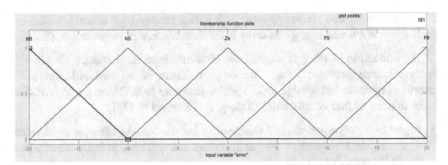

Fig. 7.5 Fuzzy representation of input error

If error (the difference between setpoint value and process value) is the input of fuzzy controller, the linguistic values can be represented like in Fig. 7.5.

The input error is represented with 5 triangular membership functions: NB—Negativ Big, NS—Negativ Small, Ze—Zero, PS—Positive Small, PB—Positive Big. If the error is −5 (crisp value) the fuzzy representation is [0 0.5 0.5 0 0] with the matching degrees of the crisp value to each membership function defined on the discourse universe.

The choice of shape and its parameters are subjective, Sozen et al. [69], reported that the fuzzification stage requires good understanding of all the variables. Jantzen [34] found a few rules for fuzzification stage:

- the noise in measurement should be take into account when establishing fuzzy set parameters;
- a certain amount of overlap is desirable (to ensure rule base continuity); otherwise the controller may run into poorly defined states, where it does not return a well defined output;
- if there is a gap between two neighbouring sets, no rule fire for values in the gap, the controller state is therefore undefined;
- the necessary and sufficient number of sets in a family depends on the width of the sets and vice versa.

In fuzzy logic an important concept is fuzzy proposition. A fuzzy proposition is a statement like "c is small", where "small" is a linguistic label defined by a fuzzy set on the universe of variable c. Fuzzy propositions connect variables with their defined fuzzy labels. Fuzzy propositions are the basis of fuzzy logic and reasoning. Fuzzy propositions can be combined using logical connectives like *and* and *or*. There are different operators proposed for and and or logical connectives presented in table.

The rules of fuzzy controller are in the IF-THEN format, the premise is on the IF side and the conclusion in the THEN side. The premise can be formed by a combination of propositions using logical connectives *and* and *or*. A fuzzy PI-like controller can have the rules:

1. If *error* is PS and *change in error* is NS then *change in command* is Ze.
2. If *error* is PS and *change in error* is Ze then *change in command* is PS.

The collection of rules is a rule base. The rule base must contain all possible fuzzy relations between inputs and outputs. There is no standard method to transform operator knowledge and expertise into rule base. There are at least four main sources to find control rules (Takagi and Sugeno in [38]):

- expert experience and control engineering knowledge. As example in literature the operation handbook of cement kiln plant that was used to build the rule base for FLS controller [26, 27].
- based on the operator's control actions. The rules derives from observing operator's control actions or from operators' intervention journal.
- based on a fuzzy model of the plant. The fuzzy rule base (that generates the control algorithm) may be viewed as an inverse model of the controlled process. Another approach is fuzzy identification [38, 59].
- based on learning. The self optimizing controller is an example of adaptive controller that generates the rule base.

A fuzzy PI-like [47] controller contains rules in linguistic IF-THEN format:

1. If *error* is NB and *change in error* is NB then *change in command* is NB.
2. If *error* is NB and *change in error* is NS then *change in command* is NB.
3. If *error* is NB and *change in error* is Ze then *change in command* is NB.
4. If *error* is NB and *change in error* is PS then *change in command* is NB.
5. If *error* is NB and *change in error* is PB then *change in command* is Ze.
6. If *error* is NS and *change in error* is NB then *change in command* is NB.
7. If *error* is NS and *change in error* is NS then *change in command* is NS.
8. If *error* is NS and *change in error* is Ze then *change in command* is NS.
9. If *error* is NS and *change in error* is PS then *change in command* is Ze.
10. If *error* is NS and *change in error* is PB then *change in command* is PB.
11. If *error* is Ze and *change in error* is NB then *change in command* is NB.
12. If *error* is Ze and *change in error* is NS then *change in command* is NS.
13. If *error* is Ze and *change in error* is Ze then *change in command* is Ze.
14. If *error* is Ze and *change in error* is PS then *change in command* is PS.
15. If *error* is Ze and *change in error* is PB then *change in command* is PB.
16. If *error* is PS and *change in error* is NB then *change in command* is NB.
17. If *error* is PS and *change in error* is NS then *change in command* is Ze.
18. If *error* is PS and *change in error* is Ze then *change in command* is PS.
19. If *error* is PS and *change in error* is PS then *change in command* is PS.
20. If *error* is PS and *change in error* is PB then *change in command* is PB.
21. If *error* is PB and *change in error* is NB then *change in command* is Ze.
22. If *error* is PB and *change in error* is NS then *change in command* is PB.
23. If *error* is PB and *change in error* is Ze then *change in command* is PS.
24. If *error* is PB and *change in error* is PS then *change in command* is PB.
25. If *error* is PB and *change in error* is PB then *change in command* is PB.

The output of controller is in this case the change in command because the controller is PI-like. The inputs of controller are the signal error and the change in error. The same set of rules is presented in relational format.

Error	Change in error	Change in command
NB	NB	NB
NB	NS	NB
NB	Ze	NB
NB	PS	NB
NB	PB	Ze
NS	NB	NB
NS	NS	NS
NS	Ze	NS
NS	PS	Ze
NS	PB	PB
Ze	NB	NB
Ze	NS	NS
Ze	Ze	Ze
Ze	PS	PS
Ze	PB	PB
PS	NB	NB
PS	NS	Ze
PS	Ze	PS
PS	PS	PS
PS	PB	PB
PB	NB	Ze
PB	NS	PB
PB	Ze	PS
PB	PS	PB
PB	PB	PB

The relational format is more compact and suitable for storing in a relational database. In this format it is assumed that the fuzzy propositions in all rules are connected with the same connective *and* or *or*. The tabular format is more compact:

		Change	in	Error		
Error		**NB**	**NS**	**Ze**	**PS**	**PB**
	NB	NB	NB	NB	NB	Ze
	NS	NB	NS	NS	Ze	PB
	Ze	NB	NS	Ze	PS	PB
	PS	NB	Ze	PS	PS	PB
	PB	Ze	PB	PS	PB	PB

The tabular form allow detecting symmetries, empty cells that indicate missing rules (this is particularly efficient when checking the rule base for continuity, completeness and consistency). The premises from IF part are usually connected with "*and*" connective. Using the "or" connective can generate a non consistent rule base.

There are two types of fuzzy control rules: Mamdani rules or Sugeno rules. The Mamdani rules were considered thus far in this chapter. The Sugeno rules have consequents that are (linear) functions of the controller inputs (*e*—error, *Δe*—change in error, *Δc*—change in command):

Mamdani   if *e* is NB and *Δe* is Ze then *Δc* is NB;
Sugeno   if *e* is PB and *Δe* is Ze then *Δc = e + 2 Δe*

The practical approach of fuzzy inference is formed by the following steps:

- matching the fuzzy propositions used in the rules' premises with numerical data (controller inputs)—fuzzification stage;
- determining the degree of fulfilment for each rule (the firing strength of the rules) using *and* and *or* connectives or a combination of them;
- determining the consequence (results) of each individual rule using implication;
- aggregation of the overall results of the individual fuzzy rules (usually *max* and *sum* are used for union operator) and defuzzification (the resulting fuzzy set conversion to single number as control signal to the process).

The most common inference method used in fuzzy control are denoted with two words: first represents the aggregation operator and the second is the implication function used (e.g. max-min, max-prod, sum-prod, etc.). The most used defuzzification methods in fuzzy control are center of gravity(cog), center of area (coa), mean of maxima (mom), fuzzy-mean (FM), extended center of area (Xcoa), center of gravity for singletons (COGS), the bisector of area (BOA). In robot control, the output must choose between left and right to avoid an obstacle, the position being the leftmost maximum (LM) or rightmost maximum (RM). The LM and RM are indifferent to the shape of fuzzy set with low computational effort.

The designer of a fuzzy controller must choose the controllers parameters related to controlled process. The design decisions refer to rule base related choices like the number of inputs and outputs, type of rules (preferably Sugeno type), universes, the number of membership functions, their overlap and parameters, the singletons for conclusions. The inference method must be chosen according to the preferred connectives, modifiers, implication operators and aggregation operators for the controlled process. Also the choice of defuzzification method between COG, COA, COGS, BOA, MOM, FM, LM or RM is a design stage. The control engineer must also establish interface connection parameters to the real process (scaling, sampling time, transformation in engineering units etc.)

The main advantages of the fuzzy controllers refers to the fact that they do not need a mathematical model of the process (shared with artificial neural networks), there is a priori knowledge within IF-THEN rules, the implementation and inter-pretation of the results that emulates human thinking is rather simple.

The main disadvantages of fuzzy controllers are:

- the process behaviour may be described by IF-THEN rules;
- there are no learning capabilities;
- there is no standard tuning method for fuzzy controller parameters;
- system adaptation to changes is difficult;
- the tuning of fuzzy parameters is trial and error method based.

The design of fuzzy controllers is the main task for control engineers. Lee [38] presents guidelines on how fuzzy controllers must be designed. An important book on fuzzy control is [59]. The theory associated to fuzzy controllers is extensively presented in [31]. The theoretical principles of fuzzy controllers are also presented in [16]. Control techniques based on fuzzy logic are covered in many aspects in [80]. Fuzzy model identification is covered in many papers one of the earliest being the paper of Takagi and Sugeno [73]. Babuska [3] offers methods of fuzzy model identification. Since hybrid methods appeared, [33] the studies have focused on developing new hybrid methods based on fuzzy logic. A complete review on fuzzy control can be found in [34]. The Fuzzy Logic Toolbox from MatLab [45] offers an efficient way to study different approaches in designing a fuzzy controller.

Industrial applications of fuzzy control are reported since the first experiments using a real cement kiln were made at F.L. Smidth & Co. cement plant in Denmark, in 1978 [34]. The presented case studies refer to home appliances (air conditioning, heating, clothes dryer, and washing machines) and process control. The main applications for process control were dedicated to nonlinear processes like reactors [2], hydrogen purification process [85] waste water—neutralization process [18] and other (decanter, incinerator, ethylene production, cooling, food processing) described in [71] and [78].

7.2.3 Adaptive Network Based Fuzzy Inference Systems

The ANFIS (Adaptive Network based Fuzzy Inference Systems) method com-pensates the disadvantages of fuzzy logic systems (high time consuming in designing and adapting a consistent, continuous and complete rule base and also trial-error methods in tuning membership functions parameters). Another disad-vantage of fuzzy based systems is the lack of standard methods for transforming human knowledge into a rule base. An efficient solution is completing fuzzy inference systems (FIS) with artificial neural networks (ANN) in order to enhance FIS capabilities with those of ANN. The result is the hybrid ANFIS with an architecture proposed by Jang [33], Fig. 7.6.

A fuzzy inference system is formed by five functional blocks: a fuzzification block that transforms a crisp value into a fuzzy set, a defuzzification block that converts a fuzzy set into a crisp value, the database block with the description of

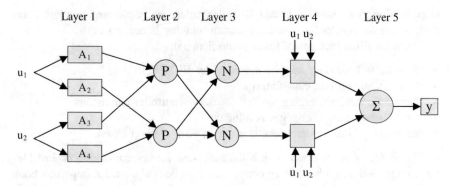

Fig. 7.6 ANFIS architecture

membership functions for input and output variables, a rule base block with the rules defined for FIS, and the decision block that performs the inference operations on the rules. The main advantages of a fuzzy system are the fact that it does not require a mathematical model of the process, the process description with rules emulates human thinking, the interpretations of the results is simple and rule based. The ANNs do not require a mathematical model of the process (this is also the case of FIS) and have learning/adapting capabilities (additional to FIS). A combination between the two architectures (neuro-fuzzy system) is capable to learn new rules or membership functions, to optimize the existing ones. The starting point is the training data in creating the rule base and membership functions. There are three possible situations:

- the rule base is empty and the ANFIS creates rules until the problem is solved. It is possible that the resulted rule base is inconsistent and oversized with negative consequences on time consuming, therefore it is better to have a minimum number of rules;
- the rule base is complete and the proper training eliminates some rules. Sometimes the result has a smaller number of rules than needed and the resulted rule base must be checked for consistency;
- the initial rule base has an fixed number of rules. Using learning mechanism old rules are replaced with new ones in order to maintain system size and time consuming.

In each situation the rule base must be checked for completeness, consistency and continuity. In Fig. 7.6 the fixed nodes are represented with circles while the adaptive nodes are squares. The proposed architecture has five layers. The Takagi Sugeno models are more efficient from computing effort perspective and results interpretation. For a two inputs $u = [u_1 \ u_2]^T$—one output y system the fuzzy rules are:

Rule 1   if u_1 is A_1 and u_2 is A_3 then $y = f_1 u_1 + g_1 u_2 + q_1$,
Rule 2   if u_1 is A_2 and u_2 is A_4 then $y = f_2 u_1 + g_2 u_2 + q_2$,

where A_i are the fuzzy sets associated to u_i inputs, and f_i, g_i and q_i are parameters that will be determined after training process.

Layer 1 The membership degree of an input variable to a fuzzy set is defined through membership functions. The i-node function is:

$$O_i^1 = \mu_{A_i}(u_j), \quad i = \overline{1,4}, j = \overline{1,2}.$$

where membership functions can have any analytical form. At this layer are formed the premise parameters (adaptive node).

Layer 2 Each node from this layer is denoted with P, the output being the product of input signals:

$$O_i^2 = w_i = \mu_{A_i}(u_1) \cdot \mu_{A_j}(u_2), \quad i = \overline{1,2} \, j = \overline{3,4}$$

Layer 3 The fixed i-node makes a normalized sum of the inputs:

$$O_i^3 = \overline{w_i} = \frac{w_i}{w_1 + w_2}, \quad i = \overline{1,2}$$

Layer 4 The adaptive i-node computes the contribution of i rule to ANFIS output:

$$O_i^4 = \overline{w_i} y_i = \overline{w_i}(f_i u_1 + g_i u_2 + q_i), \quad i = \overline{1,2}$$

The parameters from this layer are called consequence parameters.

Layer 5 The fixed node makes the summation of all inputs:

$$O_i^5 = \sum_{i=1}^{2} \overline{w_i} y_i = \frac{w_1 y_1 + w_2 y_2}{w_1 + w_2}$$

ANFIS applies a hybrid learning algorithm formed by combining gradient method used to identify premise parameters with least square method used to generate consequence parameters. The learning algorithm task is tuning the premises (the parameters associated with membership functions) and the consequences in order that ANFIS response match the training data. At feedforward propagation step from hybrid learning, the system output reaches layer 4, and the consequence parameters are formed with least square method. At backpropagation step, the error signal is fed back and the premise parameters are updated through gradient method.

The ANFIS solution is used in process control to design the controller or to model the process to be controlled. It is an efficient solution in decision support model for the safety of railway systems [12], in modelling the evaporation from a dam reservoir [15], for predicting wheat grain yield [37], in hydraulic plant generation forecasting [49], in rainfall-runoff modelling [74], in groundwater level prediction [67], in evapotranspiration modelling [72], in municipal water consumption modelling [82], in estimating mean monthly air temperatures [13], in automobile sales forecasting [79] and airborne pollution forecasting [44]. The ANFIS controllers are developed to control known nonlinear processes: distillation process [22], nonlinear integrating processes [77], fed-batch reactor [84], thermal treatment furnace [6], exothermic chemical reactor [32] and methane production in anaerobic process [23].

7.2.4 Genetic Algorithms

Genetic algorithms are evolutive algorithms applied to optimization problems solving [20]. They start a search of the solution in a population of individuals (i.e. possible solutions), that evolve in time through the application of three types of operators: selection, recombination and mutation. The best solution (i.e. global optimal solution) is found according to a fitness function, specific to the problem that is solved. Genetic algorithms can be applied to process control as a standalone technique or in a variety of combinations with other AI or non AI techniques.

There are various intelligent process control solutions based on genetic algorithms, most of them being combined with other techniques. A common solution is given by the combination of an ANN with genetic algorithms, in a so called neuro-genetic control solution. Some examples are given in [14, 42].

Swarm intelligence provides several intelligent techniques that are inspired from nature, simulating the group behaviour of ants (ACO), particles (PSO), wasps (WSP), bees (ABC) etc., when solving their specific tasks (usually, food searching). These techniques are used for solving optimization problems, and they are metaheuristics.

In the last decade, some research work performed in the domain of intelligent process control proposed solutions based on swarm intelligence. An example is given in [39] where it is presented an optimized state feedback controller based on the ACO control law for nonlinear systems described by continuous-time Takagi-Sugeno-Kang (TSK) models. Another recent example of application is described in [52] for model identification.

Hybrid intelligent techniques combine two or more techniques from the classical AI techniques and/or computational intelligence techniques, sometimes with conventional or advanced process control techniques, that are not based on artificial intelligence.

Intelligent agents were applied to industrial process control, usually, in combination with other AI techniques (ANNs, FIS, ANFIS, GAs etc.). An example of an agent-based system for industrial process control assistance is described in [63].

7.3   Knowledge Based Systems

7.3.1   Basic Notions of Knowledge-Based Systems

Intelligent systems that incorporate the knowledge from a specific expertise domain and use it to solve complex problems from that domain by reasoning mechanisms are named knowledge-based systems. The main components of a knowledge-based system (KBS) are the knowledge base (KB) and the inference engine (IE). Figure 7.7 shows the generic structure of a KBS.

The knowledge base of a KBS is a set of knowledge from the expertise domain, including methods, techniques, theories, formalisms, rules, functions, formulas that are used to solve the problems from that domain. A special type of knowledge is given by heuristics, which is knowledge derived by experts from their experience, which can provide a faster suboptimal solution, in certain contexts. Usually, heuristic knowledge is hard to be formalized and their correctness cannot be proved in all cases. They find solutions to particular problems.

The inference engine is the knowledge processing module of a KBS, which performs a reasoning process on the KB, for problem solving. The knowledge is chained forward (deductive reasoning), backward (inductive reasoning) or forward-backward/backward-forward (combined reasoning), in reasoning chains.

Knowledge representation is a key issue when developing a KBS. There are several types of methods for knowledge representation: symbolic logic (e.g. first order predicates logic), production rules (IF *<premise>* THEN *<conclusion>*), procedural methods, frames, scenarios, semantic networks, conceptual graphs and dependencies etc. These methods are used for certain knowledge representation. Uncertain knowledge can be represented by using specific methods, such as: probabilistic methods, certainty factors, fuzzy logic, Dempster-Shafer theory, Bayesian networks etc. The two classes of knowledge representation methods, for certain and uncertain knowledge are usually combined, common combinations being production rules with certainty factors and/or fuzzy logic. Certainty factors

Fig. 7.7 The generic structure of a knowledge-based system

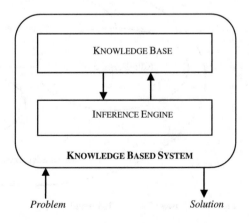

method assign to each piece of knowledge (e.g. each rule) a numerical value from the interval [0, 1] or [0, 100], in case percentage values are used, representing the knowledge certainty degree. The certainty values are given by the human experts, based on their experience.

Knowledge uncertainty is managed by the inference engine during reasoning. The inference engine can include an explanation module that provides justifications (*Why?* and *How?* explanations) of the reasoning chains.

The knowledge-based systems are a general class of intelligent systems. Particular types of KBSs are expert systems (ES), rule-based systems (RBS), case based reasoning systems (CBR) and other combinations of systems that use domain knowledge. An example of this last type of KBS is given by knowledge-based artificial neural networks (KANN), a class of artificial neural networks using expertise domain knowledge to generate the initial ANN architecture that is refined later, during ANN training [75]. Figure 7.8 presents a possible classification of KBSs.

Expert systems are knowledge-based systems that include the expertise domain knowledge with heuristic knowledge provided by human experts, and solve problems at the human expert level. Expert systems are an important subclass of knowledge-based systems [19]. In case multiple knowledge bases (specific to related expertise domains) are incorporated, the expert system is named a multi-expert system. A recent example of such system is SBC-MEDIU, a multi-expert system for environmental diagnosis presented in [57], that includes three knowledge bases: KB-Air (for air pollution analysis and dispersion assessment), KB-Water (for surface water pollution analysis), and KB-Soil (for soil erosion diagnosis).

Rule-based systems are KBSs that use knowledge under the form of production rules, i.e. IF-THEN rules. Some expert systems are rule-based systems, in case they use knowledge under the rules form. The knowledge base is composed by a rules

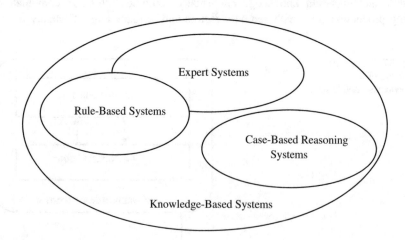

Fig. 7.8 Classes of knowledge based systems

base and a facts base. The rules base contains rules from the expertise domain, including heuristic rules and meta-rules, while the facts base contains the permanent facts, as well as, the temporary facts (i.e. the initial ones and those derived during the reasoning process performed by the IE). The temporary facts are used by the IE to fire rules during reasoning. The general form of a rule is as follows.

IF *<premise>* **THEN** *<conclusion>*

where, *<premise>* is a condition or a set of conditions linked by the AND/OR logical operators; and *<conclusion>* is a new knowledge (derived knowledge, i.e. derived fact) or an action that should be performed (e.g. by a device of a plant).

Some examples of rules, used in the monitoring and control of a specific type of boiler functioning, are giving below.

RULE 12
IF water_flow is low **THEN**
  control_valve_1 is closed; // the conclusion is a new knowledge—a fact

RULE 17
IF temperature is high **AND** water_level is high **THEN**
  pressure is high; // the conclusion is a new fact type knowledge

RULE 18
IF temperature_value > upper_temperature_limit **THEN**
  open valve_2; // the conclusion is an action

A KB can contain a special kind of rules, named meta-rules (rules about rules), that control the application of some rules. An example of meta-rule is rule 23.

RULE 23
IF alarm_parameter is closer to the limit **THEN**
  apply with priority the safety rules; // meta-rule that control other rules

Rules and meta-rules use symbolic and numerical values of the analyzed parameters. The symbolic values are generated from the numerical values or they are given by the human experts. Usually, the symbolic values are linguistic terms (i.e. fuzzy terms), that incorporate knowledge uncertainty.

The generic structure of a rule-based expert system is shown in Fig. 7.9. Rule-based systems do not perform well when solving problems with incomplete or missing data.

Case-based reasoning is a problem solving method based on human reasoning, which applies the principle that "*similar problems have similar solutions*" and uses similarity measures to retrieve similar cases. The problems and their solutions are stored as cases in a case base. CBR systems can be viewed as a subclass of KBSs, as the stored cases represent a sort of knowledge and the case base is a type of knowledge base. Case-based reasoning uses cases instead of rules, and is a proper method to solve problems with incomplete or missing data. Another advantage of storing knowledge under the form of cases is that, both tacit and implicit knowledge can be used to solve problems. CBR systems are usually modeled with the R4 model, introduced in [1], which applies four processes: retrieve, reuse, revise and retain, to the possible problem solution cases from the case base. The generic

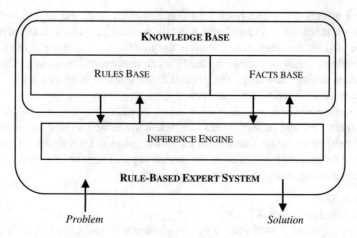

Fig. 7.9 The generic structure of a rule-based expert system

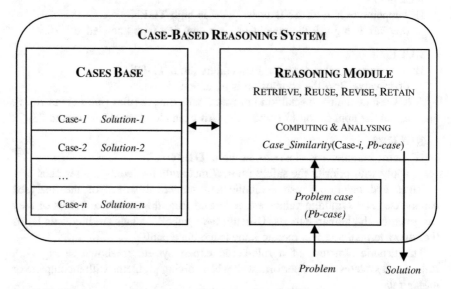

Fig. 7.10 The generic structure of a case-based reasoning system

structure of a CBR system is depicted in Fig. 7.10. The main components of a CBR system are the case base and the reasoning module that has as main goal to find a solution for the current problem that need to be solved (i.e. for a new case).

The solved problem of each case is represented as a set of features, a feature f_k having a weight of its importance, w_k. For example, a case i has the following form:

$$\left(\text{CASE} - i : \left(\left(f_1, w_1\right); \left(f_2, w_2\right); \ldots; \left(f_k, w_k\right); \ldots, \left(f_p, w_p\right)\right); \text{SOLUTION} - i\right).$$

The overall similarity between two cases is computed according to the similarity of corresponding features. The degree of similarity between the current problem case and each case from the case base is computed (according to a certain case similarity measure or metric) and it is analyzed, choosing the closest case, and performing to it the processes of retrieve, reuse, revise and retain. New cases are retained as they contain valuable knowledge, allowing a CBR system to improve its performance by learning.

The implementation of KBSs can be made by using various types of software tools: programming languages (object-oriented procedural languages such as C++, C#, Java, or AI-specific programming languages such as Prolog, Lisp), expert systems generators (such as VP-Expert, H-Expert), and specific KBS development tools (such as Jess, OPS5).

Knowledge-based systems were applied in different domains (e.g. industrial engineering, environmental protection, education, medicine, pharmacology, economy, law, business), providing good solutions for problems of monitoring, analysis, diagnosis, control, planning, quality management, training etc.

In the rest of this subchapter, we shall focus on intelligent process control, based on knowledge.

7.3.2 Knowledge-Based Process Control

Synthesizing several research work done in knowledge-based process control and real time expert control systems, reported in the literature during the last 35 years, we have identified three main types of knowledge that were used: generic and process specific control knowledge, which describe intelligent process control methods (i.e. knowledge from the expertise domain such as heuristic control rules, meta-knowledge), process knowledge (i.e. knowledge of a process—analytic knowledge, describing the process model), plant structure knowledge (e.g. connections of the process units—under a graphical form of knowledge representation). The process control knowledge is provided by engineers, process experts and plant experts. Figure 7.11 shows the main types of knowledge that are used in an intelligent process control system, and included in its KB.

The process knowledge includes analytic knowledge under the form of formulas, functions and differential equation models (material and energy balances, and reactions kinetics). This type of knowledge contains temporal knowledge, for the representation of the process dynamic behaviour.

The basic structure of a knowledge-based process control (i.e. a direct knowledge-based controller) is presented in Fig. 7.12. The KBS is used as a feedback controller (where, y is the feedback variable), that generates the control input (u) for the process/plant.

If the KBS is used as a supervisor of the controller (e.g. conventional controller, fuzzy controller, neuro-fuzzy controller), it is an indirect knowledge-based controller.

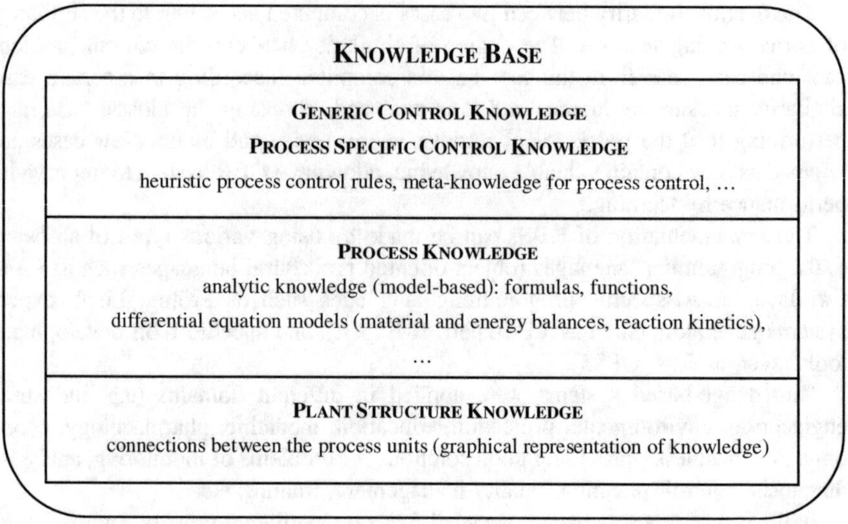

KNOWLEDGE BASE

GENERIC CONTROL KNOWLEDGE

PROCESS SPECIFIC CONTROL KNOWLEDGE

heuristic process control rules, meta-knowledge for process control, ...

PROCESS KNOWLEDGE

analytic knowledge (model-based): formulas, functions,

differential equation models (material and energy balances, reaction kinetics),

...

PLANT STRUCTURE KNOWLEDGE

connections between the process units (graphical representation of knowledge)

Fig. 7.11 The main types of knowledge from the KB of an intelligent process control system

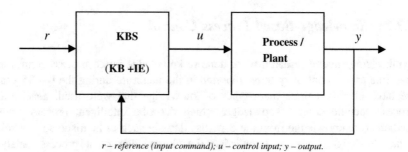

r – reference (input command); u – control input; y – output.

Fig. 7.12 The basic structure of a knowledge-based process control (direct controller)

The main limitations of knowledge-based systems applied to process control are given by:

(1) the difficulty to express the human expert knowledge for process control in an adequate manner for a KBS—the human expert knowledge is heuristic, uncertain, and enough difficult to be acquired and represented;

(2) the inefficiency of using complex KBS in real time implementations—for large and very large knowledge bases, the inference engine could not find always a solution, in real time;

(3) the KBS dependency on the individual application—the KBS is expertise domain dependent, being developed for a certain application.

A possible solution to these limitations is given by hybrid knowledge-based systems, which combine KBS with other techniques: fuzzy logic, artificial neural networks, neuro-fuzzy systems, genetic algorithms, swarm intelligence, intelligent agents etc.

In the next section it is presented a brief state of the art of intelligent process control based on KBSs and hybrid KBSs.

7.3.3 Brief State of the Art

Several KBSs were developed for process control in the period 1980–1990. One of the basic research work in the area of real time process control is described in [48], where it is presented a real time expert system. An overview of real-time industrial process control knowledge-based expert systems (KBES) reported in the period 1980–1990 is made in [70]. The main conclusion of the overview was that KBES integrates new techniques (e.g. rule-based expert systems and object oriented systems) with conventional control methods. Since those years, one of the most used software product for real time expert control of industrial processes is G2® [28], a real time KBES shell developed by Gensym Corporation, USA.

Starting with 1991 year, several combinations of knowledge-based process control were proposed for a variety of applications. We have selected some of them for chemical process control, power systems control, industrial continuous process control, geothermal plant control etc. A survey article was also selected.

An intelligent chemical process control system with supervisory KBSs is presented in [36]. The role of the KBS is to provide and to execute a corrective action, if needed, when a fault or performance change is detected. A rule-based approach for real time implementation of voltage/reactive power control is described in [17]. Two rule-based modules for voltage profile improvement and power loss minimization were implemented in Turbo Prolog. Another knowledge-based process control system is introduced in [21]. AEROLID, an abductive expert reasoning in an online industrial diagnoser, is a knowledge-based supervisory system developed for an industrial continuous process plant by using G2®.

A more recently literature survey on intelligent process monitoring, diagnosis and control systems is made in [76]. Four solutions proposed by research leading laboratories that combine data-driven, analytical and knowledge-based approaches in decision support systems are analyzed, and a comparative study is made. The conclusion was that, so far, there is no successful integration of the three approaches. The AI techniques that were used by the analyzed solutions are expert systems, ANNs, case-based reasoning and data mining.

A possible architecture for applying KBSs in real time control systems is proposed in [87], and a case study for a geothermal plant is detailed.

A hybrid knowledge-based process control system, ECSNN (Expert Control System using Neural Networks), which combines an expert system with a feed forward artificial neural network, is presented in [81].

Table 7.1 Operators used for *and* and *or* connectives [31]

and	or	Remark
min(a, b)	max(a, b)	Zadeh
max(a + b−1, 0)	min(a + b, 1)	Lukasiewicz
ab	a + b−ab	probability

Case based reasoning systems were applied to statistical process control. Such examples are described in [4, 5, 62]. The solution given in [5] designs the case base as a set of autonomous agents, where an agent is a past case.

At present, the research work is focused on developing hybrid knowledge-based process control solutions adapted to the specific process/plant, as it is not an easy task to develop a structured knowledge base that cover a broad expertise domain (Table 7.1).

7.4 Industrial Case Study: The Development of a Knowledge-Based System for Wastewater pH Control

The *pH* neutralization process from an industrial wastewater treatment plant takes place in the chemical step of the plant, in the so-called reaction-mixing tank. This process it is known for its high nonlinear behaviour, fact that raises various problems, such as: the reactant (acid—H_2SO_4 or alkaline—$Ca(OH)_2$) dosage accuracy and the strong variation of *pH* around the equivalence point (*pH* = 7, neutral *pH*) [8, 40, 41, 43, 55, 61, 68]. In literature there are many mathematical models for *pH* neutralization process [24, 25, 30, 46, 51], one of these models [30] being validated for the flowrates and volume (presented in Table 7.2) used in a studied plant from a Romanian refinery in [10].

In this section it is presented a knowledge-based system (KBSpHCTRL) which addresses the *pH* neutralization process, respectively the *pH* control process from an industrial wastewater treatment plant (WWTP). KBSpHCTRL system developed in C++ Builder programming environment, establishes the necessary reactant (calcium hydroxide—$Ca(OH)_2$ or sulphuric acid—H_2SO_4) flowrate for acid or alkaline type *pH* neutralization. The developed system uses the values of the parameters (F_1—the acid stream H_2SO_4 flowrate with concentration C_1, F_2—the alkaline stream $Ca(OH)_2$ flowrate with concentration C_2 and V—the volume of admixture-reaction tank) used by the reaction-mixing tank from the studied industrial plant, values presented in Table 7.2.

Table 7.2 Industrial case study—chemical step reactants parameters [56]

F_1 (l/hr)	C_1 (%)	F_2 (l/hr)	C_2 (%)	V [l]
25–300	95	5000–7000	10	4000

Next are presented the intermediate steps (the set rules determination—rules necessary for the system knowledge base (KB)—through the usage of a data mining (DM) technique and the KBS inference engine (IE)), which led to the KBSpHCTRL system development in C++ Builder 6 environment. Also are presented the results of the simulations achieved with this system.

7.4.1 Rule Set Determination Using a DM Technique

For rules induction, rules that compose the KBSpHCTRL system knowledge base was used a DM technique, respectively the classification and decision trees technique discussed in [10]. The data mining and the expert systems (ES) are two knowledge-based methods. From the algorithms associated with this DM technique, the C5.0 algorithm (improved version of ID3 and C5.0 algorithms) was considered more appropriate to use due to its characteristics and advantages presented in [9].

In this case was used the Windows version of C5.0 data mining tool, respectively See5 [66]. See5 is an instrument for patterns extraction and classifiers development under decision trees or rules sets form [9].

To obtain the set of rules for the system KB using See5 tool, were built two .data files and two .names files, as follows:

1. pH1.data—that contains the training data for the case of acid pH neutralization (Table 7.3), respectively the acid pH value and the associated flowrates F_2 [L/hr]. The F_2 range of values belongs to [d1, d46] interval
2. pH2.data—that contains the training data for the case of alkaline type pH neutralization (Table 7.4), respectively the alkaline pH value and the associated flow rates F1 [L/hr]. The F_1 range of values belongs to [d1, d36] interval
3. pH1.names—that describes the attributes for the acid type pH case (pH and the target attribute flow rate—F_2—as well as the associated classes—d1, d2, ..., d46)
4. pH2.names—that describes the attributes for the alkaline type pH case (pH and the target attribute—F_1—as well as the associated classes—d1, d2, ..., d36). From structural point of view pH2.names is similar to pH1.names file, but with a smaller number of classes

It must be mentioned that the training data presented in Tables 7.3 and 7.4 were obtained through a set of simulations carried out to determine the neutralization process statics characteristics (CS), characteristics presented in [8, 11].

Due to the fact that the rule sets are much easier to understand and to interpreted than decision trees (reduced number of rules, the classifiers based on rule sets are more accurate predictors than decision trees) was chosen to use the rule sets form of the classifiers (the rulesets option from Classifier Construction Option window).

By *rulesets* option (applied to the training data from Tables 7.3 and 7.4) enabling were generated two sets of rules for the KBSpHCTRL system knowledge base.

Table 7.3 Training data—acid pH neutralization (selection)

F_2		pH value
(L/hr)	Symbol	(Units)
5704	d1	2.70
5750	d2	2.75
5800	d3	2.82
5850	d4	2.89
5900	d5	2.09
6105	d10	3.75
6110	d11	3.80
6115	d12	3.86
6120	d13	3.93
6125	d14	4.01
6130	d19	4.11
6131	d20	4.13
6133	d21	4.18
6148	d28	5.14
6148.1	d29	5.16
6148.8	d33	5.38
6149.2	d35	5.59
6149.71	d38	6.22
6149.73	d40	6.27
6149.83	d45	6.84
6149.84	d46	7.00

Table 7.4 Training data—alkaline *pH* neutralization (selection)

F_1		pH value
(Liters/hr)	Symbol	(Units)
180	D1	13.42
200	D2	13.13
220	D3	11.76
220.2	D5	11.65
220.4	D7	11.51
220.6	D9	11.28
220.8	D11	10.80
220.82	D13	10.71
220.85	D15	10.50
220.89	D17	9.77
220.892	D19	9.66
220.896	D23	9.30
220.899	D26	8.06
220.8991	D27	7.88
220.8994	D30	7.43
220.8997	D33	7.17
220.8999	D35	7.05
220.9	D36	7.00

The set of rules for acid type *pH* case written under *if-then* form is the following:

1. **if** $(pH<=2.75)$ **then** flow(F_2)=d1(5704 –> 5750 [l/hr])
2. **if** $(pH<=2.98)$ **then** flow(F_2)=d1(5704 –> 5750 [l/hr])
3. **if** $(pH<=4.3)$ **then** flow (F_2)=d1(5704 –> 5750 [l/hr])
4. **if** $((pH>2.75)$and$(pH<=2.98))$ **then** flow(F_2)=d3(5800 –> 5900 [l/hr])
5. **if** $((pH>2.98)$ and $(pH<=3.4))$ **then** flow(F_2)=d6(5950 –> 6050 [l/hr])
6. **if** $((pH>3.4)$ and $(pH<=3.8))$ **then** flow(F_2)=d9(6100 –> 6110 [l/hr])
7. **if** $((pH>3.8)$ and $(pH<=4.01))$ **then** flow(F_2)=d12(6115 –> 6125 [l/hr])
8. **if** $((pH>4.01)$ and $(pH<=4.07))$ **then** flow(F_2)=d15(6126 –> 6128 [l/hr])
9. **if** $((pH>4.07)$ and $(pH<=4.13))$ **then** flow(F_2)=d18(6129 –> 6131 [l/hr])
10. **if** $((pH>4.13)$ and $(pH<=4.3))$ **then** flow(F_2)=d21(6133 –> 6137 [l/hr])
11. **if** $(pH>4.3)$ **then** flow$(F2)$=d24(6140 –> 6142 [l/hr])
12. **if** $((pH>4.51)$ and $(pH<=5.14))$ **then** flow(F_2)=d26(6144 –> 6148 [l/hr])
13. **if**$((pH>5.14)$ and $(pH<=5.24))$ **then** flow(F_2)=d29(6148.1–>6148.4 [l/hr])
14. **if** $((pH>5.24)$ and $(pH<=5.47))$ **then** flow(F_2)=d32(6148.6 –> 6149 [l/hr])
15. **if** $(pH>5.47)$ **then** flow(F_2)=d35(6149.2 –> 6149.7 [l/hr])
16. **if** $((pH>6.19)$ and $(pH<=6.27))$ **then** flow(F_2)=d38(6149.71–>6149.73[l/hr])
17. **if** $(pH>6.27)$ **then** flow(F_2)=d41(6149.75 –> 6149.79 [l/hr])
18. **if** $(pH>6.53)$ **then** flow(F_2)=d44(6149.81 –> 6149.84 [l/hr])

The set of rules for alkaline type *pH* case written under *if-then* form is the following:

1. **if** $(pH>11.71)$ **then** flow(F_1)=d1(180 –> 220 [l/hr])
2. **if** $(pH>11.59)$ **then** flow (F_1)=d1(180 –> 220 [l/hr])
3. **if** $(pH>11.11)$ **then** flow (F_1)=d1(180 –> 220 [l/hr])
4. **if**$((pH>11.59)$ and $(pH<=11.71))$**then** flow(F_1)=d4(220.1 -> 220.2 [l/hr])
5. **if**$((pH>11.41)$ and $(pH<=11.59))$**then** flow (F_1)=d6(220.3 -> 220.4 [l/hr])
6. **if**$((pH>11.11)$ and $(pH<=11.41))$**then** flow (F_1)=d8(220.5 -> 220.6 [l/hr])
7. **if**$((pH>10.71)$ and $(pH<=11.11))$**then** flow(F_1)=d10(220.7 -> 220.81 [l/hr])
8. **if**$((pH>10.5)$and$(pH<=10.71))$**then** flow(F_1)=d13(220.82 -> 220.83 [l/hr])
9. **if**$((pH>9.77)$and$(pH<=10.5))$ **then** flow(F1)=d15(220.85 -> 220.87 [l/hr])
10. **if**$((pH>9.66)$and$(pH<=9.77))$**then** flow(F1)=d17(220.89 -> 220.891 [l/hr])
11. **if**(pH<=9.66)) **then** flow (F1)=d19(220.892 –> 220.894 [l/hr])
12. **if**$((pH>9.13)$ and (pH<=9.42)) **then** flow(F1)=d22 (220.895 -> 220.896 [l/hr])
13. **if**$((pH>8.06)$ and (pH<=9.13)) **then** flow(F1)=d24 (220.897 –> 220.898 [l/hr])
14. **if**$((pH>7.71)$ and (pH<=8.06)) **then** flow(F1)=d26 (220.899 –> 220.8991 [l/hr])
15. **if**(pH<=7.71) **then** flow(F1)=d28(220.8992 –> 220.8994 [l/hr])
16. **if**$((pH>7.17)$ and (pH<=7.33)) **then** flow(F1)=d31(220.8995 –> 220.8996 [l/hr])
17. **if**(pH<=7.17) **then** flow(F1)=d33(220.8997 –> 220.8998 [l/hr])
18. **if**(pH<=7.05) **then** flow(F1)=d35(220.8999 –> 220.9 [l/hr])

The rules automatically generated using C5.0 algorithm implemented in See5 were adapted, respectively completed with a consistent number of rules developed using the knowledge (the heuristics) about the pH neutralization process from an industrial WWTP and also the research on the development of automatic systems for wastewater pH control [11]. The new set of rules composes the KBSpHCTRL system knowledge base.

A selection of the adapted and completed set of rules under if-then form for alkaline type pH is:

1. **if** ((pH>7) and (pH<=9.66)) **then** initial_F1_variation=d19(220.892 –> 220.894 [l/hr]);
2. **if** ((pH>7) and (pH<=9.66)) **then** final_F1_flow=220.9 [l/hr];
3. **if** ((pH>9.13)and(pH<=9.42)) **then** initial_F1_variation=d22(220.895 –> 220.896 [l/hr]);
4. **if** ((pH>9.13)and(pH<=9.42)) **then** final_F1_flow=220.9 [l/hr];
5. **if** ((pH>8.06)and(pH<=9.13)) **then** initial_F1_variation=d24(220.897 –> 220.898 [l/hr]);
6. **if** ((pH>8.06)and(pH<=9.13)) **then** final_F1_flow=220.9 [l/hr];
7. **if** ((pH>7.71)and(pH<=8.06)) **then** initial_F1_variation=d26(220.899 –> 220.8991 [l/hr]);
8. **if** ((pH>7.71)and(pH<=8.06)) **then** final_F1_flow=220.9 [l/hr];
9. **if** ((pH>7)and(pH<=7.71)) **then** initial_F1_variation=d28(220.8992 –> 220.8994 [l/hr]);
10. **if** ((pH>7)and(pH<=7.71)) **then** final_F1_flow=220.9 [l/hr];
11. **if** ((pH>7.17)and(pH<=7.33))**then** initial_F1_variation=d31(220.8995 –> 220.8996 [l/hr]);
12. **if** ((pH>7.17)and(pH<=7.33)) **then** final_F1_flow=220.9 [l/hr];
13. **if** ((pH>7)and(pH<=7.17)) **then** initial_F1_variation=d33(220.8997 –> 220.8998 [l/hr]
14. **if** ((pH>7)and(pH<=7.17)) **then** debit_initial_F1_variation=d35(220.8999 –> 220.9 [l/hr]);
15. **if** ((pH>7)and(pH<=7.05)) **then** final_F1_flow=220.9 [l/hr];
16. **if** (pH==7) **then** initial_F1_variation=0 (neutral pH);
17. **if** (pH==7) **then** final_F1_flow=0 (neutral pH).

A selection of adapted and completed set of rules under if-then form for acid type pH is:

1. **if**((pH>5.14)and(pH<=5.24)) **then** final_F2_flow=6149.84 [l/hr];
2. **if**((pH>5.24)and(pH<=5.47)) **then** initial_F2_variation=d32(6148.6 –> 6149 [l/hr]);
3. **if**((pH>5.24)and(pH<=5.47)) **then** final_F2_flow=6149.84 [l/hr];
4. **if**((pH>5.47)and(pH < 7)) **then** initial_F2_variation=d35(6149.2 –> 6149.7 [l/hr]);
5. **if**((pH>5.47)and(pH < 7)) **then** final_F2_flow=6149.84[l/hr];
6. **if**((pH>6.19)and(pH<=6.27)) **then** initial_F2_variation=d38(6149.71 –> 6149.73 [l/hr]);
7. **if**((pH>6.19)and(pH<=6.27)) **then** final_F2_flow=6149.84 [l/hr];

8. **if**((pH>6.27)and(pH < 7)) **then** initial_F2_variation=d41(6149.75 –> 6149.79 [l/hr]);
9. **if**((pH>6.27)and(pH < 7)) **then** final_F2_flow=6149.84 [l/hr];
10. **if**((pH>6.66)and(pH < 7)) **then** initial_F2_variation=d44(6149.81 –> 6149.84 [l/hr]);
11. **if**((pH>6.66)and(pH < 7)) **then** final_F2_flow=6149.84 [l/hr];
12. **if**(pH==7) **then** initial_F2_variation=0 (neutral pH);
13. **if**(pH==7) **then** final_F2_flow=0 (neutral pH).

All these rules compose, as it was mentioned, the system KBSpHCTRL knowledge-base. The system developed inference engine (IE) is a deductive one, presented under pseudo code form as follows:

```
if (( pH>=2.7) and (pH<=7)) then begin
    KBS maintains F1 constant
        IE control the alkaline agent pump (establishes the alkaline reactant
            flowrate)
        Near the pH equivalence point, IE deactivates alkaline agent pump
                end
if (( pH>=7) and (pH<=13.42)) then begin
    KBS maintains F2 constant
        IE control the acid agent pump (establishes the acid reactant flowrate)
        Near the pH equivalence point, IE deactivates acid agent pump
                end
```

As it can be observed, the conditions from the IE were established using as criterion the measured pH range (the pH value supplied by a simulated transducer).

7.4.2 KBSpHCTRL Development

To develop the knowledge-based system KBSpHCTRL, system whose purpose is that of establishing the reactant flowrate (H_2SO_4 or $Ca(OH)_2$) necessary to neutralize an acid or alkaline type *pH*, were used the new sets of rules (the new rules adapted by the author using its knowledge about the *pH* neutralization process from a studied industrial plant) and C++ Builder programming environment. The system interface is presented in Fig. 7.13 and the results file named *results.txt* is presented in Fig. 7.14.

The knowledge-based developed system (KBSpHCTRL) gives to the users the following facilities:

1. Depending on the pH value automatically taken from a simulated pH sensor (pHT), KBSpHCTRL system sets the reactant flowrate (F1—H_2SO_4 or F2—Ca $(OH)_2$) necessary for an acid or alkaline type pH neutralization, namely to bring the pH to the imposed reference value (SPpH). Namely, it provides the domain in which F1 and F2 initial flowrates can vary and their final value

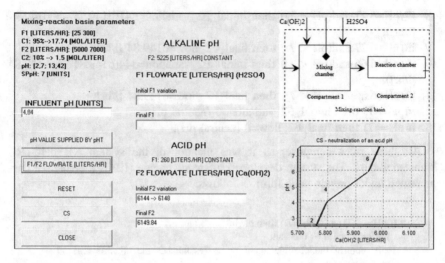

Fig. 7.13 KBSpHCTRL system interface [11]

```
            INITIAL_FLOW_VARIATION                    FINAL_FLOW
pH=11,63    (F1/F2)=220.1 --> 220.2              (F1/F2)=220.9
pH=4,77     (F1/F2)=6144 --> 6148               (F1/F2)=6149.84
pH=6,01     (F1/F2)=6149.2 --> 6149.7            (F1/F2)=6149.84
pH=8,49     (F1/F2)=220.897 --> 220.898          (F1/F2)=220.9
pH=9,01     (F1/F2)=220.897 --> 220.898          (F1/F2)=220.9
pH=8,49     (F1/F2)=220.897 --> 220.898          (F1/F2)=220.9
pH=2,92     (F1/F2)=5800 --> 5900               (F1/F2)=6149.84
pH=4,72     (F1/F2)=6144 --> 6148               (F1/F2)=6149.84 |
```

Fig. 7.14 Results.txt file [11]

2. It generates a graphical representation of pH according to the titrant (CS)
3. It store in an external file (results.txt) previous values provided by the system (the pH values recorded by the transducer—pHT from the process and the reactant flowrate necessary to its neutralization)

This information's are useful to the human operator of a wastewater treatment plant in the process of taking the best decisions regarding the wastewater pH neutralization process operating and control (the establishing of the necessary reactant flowrate for an acid or alkaline type pH neutralization).

The possibility to consult a file that keeps information's regarding the previous functioning of the process (namely the pH evolution and the reactant flowrates) it is useful in situations that requires the human operator intervention, such as: failure of the reagent dosing pumps, failure in the automatic dosing system, various disturbances in the neutralization process (variations in the reactants concentrations—C_1 and C_2).

Table 7.5 KBSpHCTRL—simulations results

No	pH (Acid)	Value (Basic)	Initial F_1 flow variation (L/hr)	Final F_1 flow (L/hr)	Initial F_2 flow variation (L/hr)	Final F_2 flow (L/hr)
1	–	11.58	220.3–220.4	220.9	–	–
2	5.95	–	–	–	6149.2–6149.7	6149.84
3	–	8.58	220.897–220.898	220.9	–	–
4	–	10.2	220.85–220.87	220.9	–	–
5	–	11.47	220.3–220.4	220.9	–	–
6	–	13.1	200–220.2	220.9	–	–
7	4.28	–	–	–	6133–6137	6149.84
8	4.9	–	–	–	6144–6148	6149.84
9	3.68	–	–	–	6100–6110	6149.84
10	4.31	–	–	–	6140–6142	6149.84
11	6.81	–			6149.81–6149.84	6149.84
12	–	9.3	220.895–220.896	220.9	–	–
13	–	10.8	220.7–220.81	220.9	–	–
14	–	13.29	180–220	220.9	–	–
15	3.61	–	–	–	6100–6110	6149.84
16	5.1	–	–	–	6144–6148	6149.84
17	3.02	–	–	–	5950–6050	6149.84
18	4.64	–	–	–	6144–6148	6149.84
19	6.15	–	–	–	6149.2–6149.7	6149.84
20	–	9.76	220.89–220.891	220.9	–	–

7.4.3   KBSpHCTRL—Results of the Simulations

In Table 7.5 are presented some results (at simulation level) supplied by KBSpHCTRL system. It must be mentioned that this knowledge-based system was designed based on the mixing-reaction tank associated parameters from the studied industrial wastewater treatment plant, parameters presented in Table 7.2.

As it can be observed in Table 7.5, the system supplies the flowrates of the reactants (F_1 or F_2) depending on the pH character (basic or acid).

7.5   Conclusions

Knowledge-based systems were intensively used in the 1980s years, and starting with the 1990s they were applied in various combinations with other AI and non AI techniques to solve real-time process control problems. Their limitations can be overcome by hybrid methods that combine KBSs with other intelligent techniques

(ANNs, ANFIS, GAs, intelligent agents, swarm intelligence etc.) or conventional/ advanced control methods that are not based on AI, thus improving the solutions to real-time process control.

The human expert knowledge with respect to the process functionality, characteristics, behaviour is essential in developing knowledge-based systems for complex industrial processes, such as the pH neutralization process from an industrial wastewater treatment plant. The lack of information's makes very difficult or even impossible the development of this type of systems. Having at disposal a comprehensive process data base and using knowledge-based methods (data mining techniques, expert systems, etc.) can be developed knowledge-based systems for any processes, industrial or not.

The system KBSpHCTRL can be improved, namely it can be connected to a data acquisition board or to an industrial pH transducer (such as an pH/ORP analyzer— model P53 [60]), the scaled output signal can be used as the control signal of two reagent dosing pumps and also the data supplied by the system can be stored in a MySQL data base.

References

1. Aamodt, A., Plaza, E.: Case-based reasoning: foundational issues, methodological variations, and system approaches. Artif. Intell. Commun. **7**, 39–59 (1994)
2. Al-Jenani, Nounou, H., Nounou, M.: Fuzzy control of a CSTR process. In: Proceedings of the 8th international symposium on mechatronics and its applications, ISMA 2012, ISBN-978-1-4673-0862-5 (2012)
3. Babuska, R.: Fuzzy Modelling for Control. Kluwer Academic Publishers (1998)
4. Behbahani, M., Saghaee, A., Noorossana, R.: A case-based reasoning system development for statistical process control: case representation and retrieval. Comput. Ind. Eng. **63**, 1107–1117 (2012)
5. Bergmann, R., Althoff, K.D., Minor, M., Reichle, M., Bach, K.: Case-based reasoning— introduction and recent developments. Kunstliche Intelligenz: Spec. Issue Case-Based Reasoning **23**, 5–11 (2009)
6. Bhogle, P.P., Patre, B.M., Waghmare, L.M., Panchade, V.M.: Neuro-fuzzy temperature controller. In: Proceedings of IEEE international conference on mechatronics and automation, Harbin, China, p. 3344–3348, ISSN 1-4244-0828-8 (2007)
7. Bloch, G., Denoeux, T.: Neural networks for process control and optimization: two industrial applications. ISA Trans. **42**(1), 39–51 (2003)
8. Cărbureanu, M., Gheorghe, C.: pH variation in the presence of the coagulants used in oil well industry wastewater treatment. Rev. Chim. **65**(12), 1498–1501 (2014)
9. Cărbureanu, M.: A system for monitoring the effluent's quality of an industrial wastewater treatment plant. In: Proceedings of the 5th international conference on manufacturing science and education-MSE 2011, pp. 409–412. Sibiu (2011)
10. Cărbureanu, M.: Sistem expert neuro-fuzzy pentru controlul proceselor de epurare a apelor uzate. Ph.D. Thesis, Petroleum-Gas University (2014a)
11. Cărbureanu, M.: The development of a neuro-fuzzy expert system for wastewater pH control. J. Control Eng. Appl. Inform. **16**(4), 30–41 (2014)

12. Ćirović, G., Pamučar, D.: Decision support model for prioritizing railway level crossings for safety improvements: application of the adaptive neuro-fuzzy system. Expert Syst. Appl. **40** (6), 2208–2223, ISSN 0957-4174 (2013)
13. Cobaner, M., Citakoglu, H. Kisi, O., Haktanir,T.: Estimation of mean monthly air temperatures in Turkey. Comput. Electron. Agricu. **109**, 71–79, ISSN 0168-1699 (2014)
14. de Canete J.F., del Saz-Orozco, P., Garcia-Cerezo A., Garcia-Moral I.: Neural and genetic control approaches in process engineering, chapter in frontiers in advances control systems. InTech 59–74 (2012)
15. Dogan, E., Gumrukcuoglu, M., Sandalci, M., Opan, M.: Modelling of evaporation from the reservoir of Yuvacik dam using adaptive neuro-fuzzy inference systems. Eng. Appl. Artif Intell. **23**(6), 961–967, ISSN 0952-1976 (2010)
16. Driankov, D., Hellendoorn, H., Reinfrank, M.: An Introduction to Fuzzy Control, 2nd edn. Springer, ISBN 978-3-662-03284-8 (1996)
17. El-Sayed, M.A.H.: Rule-based approach for real-time reactive power control in interconnected power systems. Expert Syst. Appl. **14**(3), 355–360 (1998)
18. Fuente, M.J., Robles, C., Casado, O., Syafiie, S., Tadeo, F.: Fuzzy control of a neutralization process. Eng. Appl. Artif Intell. **19**, 905–914 (2006)
19. Giarratano, J.C., Riley, G.D.: Expert Systems—Principles and Programming, 4th edn. Thomson, Boston (2005)
20. Goldberg, D.E.: Genetic Algorithms in Search, Optimization and Machine Learning. Addison-Wesley, Massachusetts (1989)
21. González, C.A., Acosta, G., Mira, J., de Prada, C.: Knowledge based process control supervision and diagnosis: the AEROLID approach. Expert Syst. Appl. **14**(3), 371–383 (1998)
22. Gupta, A. Rani, A.: Control of distillation process using neuro-fuzzy technique. Int. J. Electr. Data Commun. **1**(9) 16–20, ISSN 2320-2084 (2013)
23. Gurubel, K.J., Sanchez, E.N.: Carlos-Hernandez, neuro-fuzzy control strategy for methane production in an anaerobic process, In: Proceedings of the 2013 International Joint Conference on Neural Networks, p. 1–8, 1467361291 (2013)
24. Gustafsson, T., Waller, K.: Dynamic modelling and reaction invariant control of pH. Chem. Eng. Sci. **38**(3), 389–398 (1983)
25. Henson, M., Seborg, D.: Adaptive nonlinear control of a pH neutralization process. Control Syst. Technol. **2**(3), 169–182 (1994)
26. Holmblad, L.P., Østergaard, J.J.: The FLS application of fuzzy logic. Fuzzy Sets Syst. **70**, 135–146 (1995)
27. Holmblad, L.P., Østergaard, J.J.: Control of a cement kiln by fuzzy logic. In: Gupta, M.M., Sanchez, E.(eds.) Fuzzy Information and Decision Processes. North-Holland, Amsterdam, pp. 389–399. (Reprint in: FLS Review **67**, FLS Automation A/S, Høffdingsvej 77, DK-2500., Valby, Copenhagen, Denmark) (1982)
28. http://www.gensym.com/
29. Hussain, M.A.: Review of the applications of neural networks in chemical process control— simulation and online implementation. Artif. Intell. Eng. **13**(1), 55–68 (1999)
30. Ibrahim, R.: Practical modelling and control implementation studies on a pH neutralization process pilot plant. Ph. D. Thesis, Department of Electronics and Electrical Engineering, University of Glasgow (2008)
31. Jager, R.: Fuzzy logic in control. PhD Thesis, Technische Universiteit Delft, ISBN 90-9008318-9 (1995)
32. Jana K., Vasičkaninová, A., Dvoran, J.: Neuro-fuzzy control of exothermic chemical reactor, 2013 In: Proceedings of the IEEE International Conference on Process Control, p. 168. Slovakia, ISSN 978-1-4799-0927-8 (2013)
33. Jang, R.: ANFIS: adaptive-network-based fuzzy inference system. IEEE Trans. Syst, Man Cybern. **23**(3), 665 (1993)
34. Jantzen, J.: Foundations of Fuzzy Control. Wiley, ISBN 0-470-02963 (2007)
35. Kadam D.B., Patil A.B., Paradeshi K.P.: Neural networks based intelligent process control system. Proc. Int. Conf. Recent Trends Inf. Telecommun. Comput. 356–358 (2010)

36. Kendra, S.J., Basila, M.R., Cinar, A.: Intelligent process control with supervisory knowledge-based systems. IEEE Control Syst **14**(3), 37–47 (1994)
37. Khoshnevisan, B., Rafiee, S., Omid, M., Mousazadeh, H.: Development of an intelligent system based on ANFIS for predicting wheat grain yield on the basis of energy inputs. Inf. Process. Agric. **1**(1), 14–22, ISSN 2214-3173 (2014)
38. Lee, C.C.: Fuzzy logic in control systems: fuzzy logic controller. IEEE Trans. Syst. Man Cybern. **20**(2), 404–435 (1990)
39. Liouane, H., Douik, A., Messaoud, H.: Design of optimized state feedback controller using ACO control law for nonlinear systems described by TSK models. Stud. Inf Control **16**(3), 307–320 (2007)
40. Liteanu, C., et al: Quantitative Analytical Chemistry, pp. 178–185. Pedagogic and Didactic Press, Bucharest (1972)
41. Luca, C., et al: Analytical Chemistry and Instrumental Analysis, pp. 136–146. Pedagogic and Didactic Press, Bucharest (1983)
42. Marcolla, R.F., Machado, R.A.F., Cancelier, A., Claumann, C.A., Bolzan, A.: Modelling techniques and process control application based on neural networks with on-line adjustment using genetic algorihms. Braz. J. Chem. Eng. **26**(1), 113–126 (2009)
43. Marinoiu, V., Paraschiv, N.: Chemical Processes Automation, vol. 1, pp. 316–325. Technical Press, Bucharest (1992)
44. Martinez-Zeron, E., Aceves-Fernandez, M.A., Gorrostieta-Hurtado, E., Sotomayor-Olmedo, A., Ramos-Arreguín, J.M.: Method to improve airborne pollution forecasting by using ant colony optimization and neuro-fuzzy algorithms. Int. J. Intell. Sci. **4**, 81–90 (2014)
45. Mathworks: Fuzzy logic toolbox for use with Matlab: user's guide, version online edition. The MathWorks Inc. Available from http://www.mathworks.com/help/pdf_doc/fuzzy/fuzzy.pdf, (2014)
46. McAvoy, J., et al.: Dynamics of pH in controlled stirred tank reactor. Chem. Process Des Develop. **11**(1), 68–78 (1972)
47. Mihalache, S.F., Carbureanu, M.: Fuzzy logic controller design for tank level control. Petrol-Gas Univ. Ploiesti Bull., Technical series, **LXVI**(3) (2014)
48. Moore, R., Rosenof, H., Stanley, G.: Process control using a real time expert system. Proc. IFAC **1990**, 234–239 (1990)
49. Moreno, J.: Hydraulic plant generation forecasting in Colombian power market using ANFIS, Energy Economics **31**(3), 450–455, ISSN 0140-9883 (2009)
50. Murthy B.V., Kumar Y.V.P., Kumari U.V.R.: Application of neural networks in process control: automatic/online tuning of PID controller gains for ±10 % disturbance rejection. In: Proceedings of the IEEE International Conference on Advanced Communication Control and Computing Technology, pp. 348–352 (2012)
51. Mwembeshi, M., et al.: An approach to robust and flexible modelling and control of pH in reactors. Chem. Eng. Res. Des. **79**(3), 323–334 (2001)
52. Naitali, A., Giri, F., Radouane, A., Chaoui, F.Z.: Swarm intelligence based partitioning in local linear models identification, pp. 843–848. Proc. of IEEE Int. Symp. Intell., Control (2014)
53. Narendra, K.S., Mukhopadhyay, S.: Adaptive control of nonlinear multivariable systems using neural networks. Neural Networks **7**(5), 737–752 (1994)
54. Narendra, K.S., Parthasarathy, K.: Identification and control of dynamically systems using neural networks. IEEE Trans. Neural Networks **1**, 4–27 (1990)
55. Nenițescu, C.: General Chemistry, pp. 482–495. Pedagogic and Didactic Press, Bucharest (1972)
56. Operating manual of the ECBTAR wastewater treatment from the romanian refinery (2010)
57. Oprea, M., Dunea, D.: SBC-Mediu: A multi-expert system for environmental diagnosis. Environ. Eng. Manage. J. **9**(2), 205–213 (2010)
58. Oprea, M.: A case study of knowledge modelling in an air pollution control decision support system. AiCommunications **18**(4), 293–303 (2005)
59. Pedrycz, W.: Fuzzy control and fuzzy systems. 2nd edn. Wiley (1993)

60. pH/ORP analyzer manual (2014). www.gliint.com. Accessed 15 Sept 2014
61. Pietrzyk, D. et al: Analytical Chemistry, 2nd edn. pp.111–160. Technical Press, Bucharest (1989)
62. Rojek, G., Kusiak, J.: Case-based reasoning approach to control of industrial processes. Comput Methods Mater. Sci. **12**, 250–258 (2012)
63. Rojek G.: Agent based systems for assistance at industrial process control with experience modelling. In: Proceedings of the Federated Conference on Computer Science and Information Systems, pp. 1037–1040 (2013)
64. Rumelhart, D., Widrow, B., Leht, M.: The basic ideas in neural networks. Commun. ACM **37**, 87–92 (1994)
65. Russel S., Norvig P.: Artificial intelligence—a modern approach. Prentice Hall (2010)
66. See5 (2014). https://www.rulequest.com/See5-info.html. Accessed 10 May 2014
67. Shiri, J., Kisi, O., Yoon, H., Lee, K.K., Nazemi, A.H.: Predicting groundwater level fluctuations with meteorological effect implications—a comparative study among soft computing techniques. Comput. Geosci. **56**, 32–44, ISSN 0098-3004 (2013)
68. Skoog, D. et al: Fundamentals of analytical chemistry, pp. 182–230. Saunders College Publishing, New York (1988)
69. Sozen, A., Kurt, M., Akcayol, M.A., Ozalp, M.: Performance prediction of a solar driven ejector-absorption cycle using fuzzy logic. Renew. Energy **29**, 53–71 (2004)
70. Stanley, G.M.: Experiences using knowledge-based reasoning in online control systems. IFAC Symp. Comput. Aided Design Control Syst., Swansea, UK (1991)
71. Sugeno, M. (ed.): Industrial applications of fuzzy control, ISBN 0444878297 North-Holland (1985)
72. Tabari, H., Kisi, O., Ezani, A., Talae, P.H.: SVM, ANFIS, regression and climate based models for reference evapotranspiration modelling using limited climatic data in a semi-arid highland environment. J. Hydrol. **444–445**, pp. 78–89, ISSN 0022-1694 (2012)
73. Takagi, T., Sugeno, M.: Fuzzy identification of systems and its applications to modelling and control. IEEE Trans. Syst. Man Cybern. **15**(1), 116–132 (1985)
74. Talei, A., Chua, L.H.C., Wong, T.: Evaluation of rainfall and discharge inputs used by adaptive network-based fuzzy inference systems (ANFIS) in rainfall–runoff modelling. J. Hydrol. **391**(3–4) 248–262, ISSN 0022-1694 (2010)
75. Towell, C.G., Shavlik, J.W.: Knowledge-based artificial neural network. Artif. Intell. **70**, 119–165 (1994)
76. Uraikul, V., Chan, C.W., Tontiwachwuthikul, P.: Artificial intelligence for monitoring and supervisory control of process systems. Eng. Appl. Artif. Intell. **20**, 115–131 (2007)
77. Vasickaninová, A., Bakošová, M.: Neuro-fuzzy control of integrating processes. Acta Montanistica Slovaca **16**(1), 74–83 (2011)
78. Von Altrock, C.: Fuzzy logic and neurofuzzy applications explained. Prentice Hall (1995)
79. Wang, F.K., Chang, K.K., Tzeng, C.W.: Using adaptive network-based fuzzy inference system to forecast automobile sales, expert systems with applications, **38**(8) 10587–10593, ISSN 0957-4174 (2011)
80. Wang, L.X.: A course in fuzzy systems and control, international edition, ISBN:0-13-540882-2. Prentice Hall (1997)
81. Wu, M., She, J.H., Nakano, M.: An expert control system using neural networks for the electrolytic process in zinc hydrometallurgy. Eng. Appl. Artif. Intell. **14**(5), 589–598 (2002)
82. Yurdusev, M. A., Firat, M.: Adaptive neuro fuzzy inference system approach for municipal water consumption modelling: an application to Izmir, Turkey. J. Hydrol. **365**(3–4) 225–234, ISSN 0022-1694 (2009)
83. Zamoum Boushaki R., Chetate B., Zamoum Y.: Artificial neural networks control of the recycle compression system, Studies in Informatics and Control, 23(1), 65–76 (2014)
84. Zhang, J.: Modelling and optimal control of batch processes using recurrent neuro-fuzzy networks. IEEE Trans. Fuzzy Syst. **13**(4), 417–427, ISSN 1063-6706 (2005)

85. Zhao, Y., Xiong, W.: fuzzy intelligent control of hydrogen purification process In: Proceedings of the International Conference on Intelligent Control and Information Processing, Dalian, China (2010)
86. Zhihong, M., Yu, X.H., Wu, H.R.: An RBF neural network-based adaptive control for SISO linearisable nonlinear systems. Neural Comput. Appl. **7**(1), 71–77 (1998)
87. Zmaranda D., Silaghi H., Gabor G., Vancea C.: Issues on applying knowledge-based techniques in real-time control systems. Int. J. Comput. Commun. Control (IJCCC) **8**(1), 166–175 (2013)
88. Zorriassatine, F., Tannock, J.D.T.: A review of neural networks for statistical process control. J. Intell. Manuf. **9**, 209–224 (1998)

Chapter 8
Ciphering of Cloud Computing Environment Based New Intelligent Quantum Service

Omer K. Jasim Mohammad, El-Sayed M. El-Horbaty
and Abdel-Badeeh M. Salem

Abstract Cloud computing environment is a new approach to the intelligent control of network communication and knowledge-based systems. It drastically guarantees scalability, on-demand, and pay-as-you-go services through virtualization environments. In a cloud environment, resources are provided as services to clients over the internet in the public cloud and over the intranet in the private cloud upon request. Resources' coordination in the cloud enables clients to reach their resources anywhere and anytime. Guaranteeing the security in cloud environment plays an important role, as clients often store important files on remote trust cloud data center. However, clients are wondering about the integrity and the availability of their data in the cloud environment. So, many security issues, which pertinent to client data garbling and communication intrusion caused by attackers, are attitudinized in the host, network and data levels. In order to address these issues, this chapter introduces a new intelligent quantum cloud environment (IQCE), that entails both Intelligent Quantum Cryptography-as-a-Service (IQCaaS) and Quantum Advanced Encryption Standard (QAES). This intelligent environment poses more secured data transmission by provisioning secret key among cloud's instances and machines. It is implemented using System Center Manager (SCM) 2012-R2, which in turn, installed and configured based on bare-metal Hyper-V hypervisor. In addition, IQCaaS solves the key generation, the key distribution and the key management problems that emerged through the online negotiation between the communication parties in the cloud environment.

O.K. Jasim Mohammad (✉)
Al-Ma'arif University College, Anbar, Iraq
e-mail: omer.k.jasim@ieee.org

E.-S.M. El-Horbaty · A.-B.M. Salem
Faculty of Computer and Information Sciences,
Ain Shams University, Cairo, Egypt
e-mail: Shorbaty@cis.asu.edu.eg

A.-B.M. Salem
e-mail: absalem@cis.asu.edu.eg

© Springer International Publishing Switzerland 2016   241
K. Nakamatsu and R. Kountchev (eds.), *New Approaches in Intelligent Control*,
Intelligent Systems Reference Library 107, DOI 10.1007/978-3-319-32168-4_8

8.1 Introduction

Nowadays, in order to develop a new generation of distributed computing environment, the integration between distributed computing systems and networking systems is required [1]. This integration allows computer networks to be involved in distributed computing as full participants like other computing resources such as CPU capacity and memory/disk space. Moreover, the output of this integration represents a new emerging trend in distributed computing world. Grid computing [2], utility computing [3] and cloud computing [3], are famous trends of distributed computing technology. These emerging distributed computing technologies, with the rapid development of new networking technologies, are changing the entire computing paradigm toward a new generation of distributed computing.

Cloud computing is clearly the new generation of data centers with virtualization nodes through hypervisor technologies such as VMs, dynamically service provisioned on demand [3]. It is based on the principle of virtualization and allocation of information technology based services to worldwide distributed computers. Cloud computing widely been applied in several industrial fields such as Google, Facebook, and Amazon. It is considered a new communication technique that combines multiple disciplines such as parallel computing, distributed computing and grid computing [4].

In addition, cloud computing is a specialized form of distributed, grid, and utility computing. It takes a style of grid computing where dynamically stable and virtualized resources are available as a service over the internet. In a sense, cloud computing offers IT resources, including storage, networking, computing platform, on-demand, and pay-as-you-go basis. This evolution and maturity have already encouraged various organizations to move from classical to cloud-based services [5]. Any cloud architecture is revolving at least three primary layers, Infrastructure as a Service (IaaS), Platform as a Service (PaaS), and Software as a Service (SaaS). These services offer numerous pros, such as reducing the costs of the hardware, providing the reliability for each client and others [5, 6].

Despite the free data accessing is deliberated an advantage in the cloud; it has several weaknesses such that any cloud client can handle any file transferred through cloud communication. Consequently, many companies have explored the critical areas in a cloud environment.

CSA [7] is an example of those companies, which delivers a package that contains cloud provider, clients and considers the security model. The CSA security model can interrupt the intruder, who is responsible for destroying and interrupting the original data files and communications. However, there are a number of issues related to cloud computing that still prevent companies from moving their business onto public clouds.

The many similarities in these perceptions indicate a grave concern for crucial security and legal obstacles to cloud computing. Service availability, data confidentiality, provider lock-in and reputation fate sharing are examples of such concerns [8, 9].

This chapter discusses in detail the well-known emerging trend in distributed computing which is cloud computing and focuses on security issues in a cloud environment. In addition, this chapter introduces an intelligent cloud environment, which poses more secured data transmission by provisioning secret key among cloud's instances based on IQCaaS and new developed symmetric encryption algorithm. The IQCaaS solves the key generation, key distribution and key management problems that emerged through the online negotiation between the communication parties. Also, this intelligent environment eliminates all cloud security concerns and hypervisor vulnerabilities.

The rest of the chapter is organized as follows: Sect. 8.2 discusses the related work of cloud computing security and cryptography. Cloud computing definition and technologies are presented in Sect. 8.3. Section 8.4 describes the cloud computing security challenges like data leakage and cloud attacks. The various cloud security solutions are discussed in Sect. 8.5. The information security in the quantum world and empirical analysis presented in Sect. 8.6. Section 8.7 describes the proposed algorithm and the performance of such algorithm on proposed cloud environment. Section 8.8 explains the cryptographic cloud environment based on intelligent quantum computation; it includes the experimental cloud environment and intelligent quantum service. Section 8.9 provides the analysis of the proposed Intelligent Quantum Cloud Environment and IQCaaS and finally, Sect. 8.9 presents the conclusion and future works.

8.2 Related Works

The cryptographic process is a well-known technology for protecting sensitive data during transmission process in a cloud environment. Recently, many researchers have focused on the encryption process using many techniques entirely to hide the sensitive transmitted data and files. Most of these studies attempt to address seven security issues in the open cloud environment. Authentication and data integrity, access control, user privacy, data privacy, data confidentiality, and key management are primary security issues in such environment.

Doelitzscher et al. [10] identify abuse of cloud resources, lack of security monitoring in cloud infrastructure and defective isolation of shared resources as focal points to be managed. They also focus on the lack of flexibility of classic intrusion detection mechanisms to handle virtualized environments, suggesting the use of special security audit tools associated with the business flow modeling through security SLAs. Their analysis of the top security concerns is also based on publications from CSA, ENISA, and NIST. After a quick evaluation of issues, they focus switch to their security auditing solution, without offering a deeper quantitative compilation of security risks and areas of concern.

Lijiang et al. [11], introduced the development and present situation of cloud computing technology, In addition, operation mechanism, characteristics and parameters of trusted computing are discussed. Furthermore, this study shows the architecture and features of the intelligence security control platform based on cloud computing and trusted computing technology. This study has a significant meaning to the automation of information and popularization of national economic information network and has very high reference value to the research of network security technology. However, the authors limited in describing the architecture of cloud security system and the environment.

Wang et al. [12] proposed the anonymity based method for achieving the cloud privacy. In this approach, the anonymity algorithm will process the data, and the output will be released to the cloud provider. Anonymized output means that the content will split into different tables or parts and placed in multiple service provider's storages. If the user needs to restore the meaningful information, he/she has to get all the parts or tables. In case, one service provider aggregates the meaningful contents by collecting from all the other service providers. Then, providing the data to the clients for client's convenience there is no preservation of privacy because the aggregating service provider may read or leak the information. In case the clients have to aggregate themselves then they have to contact multiple service providers, which will reduce the efficiency of the service on-demand.

Hossein et al. [13] developed a private cloud as an intermediary based on XaaS concept and designed the Encryption as a Service (EaaS). EaaS gets rid of the security risks of cloud provider's encryption and the inefficiency of client-side encryption. Moreover, this study computes the time for six selected algorithms has been shown in the traditional client-side and the proposed EaaS. Finally, the authors illustrated that the EaaS performs more than six times better than a single computer in all algorithms. However, authors depending on only text file format only as an input sample to the EaaS cloud computing. Therefore, authors need to use different data formats such as wave, rar, and others to reveal the efficiency of the proposed cloud cryptosystem.

Esh et al. [14] propose a new method for the data security in cloud computing based on RSA algorithm. They tried to assess the cloud storage methodology and the data security in the cloud by implementing digital signature with RSA algorithm. They illustrate the Cloud Computing Open Architecture (CCOA) to enable infrastructure–level resource sharing. Unfortunately, the RSA algorithm is known by its slowing in the data transformation process. Therefore, CCOA can easily be exposed to the timing attack.

Soren et al. [15] present a security architecture that allows establishing secure client-controlled-Cryptography-as-a-Service (CaaS) in the cloud computing. CaaS enables clients to be in control of proposing and usage of their credentials and cryptographic primitives. All these operations provided based on sub inherited domain from Xen hypervisors domain. However, this model is leakage and weak for authentication attacks due to all controlling mechanisms and encryption/decryption processes assigned to the client side. Table 8.1 gives a brief summary of the relevant security issues that have been addressed in the surveyed studies in this section.

Thus, we can conclude that the cryptography and intelligent security aspects are essential fields for the cloud security environment. Moreover, due to the dynamic characteristics of cloud technology, there are different challenges in the security

Table 8.1 Comparison between different existing studies in security issues respects in cloud computing

Cloud security issues								
Existing studies	Authentication	Access control	U. privacy	D. privacy	D. integrity	Confidentiality	Key management	Key techniques
Doelitzscher et al. [10]	Y	Y		X	X	Y	X	Identify abuse of cloud resources, lack of security monitoring and defective isolation of shared resources
Lijiang et al. [11]	Y	Y	X	Y	Y	X	X	Shows the architecture and features of the intelligence cloud security control and present the automation of information and popularization of national economic information network
Wang et al. [12]	Y	Y	Y	Y	X	X	Y	Presents the anonymity based method for achieving the cloud privacy and use anonymity algorithm to process the cloud data
Hossein et al. [13]	X	X	Y	Y		X	Y	Developed the Encryption as a Service (EaaS). It gets rid of the security risks of cloud provider's encryption and the inefficiency of client-side encryption
Esh et al. [14]	Y	Y	Y	X	X	Y	Y	Illustrate the CCOA to enable infrastructure–level resource sharing based RSA algorithm
Soren et al. [15]	Y	Y	Y	Y	X	Y	Y	Present a security architecture that allows establishing secure client-controlled-CaaS in the cloud computing

Y Yes, *X* No

process. These challenges lie in the inexistence of perfectly secured model that guarantees a trusted data transformation.

8.3   Cloud Computing Technologies

Cloud Computing is a specialized form of distributed, grid, and utility computing. It takes a style of Grid computing where dynamically stable and virtualized resources are available as a service over the internet [16].

So, in order to understand in details what is the meaning of cloud computing and what computing technologies are merged to build it, this section shows an example of such merger. Cloud computing is the result of evolution and adoption of existing technologies and computing paradigms (Distribute, Grid, Utility, Cluster, and Parallel Computing). The goal of cloud computing is to allow users to take benefit from all of these technologies, without the need for deep knowledge about or expertise with each one of them, see Fig. 8.1. The Cloud aims to cut costs, and help the users focus on their core business instead of being impeded by IT obstacles [10].

As shown in Fig. 8.1, the core technologies for cloud computing are virtualization and service-oriented architecture (SOA) [11]. Virtualization abstracts the physical infrastructure, which is the most rigged component, and makes it available as a soft part that is easy to use and manage. It provides the agility required to speed up IT operations and reduces cost by increasing infrastructure utilization. While, SOA provides a cloud framework that makes cloud much easier to understand and to manage a process. In order to take advantage of cloud computing, SOA provides the interfaces and architectures that can reach out and touch cloud computing resources.

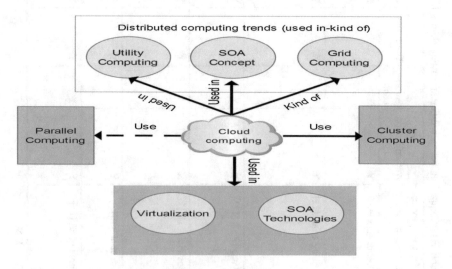

Fig. 8.1 Computing paradigms of cloud computing [6]

Additionally, cloud computing utilizes concepts from utility computing [11] in order to provide metrics for the used services, based on the benefits are gained [12]. These metrics are at the core of the pay-per-use model in public clouds. Cloud computing provides the tools and technologies to build data/compute-intensive parallel applications with many affordable prices compared to traditional parallel or cluster computing techniques. Finally, cloud computing is a kind of Grid Computing [13], it has evolved from the grid by addressing the Quality of Services (QoS) and reliability problems [14].

Accordingly, the definition of cloud computing is not simply. The complexity comes from the variety of different technologies involved, as well as to the hype surrounding the topic [15, 17]. So, cloud computing is essentially delivering computing at the Internet scale. Compute, storage, networking infrastructure, as well as development, and deployment platforms are distributed on-demand within minutes. Sophisticated applications (like Google Apps, engine, etc.) are made possibly by the abstracted, auto-scaling compute platform provided by cloud computing. However, the oft-cited definition is put forward by the National Institute of Standards (NIST), [18, 19]. As shown in Fig. 8.2, NIST defines cloud computing as" *a distributed computing system consisting of a collection of inter-connected and virtualized computers. These networks are dynamically provisioned and presented as one or more unified computing resources based on service-level agreements (SLA) established through negotiation between the service provider and clients.*

Finally, Fig. 8.2 shows that the cloud is a type of parallel, distributed, and cluster system that consisting of a collection of interconnected and virtualized nodes. These nodes are communicating based on SLA's which are established through the connection between the service provider and clients on different nodes.

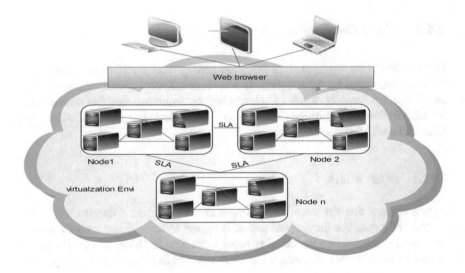

Fig. 8.2 Cloud computing architecture

8.4 Cloud Security Challenges

Data in a cloud environment are described as data transmitted, stored or processed by the cloud provider. Any client applies the same data classification used when the data are resident on the own machine or remotely data center. Trust, security, and privacy are some of the challenges existing in cloud computing, they have become significant barriers to the rapid growth of cloud computing applications. This section discusses the cloud security challenges, instantiated in the data leakage, cloud attacks, and client identity, in order to aid in providing more secure cloud environments.

8.4.1 Data Leakage

Security and privacy are the primary factors related to cloud computing. Although cloud computing environments are a multi-domain environment to sharing hardware resources. Nevertheless, the operation of placing data seem to be risky as any unauthorized person can quickly hack either accidentally or due to malicious attack [18, 20].

Therefore, a sensible strategy to ensure data security and privacy is performed based on encryption technique. Accordingly, data should be encrypted from the start, so there is no possibility of data leakage. Finally, the client should have been controlling not over the secured data, but also over the keys used for encryption/decryption process.

8.4.2 Cloud Computing Attacks

Despite cloud computing environments are getting more familiar, its main risk is the intrusion caused by the attackers. In this section, a most known communication attack mechanisms in a cloud computing environment are discussed, such as Distributed Denial-of-Service (DDoS), Side-channel, Authentication, and Man-in-the-middle attacks.

8.4.2.1 DDoS Attack

Acts a process that the users or the organization confronts a deprivation of the services of a resource they would normally expect to have. Some security professionals have argued that the cloud is more vulnerable to DDoS attacks because it shared by many consumers, which makes DDoS attacks much more damaging, see

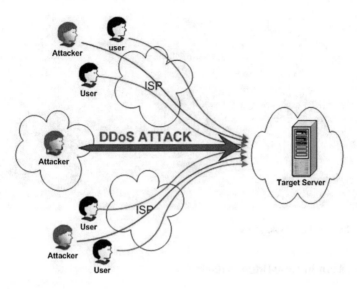

Fig. 8.3 DDoS attack mechanism

Fig. 8.3. For example, "Twitter and Saudi Aramco", suffered a devastating DoS attack during 2012 and 2013 [21]. However, our proposed system handles a new security service technique (quota) to defeat this attack.

8.4.2.2 Side-Channel Attack

An attacker could attempt to compromise the cloud by placing a malicious virtual machine in proximity to a target cloud server and then launching a side channel attack. Side-channel attacks have emerged as a kind of practical security threat, that targeting system implementation of cryptographic algorithms [22].

8.4.2.3 Authentication Attack

Authentication is a weak point in hosted and virtual services and is frequently targeted. There are many different ways to authenticate clients. The mechanisms used to secure the authentication process and the methods used are a frequent target of attackers [23]. Most clients still use a simple username and password. So, knowledge-based authentication with the exception of some financial institutions that have deployed various forms of secondary authentication. This type of knowledge increases the strength of resistance against the authentication attack [22, 23].

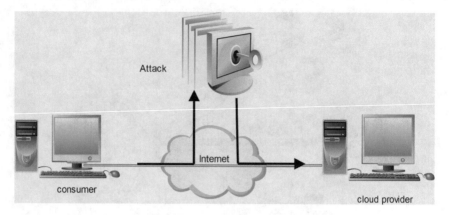

Fig. 8.4 Man-in-the-middle attack

8.4.2.4 Man-in-the-Middle Attack

This attack is carried out when an attacker places himself between Client and cloud provider at any time the attackers can place themselves into the communication's path in order to intercept and modify communications, see Fig. 8.4. Hence, some kinds of techniques are used that rely on assumptions combined with quantum cryptography concepts see [19].

8.4.3 Client Identity

Client identity is an important aspect of cloud computing. Due to this only authorized user have been writing and authority to gain access to data and to modify the contents in it. Therefore, the authorization manner with the encryption provision, give a secure environment for data resident in the cloud. Without user "id" verification, the system will not allow any of the requests made to allow some transactions, which will ultimately help in data privacy and security [20].

8.5 Cloud Security Solutions

The inferences popup from cloud security challenges shows that the each issue brings different impacts on distinct assets. Aiming to create a security model for studying security concepts in the cloud technology and for supporting decision-making. This section considers solutions for the data leakage vulnerabilities, cloud attacks, and client identification are previously illustrated and arrange them in sequence categories.

8.5.1   Services Security Solution

In a nutshell, the service layers (SaaS, PaaS, IaaS) revolving around the increasing abstraction from underlying IT infrastructure [24]. Cloud computing-based IaaS does not naturally expose actual hardware or networking layers to the renter of the service, but, these underlying resources are abstracted for the client. For instance, PaaS abstracts infrastructure to a greater extent and presents middleware containers that are tailored for categories of usage. Whereas, SaaS abstracts even further and usually exposes narrow functionality software-based services such as customer relationship management (CRM) [25].

Figure 8.5 shows that the burden of security at SaaS lies with Cloud Service Provider (CSP). Inevitably, this is because of the degree of abstraction. However, the SaaS is based on a high level of integrated functionality with minimal client control. On the other hand, the PaaS offers slightly greater extensibility and greater customer control but fewer high-level features. Predominantly, due to the relatively lower degree of abstraction, IaaS provides the more excellent tenant or client control over security than do PaaS or SaaS [26].

Moreover, at SaaS, the CSP is responsible for most security aspects, compliance, and liability. However, with these responsibilities, the CSP is more apt to change important aspects of the service or associated service contracts like SLA [27]. The service layers are influenced by the type of cloud deployment. For example, the degree of control that a tenant has a public cloud is minimal, whereas the tenant has a maximum with a private cloud. So, security concerns are more sophisticated in a public than private deployment [28].

Additionally, based on security solution at service layers, one can classify the security trends as *CSP responsibilities* and *client responsibilities*. The CSP should ensure that the architecture and the infrastructure are secure. While on the client side, should make sure that the provider has taken all measures to secure their data in the cloud. Each CSP service model has different possible implementations. However, the

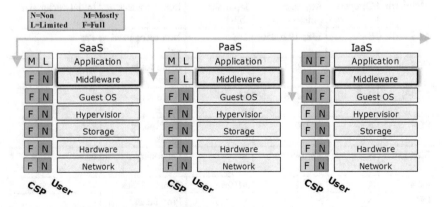

Fig. 8.5 Extent of security control at SPI service [20]

degree of control by CSP is little, and they are merely responsible for the availability of their services at the most time. However, client's responsibility is superior, and they are responsible for the confidentiality, data privacy and integrity [29].

8.5.2   Cloud Cryptographic Solution

The cloud environment's components sustain the security criteria in each layer such as authentication, integrity, and availability to guarantee the reliable communication. One of the methods to resolve these criteria is the encryption/decryption [30]. Thus, they are applying necessary security solutions to data stored, transmitted or processed by the cloud provider. The SLA cannot achieve all these solutions it must be made by an efficient cryptographic algorithm and authentication function such as AES, Kerberos, SHA-256, etc. [31]. Once data is safely transmitted to a CSP, it should be stored, transmitted and processed in a secure way [32]. Therefore, employing an appropriate strength encryption is essential. High encryption is preferable when data at rest have continuing value for an extended period. Many encryption algorithms have been developed and implemented to provide more secured data transmission process in the cloud computing environment. Data Encryption Standard (DES), AES, Rivest Cipher 4 (RC4), Blowfish, and Triple-DES (3-DES) are symmetric category, While, Rivest-Shamir-Adelman (RSA) and Diffie-Hillman (DH) are asymmetric category [31, 32]. Table 8.2 summarizes the characteristics of the encryption algorithms, both symmetric and asymmetric.

In [33], authors implemented the mentioned symmetric and asymmetric algorithms in cloud computing environment in order to ensure the data security. In addition, it examines the performance of such algorithms, considering the time of the encryption/decryption process and the size of the output encrypted files.

Table 8.2 Encryption algorithms characteristics

Algorithm	Category	Key size (bits)	Input size (bits)	Intruder attack	Initial vector size (bits)
AES	S	128; 192; 256	128	Non recognized yet	128
DES	S	56	64	Brute force attack	64
3-DES	A	112–168	64	Brute force attack	64
RC4	S	256	256	Bit flipping attack	256
Blowfish	S	32–448	32	Dictionary attack	64
RSA	A	>1024	>1024	Timing attack	–
DH	A	n	n	Ping attack	–

S symmetric, A asymmetric, n length in bit

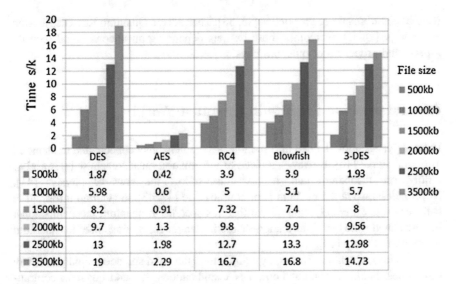

File size	DES	AES	RC4	Blowfish	3-DES
500kb	1.87	0.42	3.9	3.9	1.93
1000kb	5.98	0.6	5	5.1	5.7
1500kb	8.2	0.91	7.32	7.4	8
2000kb	9.7	1.3	9.8	9.9	9.56
2500kb	13	1.98	12.7	13.3	12.98
3500kb	19	2.29	16.7	16.8	14.73

Fig. 8.6 An efficient of modern symmetric encryption algorithms in cloud environment

Figure 8.6 shows that the time of encryption using several input file sizes: 500, 1000, 1500, 2000, 2500, and 3500 kb and the running time is calculated in milliseconds.

Moreover, Fig. 8.6 displays that the AES—symmetric encryption techniques are faster than the others. In addition, most algorithms in both categories (symmetric and asymmetric) enjoy the direct proportion relation between the running time and the input file size. Finally, after the encryption process the symmetric techniques always change the input file size to yield the encrypted file (for more details see [33]).

Despite the encryption process uses complex techniques for random key generation based on mathematical models and computations, its encryption strategy considered vulnerable. So, if the intruder is good enough in the mathematical computation field such quantum attack, he/she can easily decrypt the cipher and retrieve the original transmitted or stored documents. Furthermore, a key distribution is another critical issue that noticed in most modern encryption algorithms. It arises from the fact that communicating parties must somehow share a secret key before any secure communication can be initiated, and both parties must then ensure that the key remains secret. Of course, direct key distribution is not always feasible due to risk, inconvenience, and cost factors [34, 35].

In some situations, the direct key exchange is possible through a secure communication channel. However, this security can never guarantee. A fundamental problem remains because, in principle, any classical private channel can be monitored passively, without the sender or receiver knowing that the eavesdropper has taken place [36, 37]. This problem comes from the nature of classical physics. In general, the theory of ordinary-scale bodies and phenomena such as magnetic tapes and radio signals- allows all physical properties of an object to be measured without disturbing those proprieties [33, 38]. Therefore, a Quantum Key Distribution technology (QKD) overcomes these barriers depending on unconditional security aspects and properties of quantum physics [39].

8.6 Information Security in Quantum World

Previously, Shannon defined information as a mathematical concept and transmitted as digital signal [40]. However, a piece of information must somehow be stored or written on a medium and, hence, must follow the laws of physics [41]. Nevertheless, when the medium is of atomic scale [42, 43], the carried information behaves quite differently, and all the features specific to quantum mechanics must be translated into an information-theoretic language, giving rise to quantum information theory [43].

Nowadays, Quantum Technology (QT) flourished and spreaded rapidly in many disciplines, banking, computations, DNA applications, AI, and quantum communications are examples of such disciplines. The importance of QT laying in, its ability to solve the key challenges associated with distributed computing environments, as conserving data privacy during users' interactions with computing centers remotely [42, 43].

Therefore, the first trend of QT is presented by Wiesner in the late sixties [44] when he proposed a Quantum Banking Note (QBN). This trend uses the spin of particles to make unforgivable bank notes by encoding identification information on bank notes in a clever way using elementary particles. Consequently, a bank can verify their authenticity by later checking the consistency of this identification information. The second trend is a quantum computer, that is, a computer that uses quantum principles instead of the usual classical principles, can solve some problems much faster than the traditional computer [39]. In a classical computer, every computation is a combination of zeros and ones (i.e., bits). At a given time, a bit can either be zero or one. In contrast, a qubit, the quantum equivalent of a bit, can be a zero and a one at the same time [39]. Finally, the third trend is QKD; it enables secret quantum keys exchanging between two different parties through a two communication channel, classical and quantum channel [45]. This technology is an alternative to the classical encryption techniques and it is used to solve the common current scheme problems, such as key distribution, management, and defeating attack [46].

8.6.1 QKD Protocols

In general, data in quantum information theory are encoded based on prepared states, which are known as photons polarizations [44]. The advantage of using such states in data transmission lays in the no-cloning theorem (the quantum state of a single photon cannot be copied) [39]. These states sent as keys that are used for encryption/decryption process [45]. Therefore, QKD trend provides a securely transmitted keys between two communication parties. These keys send to trusted users, then, it integrates with different modern cryptographic algorithms.

So, in order to achieve the key distribution between two communication parties, many QKD protocols are implemented and developed. Bennet and Brassard-1984 (BB84), Ekert-1991(E91), and Einstein- Podolsk-Rosen Based Protocol (EPR) [45, 46] are well-known QKD protocols. Diversity in the protocols came from the number of states are deployed in each one. This chapter stimulates and develops the BB84-QKD protocol because of:

- Many practical communication environments are implemented BB84 protocol like as Geneva quantum network [47] Center of quantum technology [48], etc.
- Depending on the approved photon basis (rectilinear and diagonal).
- Easy to implementation, scalable, and capable to integrate with different security applications.

Accordingly, four primary phases are depended by BB84 in order to produce and distribute a final secret key. These phases are [47]:

1. **Raw Key Extraction (RKE)**: the main purpose of raw key extraction phase is to eliminate all possible errors occurred during the bits discussion (generation and transmission) over the quantum channel. Negotiated parties compare their filter types used for each photon, unmatched polarization is eliminated otherwise, bits are considered.
2. **Error Estimation (EE)**: The negotiation process might occur over a noisy quantum and unsecured (public) classical channel. Such channel can cause a partial key damage or physical noise of the transmission medium. In order to avoid such problems, both sides determine an error threshold value "E_{max}" when they are sure that there is no eavesdropping on a transmission medium. So as to calculate the error percentage (E), the raw bits are monitored and compared with the E_{max}, if $E > E_{max}$, then it is probably either unexpected noise or eavesdroppers activity.
3. **Key Reconciliation (KR)**: It is implemented to minimize the mentioned errors associated with the key as much as possible. It divides the raw key into blocks of K bits; then parity bit calculations are done for each block. Both blocks and bit calculations are performed for N-rounds that depend on the length of the raw key, where different parties wholly negotiate based on the value of N.
4. **Privacy Amplification (PA)**: It is the final step in the quantum key extraction. It is applied to minimize the number of bits that an eavesdropper might know. Sending and receiving parties use a shrinking method to their bit sequences in order to obscure the eavesdropper ability to capture bit sequence. Finally, figure summarizes the primary phases of BB84 and illustrates the procedure of the simulation it.

8.6.2 QKD-BB84: Simulation and Experimental Environment

In a sense, BB84 protocol depends on a single-photon polarization (rectilinear and diagonal) to generated key [47]. Nevertheless, both sender and receiver must use device or simulator that generate and detect light pulses with dissimilar polarizations (for more details see [45, 46]). In order to create an innovative system capable to simulate the processes of QKD system and quantum computation, the illustration and description of physical processes must be achieved. Any process of quantum mechanics can be depicted as an operator (operator is formalized as multiplication by a matrix). Consequently, the logic gates (circuits), embrace the operator as a square matrix, are given in matrices of vectors. Drastically, matrices illustration is a delightful topic in scientific quantum computing. In addition, many tools and random algorithms are considered in the simulation of QKD.

Usually, two communication parties must be guaranteed an acceptance secure communication without the presence of the attack. Then, they are going to be executed the stages of the QKD-BB84 (RK, EE, KR, and PA). The main hardware requirements for the deployment of the BB84 simulation are at least two computers machines connected by switches. Each machine has a static IP (computer join to the domain) to communicate over the switch and on each machine will implement assigned the software. The simulation is performed using a Core i5 (4.8 GHz) processor associated with 16 GB of RAM as a central server hardware. A Core i3 (2.4 GHz) processor associated with 4 GB of RAM as a client hardware [49]. Finally, the procedure of the BB84 protocol simulation is illustrated as follows:

Algorithm: key generation based BB84 protocol.
Input: a stream of photons states.
Output: secret quantum key (qubits).

Step 1 *Server agent sends to client a sequence of random photon states (independently chosen),*

$$P_s = \{p_0, p_1, \ldots, pn\} \quad \text{where } P_s \in \leftarrow, \uparrow.$$

Step 2 *For each photon deployed bases (P_b) client randomly selected one of two measure of*

$$P_b = (\mathbf{X}, +)$$

Step 3 *Server agent and client eliminate all non-valent bases.*
Step 4 *Server agent and client compute the error percentage (E_r) and compare with threshold (E_t). If $E_r \leq E_t$, go to the next step, otherwise, aborting negotiation.*

Step 5 *Server agent and client obtained a series of an initial qubit, according to photon measurement and coding.*

$$S_n = \{inqubit_0, inqubit_1, \ldots, inqubit_n\}$$

Step 6 *Server agent and client perform. (KR) phase, through*

- *Divide the intimal series into number of blocks (b)*

$$B_n = \{b_0, b_1, \ldots, b_n\}$$

- *Compute the parity bit for each two block that has even location or odd location*

$$P_{i=0}^{i=n} = b[i]$$

Step 7 *Server agent and client perform PA, in order to generate a series of final secret key (Sk).*

$$S_k = \{sk_0, sk_1, \ldots, sk_n\}$$

Finally, for more details about BB84 simulation environment, see [49].

8.6.3   QKD-BB84: Empirical Analysis

A practical comparison between the real environment of BB84-QKD and our BB84-QKD simulator is presented. The actual environment implemented in the Center for Quantum Technology (CQT) laboratories [50], while, our BB84-QKD simulated based on Visual Studio 2012-VC#. Both environments using BB84 protocol (4-bases) to achieve the negotiation and generate a final secret key.

Figure 8.7 shows the length of secret keys obtained with different configurations such as noise levels and eavesdropper influence. For example, when 1000-qubits are sent as an initial value in the actual environment, it produces 181-qubits as the final secret key, whereas, 261-qubits are produced in the implemented simulator. Two different results are obtained through 0.05 GHz as noise level and 0.5 as an Eve influence. In general, the Quantum Bit Error Rate (QBER) between the initial qubits are pumped, and the final secret key is obtained, 62.1 % in the actual environment and 18.1 % in the simulator environment.

Moreover, Fig. 8.8 reveals that any change in the eavesdropper activity or noise level, it directly impacts on the length of the secret key and the time of key generation.

More clarification, in order to get 200 qubits from 500 qubits are pumped under the impact of noise 0.05 GHz, and there is no Eve influence, we need 0.23 ms to

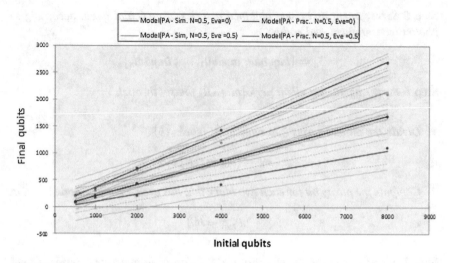

Fig. 8.7 Final secret key in the practical and simulation environment

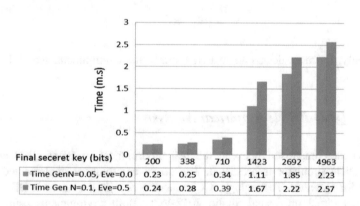

Fig. 8.8 Final key and time obtained with different configuration levels

generate it. Usually, this time simultaneously grows with the increasing of the noise or Eve power. However, the practical environment is faster than the simulator environment, due to the light nature [41].

8.7 Proposed Encryption Algorithm

Data transformation through an open communication environment needs highly secured levels, so, many cryptographic encryption algorithms rely on the unpredictable complex encryption key. To assure the strength of such keys, QKD has been integrated and QAES, a new version of the AES, has been developed by the present authors [51]. Furthermore, QAES is a private key block cipher that processes data blocks of 128 bits with a key length of 128, 192, or 256 bits.

The QAES algorithm incorporates both the QKD and the classical AES algorithm in order to provide an unconditional security level [44]. The AES enhanced version exploits the generated key based QKD in the encryption/decryption process. Since the unconditional security depends on the Heisenberg uncertainty principle [46, 47] instead of the complex mathematical model in key generation. Consequently, a truly randomness characteristic associated with quantum key generation [52]. These properties led to more attack resistance is assured, and the cipher system is hard to be attacked. Furthermore, the randomness characteristic helps to adopt the QT as a source to generate random numbers that are used with various encryption algorithms.

Cryptographic processes (encryption/decryption) in the proposed algorithm are provided by 10; 12 or 14 rounds depending on the key length. Table 8.3 shows that the key length (K_l) determine the number of rounds (N_r); QAES-128 applies the round function 10 times, QAES-192—12 times and QAES-256—14 times.

In addition, the difference between the QAES and the AES can be summarized in:

- QAES utilizes the Dynamic Quantum S-Box (DQS-Box) that is generated depending on key generation from the quantum cipher.
- QAES exploits the generated key from QKD in the encryption/decryption process. Nevertheless, AES utilizes a static S-box and ultimately depends on the mathematical model in the key generation and management.

Accordingly, the round key session enjoys the dynamic mechanism, in which the contents of each key session changes consequently in each round with the change of the key generation. Such dynamic mechanism aids in solving the problems of avoiding the off-line analysis attack, and resistance to the quantum attack.

As shown in Fig. 8.9, the performance of proposed algorithm is examined on IQCE, considering the time of the encryption/decryption process and the size of the output encrypted files. This examination implemented using several input file sizes: 500, 1000, 1500, 2000, 3000, and 4000 kb and the running time is calculated in milliseconds.

Comparing the QAES with other encryption algorithms reflects a higher security level. However, QAES takes time more than others due to the time required for a quantum key generation, for more elaborates and understanding the architecture of QAES see [51].

Table 8.3 QAES key and rounds

Cipher	K_r (bytes)	Nr
QAES-128	16	10
QAES-192	24	12
QAES-256	32	14

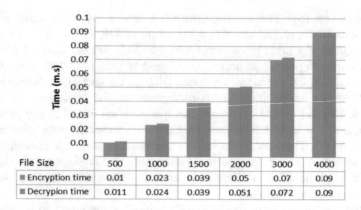

Fig. 8.9 An efficient of QAES in IQCE

8.8 Cryptographic Cloud Using Intelligent Quantum Computation

In order to solve problems associated with classical cryptographic algorithms (mentioned above). This section illustrates an intelligent quantum computing technology that eliminates all above problems by deploying an intelligent unconditional security service in a cloud environment. This technology offers: (i) Simple, low-cost for the data protection, (ii) Tools and security services integration, and (iii) An efficient disaster recovery.

8.8.1 Experimental Cloud Environment

Drastically, many operations and features are realized in cloud computing environment. These features directly depend on IaaS layer like the number of VMs, quality of services (QoS), storage capacity and other features. So, the bare-metal Hyper-V hypervisor and the SCM 2012 components are elucidated and implemented [53]. These components are System Center Virtual Machine Manager (SCVMM), System Center Operation Manager (SCOM), Application Controller (AC), Operation Services Manager (OSM), Data Protection Manager (DPM), and orchestrator (OC). The host server (*cloud providers*) utilizes the Core i5 (4.8 GHz) with 16 GB of RAM with 2 TB-HDD as the central hardware. As shown in Fig. 8.10, IQCE includes:

- Domain controller (*qcloud.net*) server: It is the central server that manages the single sign in the process and offers a response multiple cloud services. It deploys a credential account for each trusted client and provides a trust connection among clients and cloud instances. Moreover, it helps to defeat bot the

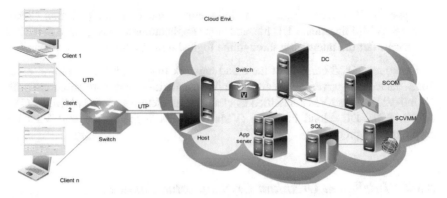

Fig. 8.10   Intelligent quantum cloud environment

insider attack and denial of service attack (DoS). Finally, it holds the right to access hardware resources.

- *SQL* Server: Each System Center 2012 site database can be installed on either the default instance or a named instance of an SQL Server installation. The SQL Server instance can be co-located with the site server, or on a remote computer. Our cloud environment, a remote SQL Server is implemented and configured as a failover cluster in either a single instance or a multiple instance configuration. It is responsible for keeping the credential of *SCVMM* and *SCOM* services and creating the report viewer for all sending/receiving request operations.
- *SCVMM* server: It manages the virtualized data center, visualization host, and storage resources. It can integrate with other system center components (mentioned above), deploy a management console operations and cloud configuration wizards. Moreover, it creates VM templates, capability profiles, ISO images, private cloud objects, and self-service user roles.
- *APPC* server: It mainly depends on the SCVMM to manage applications and services, which in turn are deployed in private or public cloud infrastructures. It provides a unified self-service portal that helps to configure, deploy, and manage virtual machines (VMs) and services on predefined templates. Although some administrator tasks can be performed via the *APPC* console, the users for APPC cannot be considered as administrators. (the URL for the implemented *APPC* is, http://scvmm.qcloud.net/#/Shell/VmsManager—using our cloud network)
- *SCOM*: A robust, comprehensive monitoring tool which provides the infrastructure that is flexible and cost-effective, helps ensure the predictable performance and availability of vital applications, and offers comprehensive monitoring for datacenter and cloud.
- *Star Wind* server: It replaces the expensive SAN and NAS physical shared storage since it is a software-based hypervisor-centric virtual machine that provides a fully fault-tolerant and high-performing storage for the virtualization platform. It could seamlessly be integrated into the hypervisor platforms, such as

Hyper-V, Xen Server, Linux and UNIX environments. In the *Quantum Cloud*, the SCVMM libraries, VHD file, and cloud applications are assigned to *StarWind* server after the integration through the logical unit (LUN) at the fabric panel.

Moreover, IQCE consists of the cloud network that entails the windows server 2012 datacenter server and the Hyper-V installations and configurations with N- full-VMs. These VMs classified as cloud infrastructure such SCVMM, SCOM, APPC, SQL, DC, cloud instances (VMs rented from the client), and VMs for cryptographic processes.

8.8.2 Intelligent Quantum Cryptographic Service

In this section, a new intelligent cryptographic service layer in the IQCE is presented. This service provides the secret key provisioning to VMs' clients, separating both clients' cryptographic primitive and credential accounts based on *qcloud.net* domain. It is applied to the multiple trusted clients (which renting the VMs) concurrently. Integrating such service achieves both confidentiality and integrity aspects.

In a sense, the Intelligent Quantum Cryptography-as-a-Service (IQCaaS) has mini-OS directly connected with the cloud platform and isolated from the cloud instances (see Fig. 8.11). Consequently, it assures both the appropriate load for

Fig. 8.11 Single IQCaaS architecture

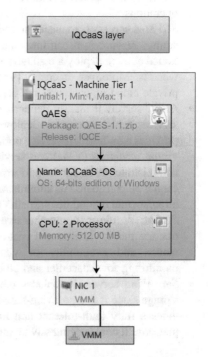

cloud performance optimization and the client controlling activities (client prevent the cloud administrator from gained or preserve his data). Accordingly, a secured environment for each client's VMs, with no possibility for insiders or external attackers, is guaranteed.

To sum up, after the signing in verification and the VM renting, IQCaaS deploys the client wizard and the CSP wizard to achieve the encryption/decryption processes and connect to the IQCE. Finally, IQCaaS includes the proposed applied algorithm (QAES) which are briefly explained in Sect. 8.7.

8.8.2.1 IQCaaS Methodology

In general, IQCaaS generates the encryption keys based on quantum mechanics instead of mathematics and computations, which in turn, provides unbroken key and eavesdropper detection. Most definite criteria associated with IQCaaS came from the nature of the quantum mechanism. IQCaaS encrypts cloud data based applied QAES using online negotiation mode. In this mode, clients directly negotiate with IQCaaS using BB84 protocol, to get a final secret key, which used with QAES algorithm to encrypt files. However, the client and IQCaaS are not able to know the secret key until the negotiations of their bases are finished (privacy amplification).

Finally, IQCE aims to: (i) Improve the availability and the reliability of the cloud computing cryptographic mechanisms by deploying both key generation and key management techniques based on IQCaaS layer, (ii) Manipulate massive computing processes that cannot be executed using personal computer only.

8.8.2.2 IQC Stages

This subsection discusses the main stages of IQCE like registration and verification, encryption/decryption, and uploading/downloading stages.

Registration and Verification Stage

At this stage, the client registers as a client to the CSP. The CSP after then verifies the MAC address assigns and generates a CA to authorize the client via Kerberos authentication function. Secondly, the client determined which his resources are needed. Thirdly, the CSP checks the resources availability and picks up, depended on the Microsoft load balancing (MLB) [54], the lowest load VM among the others. Finally, the VM-IP address is assigned to the client. Accordingly, due to such assigned process, when the client needs to reconnect his VM is assigned directly after the authentication achieved, see Fig. 8.12.

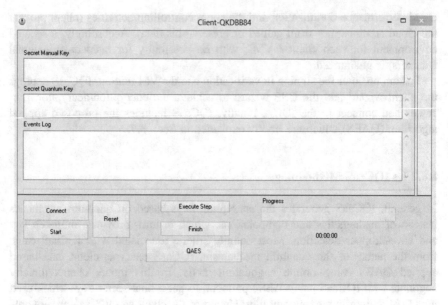

Fig. 8.12 Authenticated client wizard

Cryptographic Stage

The encryption/decryption phase entails online mode. It means the final secret key (K_q) is generated as a result of the negotiation between the client and the IQCaaS agent. In order to gain a secured communication for file transmission, the QAES-256 is used during the encryption/decryption process. Accordingly, the main procedure for the encryption/decryption process at IQCE depicted as follow:

i. **Client:** $E_n (P, K_q) \leftarrow P'$ //*trusted client encrypts file (P) on own machine using QAES.*
ii. **IQCaaS:** $D (P', K_q) \leftarrow P$ //*intelligent service decrypts an encrypted file (P') on IQCE usig QAES.*

Uploading/Downloading Stage

In this stage, necessary steps for uploading/downloading files in the IQC environment are illustrated. The steps for the uploading process are:

- The Client sends a request to CSP, (http://sharepoint2:8088/scvm_layout/request), for authentication.
- CSP sends the corresponding registered VM-IP successful authentication.
- CSP assigns an IQCaaS layer and deploys a console wizard to the client, see Fig. 8.13.

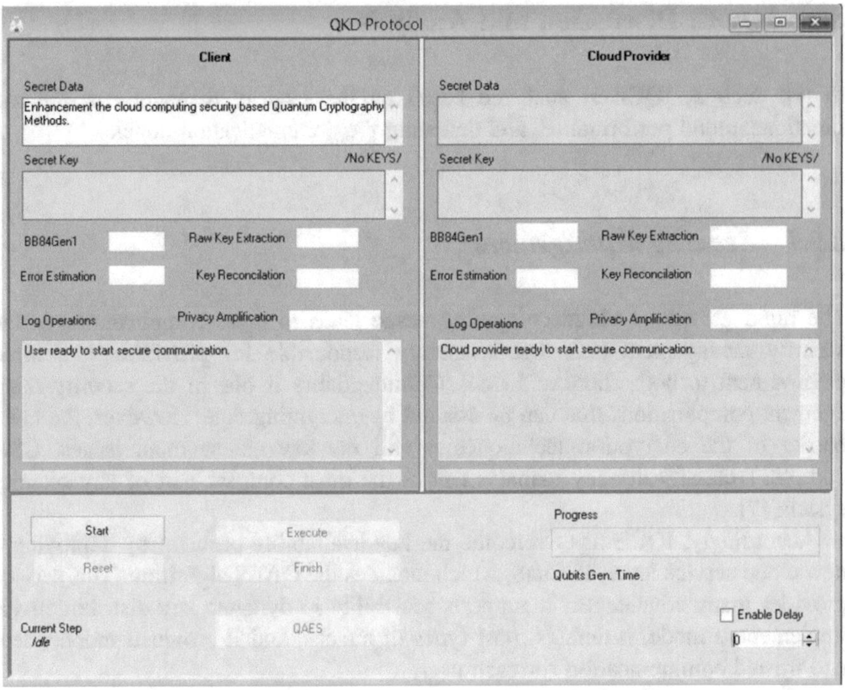

Fig. 8.13 CSP negotiation wizard

- The Client starts the negotiation with IQCaaS service.
- The Client invokes the CSP, which QAES selected length.
- Client encrypts file on own machine by QAES(128;192;256)
- The Client sends the file to the cloud environment via secure web-services.
- IQCaaS decrypts the sending file and saves it.

  The steps for downloading process are:

- The Client sends a request to CSP (http://sharepoint2:8088/scvm_layout/request), for authentication.
- CSP sends the corresponding registered VM-IP successful authentication.
- The Client starts the negotiation with IQCaaS service.
- The Client invokes the CSP, which QAES selected length.
- The Client determines the file wanted to view or download.
- IQCaaS encrypted file using selected QAES (128; 192; 256 bits)
- The Client decrypted the file, using QAES.
- Client download and view file on the own machine.

8.9 IQCE: Discussion and Analysis

In this section, IQCE is analyzed based on the security management, IQCaaS functions, cloud performance, and defeating the communication attacks.

8.9.1 Security Management

The rapid growth of cloud computing usage leads to more complications in the security management task that is mainly responsible for providing a secured environment to both client and CSP. Confidentiality is one of the security management complications that can be assured by encrypting data. However, the main barrier of the encryption techniques is still the key management issues. CSA and NIST classify the key management as the most complex part of any security system [7].

Accordingly, IQCE has overcome the key availability problem by deploying a new cloud service layer, IQCaaS, which includes the QAES algorithm. This service provides many advantages: It supports scalability in dynamic key distribution via implemented mode, it defeats most types of attacks, and it provided independent and trusted communication for each user.

8.9.2 IQCaaS Main Roles

IQCaaS protects the client's cryptographic key and file through the communication. Moreover, due to the isolation criteria for the resources, IQCaaS prevents an attacker or malicious from information extraction through the cloud. Additionally, IQCaaS provides:

- Securing the client: IQCaaS provides the encryption/decryption process by cooperating both client's machine and cloud servers, this corporation defeats two types of attacks (man-in-the middle attack and authentication attack).
- Client Encryption Permissions: IQCaaS helps the client for encrypting the flying data, which in turn, provides a higher level of security.
- Key Protection: Key generation and key distribution processes are critical in any cloud storage environment; therefore, keys must be carefully generated and protected. IQCaaS achieves these processes by dynamic key generation based QKD (for more details see [40, 47]). After then, keys are expired as soon as the sending or receiving files process completed.

8.9.3 Cloud Performance

In this section, the performance of IQCE is evaluated based on SCOM monitoring wizard, this evaluation focuses on processor efficiency and memory consumption for host cloud server. The evolution computations are calculated when 6-client machines are connected (Windows 8 operating systems installed) to the IQCE at the same time. Two of these machines implementing the download process, whereas, the others implementing the uploading process using different files size (10–150 MB). In addition, Figs. 8.14 and 8.15 illustrate the performance of IQCE based on memory consumption and processor efficiency.

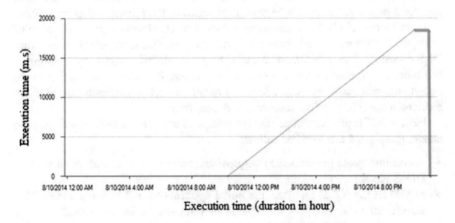

Fig. 8.14 Processor efficiency through 4 h of running host (6 connected VMs)

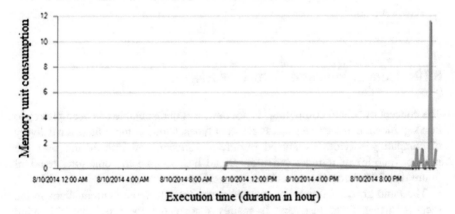

Fig. 8.15 Memory consumption through 4 h of running host (6 connected VMs)

According to above Figs. 8.14 and 8.15, IQCaaS does not affect the performance of IQCE and a normal state of the processor and memory registered. The isolation resources of cryptographic service are helped to guarantee the efficiency of the cloud.

8.9.4 Defeating the Communication Attacks

This section discusses the defeating of communication attacks are mentioned in Sect. 8.4.2, through the creation of a cloud connection and its prevention measures, through the client on the cloud server. IQCE can be used for preventing attacks in the cloud network, and it can be useful for securing the private and public cloud.

Furthermore, IQCaaS mitigates the influence of attacks through the isolation of host physical resources and domain controller connection (computer join to center DNS-domain). The *qcloud.net* domain helps to defeat attacks by creating a sub-inherited domain (forest) for each trusted client. In addition, IQCaaS helps to assign standard quota for each client and based on such quota each trusted client received a specific physical resources and time from CSP.

Finally, IQCE provides three necessary steps to mitigate a cloud communication attack; these steps are listed as follow:

- A costume quota prevents the overprovision bandwidth to absorb the impact of DDoS attack.
- SCOM in the IQCE monitors the portal applications and network traffic. This feature determines if poor application performance is due to service provider outages or an attack influence.
- The *qcloud.net* domain achieves many roles such as detects and stop malicious users, detect and stop malicious requests based on Kerberos authentication function [53].

8.10   Conclusions and Future Works

The concept of "cloud computing" is not new, it is undisputable has been proven to be a significant commercial success at recent times. Cloud computing is the delivery of computing services by shared resources, software and information over the internet (public) or intranet (private). Cloud has become the emerging trend of future distributed computing technology.

The rapid growth of cloud computing usage leads to more complications in the security management task that is mainly responsible for providing a secured environment to both client and CSP. Confidentiality is one of the security management complications that can be assured by encrypting data. However, the primary barrier of the encryption techniques is still the key management issues.

CSA and NIST classify the key management as the most complex part of any security system.

This chapter solves the key availability problem by deploying a new intelligent cloud service layer (IQCaaS). This service provides many roles such as supports scalability in dynamic key distribution, defeats the most types of attacks and provides independent and trusted communication for each user.

IQCaaS protects the client's cryptographic key and file through the communication and prevents an attacker or malicious from information extraction through the cloud. Finally, this service helps to build a secure, intelligent cloud environment based on quantum criteria. IQCE helps to improve the availability and the reliability of the cloud computing cryptographic mechanisms by deploying both key generation and key management techniques and manipulate massive computing processes that cannot be executed using the personal computer only.

In the future work, a quantum cipher cloud should be added between cloud environment and client enterprise. It is responsible for key generation and key deploying. Moreover, build a secure cloud computing system that mainly depend on randomness ratio associated with the quantum system.

References

1. Nelson, G., Charles, M., Fernando, R., Marco, S., Teresa, C., Mats, N., Makan, P.: A quantitative analysis of current security concerns and solutions for cloud computing. J. Cloud. Comput. Adv. Syst. Appl. **1**(11), 34–51 (2014)
2. Sunita, M., Seema, S.: Distributed Computing, 2nd edn. Oxford University Press, New York (2013)
3. Chander, K., Yogesh, S.: Enhanced security architecture for cloud data security. Int. J. Adv. Res. Comput. Sci. Softw. Eng. **3**(5), 67–78 (2013)
4. Mohammad, A., John, B., Ingo, T.: An analysis of the cloud computing security problem. In: APSE 2010 Proceeding—Cloud Workshop, Sydney, Australia, pp. 56–64 (2010)
5. Faiza, F.: Management of symmetric cryptographic keys in a cloud-based environment. In: A 2nd IEEE International Conference on Cloud Computing Technology and Science, pp. 90–98 (2012)
6. Mather, T., Kumaraswamy, S.: Cloud Security, and Privacy; An Enterprise Perspective on Risks and Compliance, 1st edn. O'Reilly Media, Sebastopol (2013)
7. CSA: Security a-a-services guidance for critical areas in cloud computing. Category **8**, 1–21 (2012)
8. Jasim, O.K., Abbas, S., El-Horbaty, E.-S.M., Salem, A.-B.M.: Cryptographic cloud computing environment as a more trusted communication environment. Int. J. Grid. High. Perform. Comput. (IJGHPC) **6**(2), 45–60 (2014)
9. Sekar, A., Radhika, S., Anand, K.: Secure communication using 512 bit key. Eur. J. Sci. Res. **52**(1), 61–65, 90–104 (2012)
10. Doelitzscher, F., Reich, C., Kahl, M., Clarke, N.: An autonomous agent-based incident detection system for cloud computing. In: 3rd IEEE International on Cloud Computing Technology and Sciences, CloudCom 2011, CPS. pp. 197–204 (2011)
11. Lejiang, G., XiangPing, C., Chao, G., Wang, D.P.: The intelligence security control platform based on cloud and trusted computing technology. In: International Conference on Electrical and Computer Engineering, Advances in Biomedical Engineering, vol. 11, pp. 221–229 (2012)

12. Wang, Q., Cong, X., Min, S.: Protecting privacy by multi-dimensional K-anonymity. J. Softw. 7(8), 87–90 (2012)
13. Hossein, R., Elankovan, S., Zulkarnain, A., Abdullah, M.: Encryption as a service (EaaS) as a solution for cryptography in cloud. In: 4th International Conference on Electrical Engineering and Informatics, ICEEI'13, vol. 11, pp. 1202–1210 (2013)
14. Esh, N., Mohit, P., Aman, B., Prem, H.: To enhance the data security of cloud in cloud computing using RSA algorithm. Int. J. Softw. Eng. 1(1), 56–64 (2012)
15. Soren, B., Sven, B., Hugo, I.: Client–controlled cryptography-as-a-service in the cloud. In: Proceedings of 11th International Conference, ACNS 2013, vol. 7954, pp. 74–82. Springer (2013)
16. Mandeep, K., Manish, G.: Implementing various encryption algorithms to enhance the data security of cloud in cloud computing. Int. J. Comput. Sci. Inf. Technol. 2(10), 156–204 (2012)
17. Dabrowski, C., Mills, K.: VM leakage and orphan control in open source cloud. In: 3rd IEEE International on Cloud Computing Technology and Sciences, CloudCom 2011, CPS, pp. 554–559 (2011)
18. Emmanuel, S., Navdeep, A., Prashant, T., Bhanu, P.: Cloud computing: data storage security analysis and its challenges. Int. J. Comput. Appl. 70(24), 98–105 (2013). ISSN 0975-8887
19. William, S.: Cryptography and Network Security, a Library of Congress Cataloging, 5th edn. Prentice Hall (2012). ISBN 13; 978-0-13609704-4
20. Winkler, J.: Securing the Cloud: Cloud Computer Security Techniques and Tactics, a Library of Congress Cataloging. Elsevier (2011). ISBN 13; 978-1-59749-592-9
21. Padmapriya, A., Subhas, P.: Cloud computing: security challenges and encryption practices. Int. J. Adv. Res. Comput. Sci. Softw. Eng. 3(3), 23–31 (2013)
22. Manpreet, R., Rajbir, V.: Implementing encryption algorithms to enhance data security of cloud in cloud computing. Int. J. Comput. Appl. 70(18) (2013)
23. Jensen, M., Schwenk, J., Hauschka, N., Iacono, L.: On technical security issues in cloud computing. In: IEEE ICCC, Bangalore, pp. 109–116 (2009)
24. Gawain, B., Waugh, R., Patil, S.: Enhancement of data security in cloud computing environment. Int. J. Internet Comput. 1(3), 123–132 (2012)
25. Farhad, A., Seyed, S., Athula, G.: Cloud computing: security and reliability issues. Commun. IBIMA 1(655710) (2013)
26. Eman, M., Hatem, S., Sherif, E.: Enhanced data security model for cloud computing. In: 8th International Conference on Informatics and Systems (INFO2012), Egypt, pp. 12–17 (2012)
27. Sanjo, S., Jasmeet, S.: Implementing cloud data security by encryption using Rijndael. Global J. Comput. Sci. Technol. Cloud Distrib. 13(4), 56–64 (2013)
28. Syam, P., Subramanian, R., Thamizh, D.: Ensuring data security in cloud computing using sobol sequence. In: Proceeding of 1st International Conference on Parallel, Distributed and Grid Computing (PDGC) (2010). IEEE 978-1-4244-7674-9/10
29. Wesam, D., Ibrahim, T., Christoph, M.: Infrastructure as a service security: challenges and solution. In: The Proceeding of 7th International Conference on Informatics and Systems (INFOS), Egypt, pp. 89–95 (2010)
30. Arif, S., Olariu, S., Wang, J., Yan, G., Yang, W., Khalil, I.: Datacenter at the airport: reasoning about time-dependent parking lot occupancy. IEEE Trans. Parallel Distrib. Syst. 3, 167–177 (2011)
31. Subashini, S., Kavitha, V.: A survey on security issues in service delivery models of cloud computing. J. Netw. Comput. Appl. 98–105 (2010)
32. Klems, K., Lenk, K., Nimis, J., Sandholm, T., Tai, T.: What's inside the cloud? An architectural map of the cloud landscape. In: IEEE Xplore, pp. 23–31 (2009)
33. Jasim, O.K., Abbas, S., El-Horbaty, E.-S.M., Salem, A.-B.M.: A comparative study of modern encryption algorithms based on cloud computing environment. In: 8th International Conference for Internet Technology and Secured Transactions (ICITST-2013), UK, pp. 536–541 (2013)

34. Kraska, F.: Building database applications in the cloud, Swiss federal institute of technology (2010)
35. Global Net optex Incorporated (2009) Demystifying the cloud, important opportunities and crucial choices, http://www.gni.com, pp. 4–14
36. Gongjun, Y., Ding, W., Stephan, O., Michele, C.: Security challenges in vehicular cloud computing. IEEE Trans. Intell. Transp. Syst. **14**(1), 67–73 (2013)
37. Patil, D., Akshay, R.: Data security over cloud emerging trends in computer science and information technology. In: Proceeding Published in International Journal of Computer Applications, pp. 123–147 (2012)
38. Itani, W., Kayassi, A., Chehab, A.: Energy-efficient incremental integrity for securing storage in mobile cloud computing. In: International Conference on Energy Aware Computing (ICEAC10), Egypt, pp. 234–241 (2012)
39. Solange, G., Mohammed, A.: Applying QKD to reach unconditional security in communications, European research project SECOQC, www.secoqc.net (2014)
40. Yau, Stephen S., Ho, G.An.: Confidentiality protection in cloud computing systems. Int. J. Softw. Inform. **4**(4), 351–363 (2010)
41. Bethencourt, J., Sahai, A., Waters, B.: Ciphertext-policy attribute-based encryption. In: Proceedings of the IEEE Symposium on Security and Privacy (SP '07), pp. 321–334 (2011)
42. Cutillo, A., Molva, R., Strufe, T.: A privacy-preserving online social network leveraging on real-life trust. IEEE Commun. Mag. (2012)
43. Trugenberger, C., Diamantini, C.: Quantum associative pattern retrieval. Stud. Comput. Intell. (SCI) **121**, 103–113 (2008)
44. Christain, K., Mario, P.: Applied Quantum Cryptography, Lecture Notes in Physics, vol. 797. Springer, Berlin (2010). doi:10.1007/978-3-642-04831-9
45. Ammar, O., Khaled, E., Muneer, A., Eman, A.: Quantum key distribution by using RSA. In: Proceedings of 3rd International Conference on Innovative Computing Technology (INTECH), pp. 123–128 (2013)
46. Jasim, O.K., Anas, A.: The goals of parity bits in quantum key distribution system. Int. J. Comput. Appl. **56**(18), 5–9 (2012)
47. Dani, G.: An experimental implementation of oblivious transfer in the noisy storage model. Nat. Commun. J. **5**(3418), 56–64 (2014)
48. Rawal, V., Dhamija, A., Sharma, S.: Revealing new concepts in cryptography & clouds. Int. J. Sci. Technol. Res. **1**(7), 23–33 (2012)
49. Jasim, O.K., Abbas, S., El-Horbaty, E.-S.M., Salem, A.-B.M.: Quantum key distribution: simulation and characterizations. In: International Conference on Communication, Management and Information Technology (ICCMIT 2015), Procedia Computer Science, pp. 78–88, Prague (2015)
50. Matthew, P., Caleb, H., Lamas, L., Christian, K.: Daylight Operation of a Free Space, Entanglement-Based Quantum Key Distribution System. National University of Singapore, Centre for Quantum Technologies, Singapore (2008)
51. Jasim, O.K., Abbas, S., El-Horbaty, E.-S.M., Salem, A.-B.M.: Statistical analysis of random bits generation of quantum key distribution. In: 3rd IEEE-International Conference on Cyber Security, Cyber Warfare, and Digital Forensic (CyberSec2014), pp. 45–52 (2014)
52. Matthew, G.: Statistical tests of randomness on QKD through a free-space channel coupled to daylight noise. J. Lightwave Technol. **3**(23), 56–68 (2013)
53. Aidan, F., Hans, V., Patrick, L., Damian, F.: (2012) Microsoft Private Cloud Computing. Wiley, ISBN: 978-1-118-25147-8
54. Sharif, M.S.: Quantum cryptography: a new generation of information technology security system. In: 6th International Conference on Information Technology: New Generations, pp. 1644–1648 (2009)
55. Woolley, R., Fletcher, D.: The hybrid cloud: bringing cloud-based IT services to state government (2009)

56. Raheel, A., Stuart, B., Elizabeth, C., Kavitha, G., James, E.: Quantum simultaneous recurrent networks for content addressable memory. Stud. Comput. Intell. (SCI) **121**, 57–74 (2008)
57. Center for Quantum Technology, http://www.quantumlah.org/research/topic/qcrypto. Accessed 1 Sept 2014
58. Jasim, O.K., Abbas, S., El-Horbaty, E.-S.M., Salem, A.-B.M.: Advanced encryption standard development based quantum key distribution. In: 9th International Conference for Internet Technology and Secured Transactions (ICITST-2014), UK, pp. 343–354 (2014)

Chapter 9
Paraconsistent Logics and Applications

Jair Minoro Abe

Abstract In this work we summarize some of the applications of so-called Paraconsistent logics, mainly one class of them, the paraconsistent annotated logics. Roughly speaking such systems allow inconsistencies in a non-trivial manner in its interior; so it is suitable to handle themes in which inconsistencies become a central issue, like pattern recognition, non-monotonic reasoning, defeasible reasoning, deontic reasoning, multi-agent systems including distributed systems, collective computation, among a variety of themes.

Keywords Paraconsistency and AI · Applications of paraconsistent systems · Paraconsistent logics and informatics

9.1 Introduction

The exuberant advances of computer applications in several branches such as Informatics, Automation and Robotics, Information Systems, and correlated areas over the past century and this, has changed our day-by-day lives and our world vision. Since more and more concern is necessary to these new techniques and theories, there is the need of more formal and generic tools to understand and to deal with them. Therefore, step by step the classical logic was being complemented and/or substituted partially or totally by new logics, which are commonly referred as non-classical logics. So, the appearance the non-classical logics over the past century and this is one of the most notable contributions to overcome classical logic in many ways: many-valued logics, modal logics, tense logics, fuzzy set theory, non-monotonic logics, linear logics, and so on.

Another topic of interest is of the contradiction and paracompleteness. When we face with contradictory and/or paracomplete data, the usual logical systems are not

J.M. Abe (✉)
Graduate Program in Production Engineering, ICET - Paulista University,
Rua. Dr. Bacelar, 1212, São Paulo, SP CEP 04026-002, Brazil
e-mail: jairabe@uol.com.br

© Springer International Publishing Switzerland 2016   273
K. Nakamatsu and R. Kountchev (eds.), *New Approaches in Intelligent Control*,
Intelligent Systems Reference Library 107, DOI 10.1007/978-3-319-32168-4_9

able to deal with them, at least directly. So, we need a new kind of logic, namely, the paraconsistent logics. Problems of various kinds give rise to these non-classical logics: for instance, the paradoxes of set theory, the semantic antinomies, some issues originating in dialectics, in Meinong's theory of objects, in the theory of fuzziness, and in the theory of constructivity.

However, nowadays paraconsistent logic has found various applications in Artificial Intelligence (AI), logic programming, Biomedicine, Engineering, Foundation of Physics, psychoanalysis, ultimately in almost all branches of human knowledge, showing itself to be of basic significance for Science in general.

In this work we summarize of some significant applications obtained recently to Computer Science (ParaLog—a paraconsistent logic programming language), multi-agents systems, knowledge representation (Frames), new framework for Computer Science based on paraconsistent annotated systems, implementation of paraconsistent electronic circuits, artificial neural networks, automation, and robotics.

Obviously this work is necessarily incomplete, so paraconsistent logic (theo-retical and application aspects) is a theme cultivated worldwide.

9.2 Imprecision and Actual World

Almost all concepts regarding actual world encompasses a certain imprecision degree. Let us take colors fulfilling a rainbow. Let us suppose, for instance that the rainbow begins with yellow band and ends with red. If we consider the statement, "This point of the rainbow is yellow", surely it is true if the point is on the first band and false if it is on the last band. However, if the point ranges between the extremes, there are points in which the statement is neither true nor false; or it can be both true and false. This is not because the particular instruments that we use nor it is lack of our vocabulary. The vagueness of the terms and concepts of real science has no subjective nature, arising from causes inherent to the observer, nor objective, in the sense that the reality is indeed imprecise or vague. Such a condition is imposed on us by our relationship with reality, how we are constituted psycho-physiologically to grasp it, and also by the nature of the universe.

9.3 Paraconsistent, Paracomplete, and Non-alethic Logics

A deductive theory is said to be consistent if it has no theorem, one of which is the negation of the other; otherwise it is called inconsistent (or contradictory). A theory is called trivial if all formulas (or sentences) of its language are provable; otherwise it is called non-trivial. Analogously, the same definition above applies to propo-sitional systems, sets of information, etc. (taking into account their set of consequences).

If the underlying logic of a theory T is classical logic or most of the extent logics, T is trivial iff it is inconsistent. Therefore, if we want to handle logically inconsistent but non-trivial theories or information systems, we have to use a new kind of logic.

Paraconsistent logic is a logic that can be the basis of inconsistent but non-trivial theories. Issues such as those described above have been appreciated by many logicians. In 1910, the Russian logician Nikolai A. Vasiliev (1880–1940) [27] and the Polish logician Jan Łukasiewicz (1878–1956) [9] independently glimpsed the possibility of developing such logics. Nevertheless, Stanisław Jaśkowski (1906–1965) was in 1948 effectively the first logician to develop a paraconsistent system, at the propositional level. His system is known as 'discussive propositional calculus'. Independently, some years later, the Brazilian logician Newton C.A. da Costa (1929–) constructed for the first time hierarchies of paraconsistent propositional calculi Ci, $1 \leq i \leq \omega$ of paraconsistent first-order predicate calculi (with and without equality), of paraconsistent description calculi, and paraconsistent higher-order logics (systems NFi, $1 \leq i \leq \omega$) [10]. Also, independently of Da Costa, the American mathematician and logician Nels David Nelson (1918–2003) has considered a paraconsistent logic as a version of his known as constructive logics with strong negation [22]. Relevance logics (or relevant logics): these systems were developed as attempts to avoid the paradoxes of material and strict implication. Relevant logic begins with Ackermann, and was properly developed in the work of Anderson and Belnap [7]. Many of the founders of relevant logic, such as Robert Meyer and Richard Routley, have also been directly concerned with paraconsistency [24].

Another important class of non-classical logics is the paracomplete logics. A logical system is called paracomplete if it can function as the underlying logic of theories in which there are formulas such that these formulas and their negations are simultaneously false. Intuitionistic logic and several systems of many-valued logics are paracomplete in this sense (and the dual of intuitionistic logic, Brouwerian logic, is therefore paraconsistent).

As a consequence, paraconsistent theories do not satisfy the principle of non-contradiction, which can be stated as follows: of two contradictory propositions, i.e., one of which is the negation of the other, one must be false. And, paracomplete theories do not satisfy the principle of the excluded middle, formulated in the following form: of two contradictory propositions, one must be true.

Finally, logics which are simultaneously paraconsistent and paracomplete are called non-alethic logics.

9.4   Paraconsistent Annotated Logics

Nowadays it is known infinite paraconsistent systems. Annotated logics are a family of non-classical logics initially proposed in logic programming by Subrahmanian [25]. In view of the applicability of annotated logics to these differing formalisms in computer science, it has become essential to study these logics more carefully, mainly from the foundational point of view. Thus, they were studied from a foundational viewpoint by J.M. Abe, S. Akama, N.C.A. da Costa and others Abe [1, 9]. The annotated systems initially have been applied to the development of a declarative semantics for inheritance networks, and object-oriented databases. Other applications are summarized in Abe [2].

In general, annotated logics are a kind of paraconsistent, paracomplete, and non-alethic logic. The latter systems are among the most original and imaginative systems of non-classical logic developed in last and present century. Nowadays, paraconsistent logic has established a distinctive position in a variety of fields of knowledge.

9.5   A Paraconsistent Logic Programming—ParaLog

Inconsistency is a natural phenomenon arising from the description of the real world. This phenomenon may be found in several contexts. Nevertheless, human beings are capable of reasoning adequately. The automation of such reasoning requires the development of formal theories.

The employment of logic systems allowing reasoning about inconsistent infor-mation is an area of growing importance in Computer Science, Data Base Theory and AI. For instance, if a knowledge engineer is designing a knowledge base KB, related to a domain D, he may consult n experts in that domain. For each expert ei, $1 \leq i \leq n$, of domain D, he will obtain some information and will present it in some logic such as a set of sentences KB_i, for $1 \leq i \leq n$. A simple way of combining the knowledge amassed from all experts in a single knowledge base KB is:

$$KB = KB_1 \cup KB_2 \cup \cdots \cup KB_n (1 \leq n)$$

However, certain KB_i and KB_j bases may contain conflicting propositions—p and ¬p (the negation of p). In such case, p might be a logical consequence of KB_i, while ¬p might be a logical consequence of KB_j. Therefore, KB is inconsistent and consequently meaningless, because of the lack of models. However, the knowledge base KB is not a useless set of information.

There are some arguments favoring this standpoint, as follows:

Certain subsets of KB may be inconsistent and express significant information. Such information cannot be disregarded;

The disagreement among specialists in a given domain may be significant. For instance, if physician M_1 concludes patient X suffers from a fatal cancer, while physician M_2 concludes that same patient suffers from cancer, but a benign one, the patient will probably want to know the causes of such disagreement. This disagreement is significant because it may lead patient X to take appropriate decisions—for instance, to get the opinion of a third physician.

The reasoning for the last item is that it is not always advisable to find ways to exclude formulas identified as causing inconsistency(ies) in KB, because many times important information may be removed. In such cases, the very existence of inconsistency is important.

Though inconsistency is an increasingly common phenomenon in programming environments—especially in those possessing a certain degree of distribution—it cannot be handled, at least directly, by classical logic, on which most of the current logic programming languages are based. Thus, one has to resort to alternatives to classical logic; it is therefore necessary to search for programming languages based on such new logics.

Paraconsistent Logic, despite having been initially developed from the purely theoretical standpoint, found in recent years extremely fertile applications in Computer Science, thus solving the problem of justifying such logic systems from the practical standpoint.

In Da Costa et al. [13] it was proposed a variation of the logic programming language Prolog—ParaLog—which allows inconsistency to be handled directly. This implementation was made independently of results by Subrahmanian and colleagues.

In the sequel we sketch a paraconsistent logic programming—ParaLog.

The development of computationally efficient programs in Paralog must exploit two aspects in this language:

1. the declarative aspect that describes the logic structure of the problem, and
2. the procedural aspect that describes how the computer solves the problem.

However, it is not always an easy task to conciliate both aspects. Therefore, programs to be implemented in Paralog should be well defined to evidence both the declarative aspect and the procedural aspect of the language.

It must be pointed out that programs in Paralog, like programs in standard Prolog, may be easily understood or reduced—when well defined—by means of addition or elimination of clauses, respectively.

A small knowledge base in the domain of Medicine is presented as a Paralog program. The development of this small knowledge base was subsidized by the information provided by three experts in Medicine. The first two specialists—clinicians—provided six[1] diagnosis rules for two diseases: disease1 and disease2. The last specialist—a pathologist—provided information on four symptoms: symptom1, symptom2, symptom3 and symptom4. This example was adapted from da Costa and Subrahmanian's work [11].

---

[1]The first four diagnosis rules were supplied by the first expert clinician and the two remaining diagnosis rules were provided by the second expert clinician.

```
disease1(X): [1.0,  0.0] <--
symptom1(X): [1.0,0.0] &
symptom2(X): [1.0,0.0]
disease2(X): [1.0,0.0] <--
symptom1(X): [1.0,0.0] &
symptom3(X): [1.0,0.0]
disease1(X): [0.0,1.0] <--
disease2(X): [1.0,0.0].
disease2(X): [0.0,1.0] <--
disease1(X): [1.0,0.0].
disease1(X): [1.0,0.0] <--
symptom1(X): [1.0,0.0] &
symptom4(X): [1.0,0.0]  .
disease2(X): [1.0,0.0] <--
symptom1(X): [0.0,1.0] &
symptom3(X): [1.0,0.0]  .
symptom1(john)  : [1.0,0.0].
symptom1(bill): [0.0,1.0].
symptom2(john)  : [0.0,1.0].
symptom2(bill): [0.0,1.0].
symptom3(john): [1.0,0.0].
symptom3(bill): [1.0,0.0].
symptom4(john): [1.0,0.0].
symptom4(bill): [0.0,1.0].
```

Example 9.1 A small knowledge base in Medicine implemented in Paralog

In this example, several types of queries can be performed. Table 9.1 below shows some query types, the evidences provided as answers by the Paralog inference engine and their respective meaning.

The knowledge base implemented in Example 9.1 may also be implemented in standard Prolog, as shown in Example 9.2.

Table 9.1 Query and answer forms in Paralog

Item	Query and answer form		Meaning
1	Query	Disease1(bill):[1.0, 0.0]	Does Bill have disease 1?
	Evidence	[0.0, 0.0]	The information on Bill's disease1 is unknown
2	Query	Disease2(bill):[1.0, 0.0]	Does Bill have disease 2?
	Evidence	[1.0, 0.0]	Bill has disease2
3	Query	Disease1(john):[1.0, 0.0]	Does John have disease 1?
	Evidence	[1.0, 1.0]	The information on John's disease1 is inconsistent
4	Query	Disease2(john):[1.0, 0.0]	Does John have disease 2?
	Evidence	[1.0, 1.0]	The information on John's disease2 is inconsistent
5	Query	Disease1(bob):[1.0, 0.0]	Does Bob have disease 1?
	Evidence	[0.0, 0.0]	The information on Bob's disease1 is unknown

Example 9.2 Knowledge base of Example 9.1 implemented in standard Prolog

```
disease1(X)  :-
symptom1(X),
symptom2(X).
disease2(X):-
symptom1(X),
symptom3(X).
disease1 (X):-
not disease2(X).
disease2(X)  :-
not disease1(X).
disease1(X)  :-
symptom1(X),
symptom4(X).
disease2(X)  :-
not symptom1(X),
symptom3(X).
symptom1 (john)  .
symptom3( john)  .
symptom3(bill).
symptom4(john)
```

In this example, several types of queries can be performed as well. Table 9.2 shows some query types provided as answers by the standard Prolog and their respective meaning.

Starting from Examples 9.1 and 9.2 it can be seen that there are different characteristics between implementing and consulting in Paralog and standard Prolog. Among these characteristics, the most important are:

Table 9.2 Query and answer forms in standard Prolog

Item	Query and answer form		Meaning
1	Query	Disease1(bill)	Does Bill have disease 1?
	Answer	Loop	System enters into an infinite loop
2	Query	Disease2(bill)	Does Bill have disease 2?
	Answer	Loop	System enters into an infinite loop
3	Query	Disease1(john)	Does John have disease 1?
	Answer	Yes	John has disease1
4	Query	Disease2(john)	Does John have disease 2?
	Answer	Yes	John has disease2
5	Query	Disease1(bob)	Does Bob have disease 1?
	Answer	No	Bob does not have disease1

1. the semantic characteristic; and
2. the execution control characteristic.

The first characteristic may be intuitively observed when the program codes in Examples 9.1 and 9.2 are placed side by side. That is, when compared to Paralog, the standard Prolog representation causes loss of semantic information on facts and rules. This is due to the fact that standard Prolog cannot directly represent the negation of facts and rules.

In Example 9.1, Paralog program presents a four-valued evidence representation. However, the information loss may be greater for a standard Prolog program, if the facts and rules of Paralog use the intermediate evidence of lattice $\tau = \{x \in \Re \mid 0 \leq x \leq 1\} \times \{x \in \Re \mid 0 \leq x \leq 1\}$. This last characteristic may be observed in Tables 9.1 and 9.2. These two tables show five queries and answers, presented and obtained both in Paralog and standard Prolog program.

The answers obtained from the two approaches present major differences. That is, to the first query: "Does Bill have disease1?", Paralog answers that the information on Bill's disease1 is unknown, while the standard Prolog enters into a loop. This happens because the standard Prolog inference engine depends on the ordination of facts and rules to reach deductions. This, for standard Prolog to be able to deduct an answer similar to Paralog, the facts and rules in Example 7.2 should be reordered. On the other hand, as the Paralog inference engine does not depend on reordering facts and rules, such reordering becomes unnecessary.

In the second query: "Does Bill have disease2?", Paralog answers that "Bill has disease2", while the standard Prolog enters into a loop. This happens for the same reasons explained in the foregoing item.

In the third query: "Does John have disease1?", Paralog answers that the information on John's disease1 is inconsistent, while the standard Prolog answers that "John has disease1". This happens because the standard Prolog inference engine, after reaching the conclusion that "John has disease1" does not check whether there are other conclusions leading to a contraction. On the other hand, Paralog performs such check, leading to more appropriate conclusions.

In the fourth query: "Does John have disease2?", Paralog answers that the information on John's disease2 is inconsistent, while the standard Prolog answers that "John has disease2". This happens for the same reasons explained in the foregoing item.

In the last query: "Does Bob have disease1", Paralog e answers that the information on Bob's disease1 is unknown, while the standard Prolog answers that "Bob does not have disease1". This happens because the standard Prolog inference engine does not distinguish the two possible interpretations for the answer not. On the other hand, the Paralog inference engine, being based on an infinitely valued paraconsistent evidential logic, allows the distinction to be made.

In view of the above, it is demonstrated that the use of the Paralog language may handle several Computer Science questions more naturally.

9.6 Paraconsistent Knowledge in Distributed Systems

Multi-agents systems are an important topic in AI. The use of modal systems for modeling knowledge and belief has been largely considered in Artificial Intelligence. One interesting approach is due Fagin et al. [17].

The essential ideas underlying the systems proposed by Fagin et al. [17] can be summarized as follows: iA can be read agent i knows A, i = 1, ..., n. Common knowledge and Distributed knowledge are also defined in terms of additional modal operators: $[]_G$, ("everyone in the group G knows"), $[]_G^C$ ("it is common knowledge among agents in G"), and $[]_G^D$ ("it is distributed knowledge among agents in G") for every nonempty subset G of {1, ..., n}.

Nevertheless, the most of those proposals use extensions of classical logic or at least part of it, keeping as much as fundamental characteristics of classical logic. When it is taken questions of logical omniscience, one relevant concept that appears is that of contradiction.

The attractiveness of admitting paraconsistency and paracompleteness in the system becomes evident if we observe that some agents can actually lie or be ignorant about certain propositions: an agent may state both A and ¬A (the negation of A) hold (or that none of A and ¬A hold).

Abe and Nakamatsu [5] has presented a class of multimodal annotated system Mτ which is paraconsistent and, in general, paracomplete and non-alethic logics. They ay constitute, for instance, a framework for modeling paraconsistent knowledge. In what follows, we sketch an axiomatic systems of them.

The symbol $\tau = <|\tau|, \leq, \sim>$ indicates some finite lattice with operator called the *lattice of truth-values*. We use the symbol $\leq$ to denote the ordering under which τ is a complete lattice, $\perp$ and $\top$ to denote, respectively, the bottom element and the top element of τ. Also, $\wedge$ and $\vee$ denote, respectively, the greatest lower bound and least upper bound operators with respect to subsets of $|\tau|$. The operator $\sim : |\tau| \rightarrow |\tau|$ will work as the "meaning" of the negation of the system Mτ.

The language of Mτ has the following primitive symbols:

1. Individual variables: a denumerable infinite set of variable symbols: $x_1, x_2, ...$
2. Logical connectives: ¬ (negation), $\wedge$ (conjunction), $\vee$ (disjunction), and $\rightarrow$ (implication).
3. For each n, zero or more n-ary function symbols (n is a natural number).
4. For each $n \neq 0$, n-ary predicate symbols.
5. The equality symbol: =
6. Annotational constants: each member of τ is called an annotational constant.
7. Modal operators: $[]_1, []_2, ..., []_n, (n \geq 1), []_G, []_G^C, []_G^D$ (for every nonempty subset G of {1, ..., n}).
8. Quantifiers: $\forall$ (for all) and $\exists$ (there exists).
9. Auxiliary symbols: parentheses and comma.

For each n the number of n-ary function symbols may be zero or non-zero, finite or infinite. A 0-ary function symbol is called a *constant*. Also, for each $n \geq 1$, the number of n-ary predicate symbols may be finite or infinite. We suppose that $M\tau$ possesses at least one predicate symbol.

We define the notion of *term* as usual. Given a predicate symbol p of arity n and n terms $t_1, \ldots, t_n$, a *basic formula* is an expression of the form $p(t_1, \ldots, t_n)$. An *annotated atomic formula* is an expression of the form $p_\lambda(t_1, \ldots, t_n)$, where λ is an annotational constant. We introduce the general concept of (*annotated*) *formula* in the standard way. For instance, if A is a formula, then $[]_1A, []_2A, \ldots, []_nA, []_GA, []_G^C A$, and $[]_G^D A$ are also formulas.

Among several intuitive readings, an atomic annotated formula $p_\lambda(t_1, \ldots, t_n)$ can be read: *it is believed that $p(t_1, \ldots, t_n)$'s truth-value is at least λ.*

Definition 9.1 Let A and B be formulas. We put

$A \leftrightarrow B =_{\text{Def.}} (A \rightarrow B) \wedge (B \rightarrow A)$ (equivalence)

$\daleth A =_{\text{Def.}} A \rightarrow ((A \rightarrow A) \wedge \neg(A \rightarrow A))$ (strong negation)

The symbol '$\leftrightarrow$' is called *biconditional*, '$\daleth$' is called *strong negation*.

Let A be a formula. Then: $\neg^0 A$ indicates A, $\neg^1 A$ indicates $\neg A$, and $\neg^k A$ indicates $\neg(\neg^{k-1}A)$, $(k \in N, k > 0$, N is the set of natural numbers). Also, if $\mu \in \tau$, $\sim^0\mu$ indicates μ, $\sim^1\mu$ indicates $\sim\mu$, and $\sim^k\mu$ indicates $\sim(\sim^{k-1}\mu)$, $(k \in N, k > 0)$. If A is an atomic formula $p_\lambda(t_1, \ldots, t_n)$, then a formula of the form $\neg^k p_\lambda(t_1, \ldots, t_n)$ $(k \geq 0)$ is called a *hyper-literal*. A formula other than hyper-literals is called a *complex* formula.

The postulates (axiom schemata and primitive rules of inference) of $M\tau$ are the same of the logics $Q\tau$ plus the following listed below, where A, B, and C are any formulas whatsoever, $p(t_1, \ldots, t_n)$ is a basic formula, and λ, μ, μ_j are annotational constants.

(M1) $[]_i(A \rightarrow B) \rightarrow ([]_iA \rightarrow []_iB)$, $i = 1, 2, \ldots, n$

(M2) $[]_iA \rightarrow []_i[]_iA$, $i = 1, 2, \ldots, n$

(M3) $\daleth []_iA \rightarrow []_i\daleth []_iA$, $i = 1, 2, \ldots, n$

(M4) $[]_iA \rightarrow A$, $i = 1, 2, \ldots, n$

(M5) $\frac{A}{[]_iA}$, $i = 1, 2, \ldots, n$

(M6) $[]_G A \leftrightarrow \wedge_{i \in G}[]_iA$

(M7) $[]_G^C A \rightarrow []_G(A \wedge []_G^C A)$

(M8) $[]_{\{i\}}^D A \leftrightarrow []_iA$, $i = 1, 2, \ldots, n$

(M9) $[]_G^D A \rightarrow []_{G'}^D A$ if $G' \subseteq G$

(M10) $\frac{A \rightarrow []_G(B \wedge A)}{A \rightarrow []_G^B B}$

(M11) $\forall x[]_iA \rightarrow []_i\forall xA$, $i = 1, 2, \ldots, n$

(M12) $\daleth (x = y) \rightarrow []_i\daleth (x = y)$, $i = 1, 2, \ldots, n$

with the usual restrictions.

$M\tau$ is an extension of the logic $Q\tau$. As $Q\tau$ contains classical predicate logic, $M\tau$ contains classical modal logic S5, as well as the multimodal system studied in [14] in at least two directions. So, usual all valid schemes and rules of classical positive propositional logic are true. In particular, the deduction theorem is valid in $M\tau$ ant it contains intuitionistic positive logic.

Theorem 9.2 *Mτ is non-trivial.*

Mτ admits a Kripke style model and Mτ is sound and complete when the lattice τ is finite.

9.7   A Multi-agent Paraconsistent Framework

In Prado [23] it was described a specification and prototype of an annotated paraconsistent logic-based architecture, which integrates various computing systems—planners, databases, vision systems, etc.—of a manufacture cell. Such systems will be referred to as agents.

In application domains such as robot control and flexible manufacture cells, the complexity of the control task grows proportionally with the increase and variety of stimuli coming from the external world to the system.

To deal with such complexity and to adequate to those stimuli within the time constraints imposed by the application domain, the control task should not be centralized. However, control decentralization is not easy to implement: paradoxically, it can lead to an increase in the time required to solve the problem, since it can interfere with the coherence of the resolution process. To avoid this phenomenon the architecture specifies:

- How each agent is going to use its knowledge, plans, goals and skills in the resolution process.
- How each agent is going to behave when faced with imprecise and inconsistent information
- How, and when, each agent is going to pass on to the other agents its plans, goals, skills and beliefs.
- How each agent is going to internally represent the information received from the other agents and its belief in this information.

Finally, the proposed architecture is able to "encapsulate" the existing computing systems, as well as hide from these systems the mechanisms of co-operation, co-ordination and inconsistency handling. This reduces the effort necessary to integrate the systems.

Gathering concepts and techniques from Distributed Artificial Intelligence and Annotated Paraconsistent Logic, the architecture has also enabled the agents to work in a co-operative fashion, even in the presence of inconsistent data and results, in order to achieve common or distinct interactive goals.

In Distributed Artificial Intelligence Systems the agents are members of a network, and each of them only possesses its own local perception of the problem to be solved. In a traditional Distributed Processing approach an intense message exchange among the nodes of the network is necessary, so as to supply the nodes with the information necessary to the processing and local control of each node. The result of this intense communication is a drop in the performance of the entire system, and a high level of synchronism in the agents' processing.

One possible manner to reduce the communication and synchronization rates among agents is to let them produce partial (candidate), incomplete, or incorrect results. Or, even, inconsistent and/or paracomplete results in comparison with the partial results produced by other agents.

This kind of processing requires a resolution problem architecture, which allows the co-operation among agents, in such a way that the partial results of each agent can be revised and enhanced from the information obtained during the interaction with the other agents.

For the past two decades, some Distributed Artificial Intelligence architectures have been proposed in the most varied fields, ranging from signal integration to industrial applications. However, such frameworks do not deal with the inconsistency phenomenon. In the majority of them, only the most recent data are considered during the resolution process. The earlier data (regardless their origin), which may lead to inconsistency, are not taken into account. Despite its importance, the inconsistency phenomenon is a research field in Distributed Artificial Intelligence, which has not received enough attention.

One possible reason for the current situation is that the inconsistency phenomenon and/or paracompleteness cannot be directly dealt with through the Classic Logic. Therefore, in order to tackle inconsistencies and paracompleteness directly, one should employ logic other than the classic one. In this work, we have employed the paraconsistent annotated logic to deal with the systems' inconsistencies.

In order to make possible the use of such logic in complex application domains (intense information input and critical agent response time), like the manufacture cells, it has been necessary to extend and refine the techniques and concepts of the paraconsistent annotated evidential logic programming and the Amalgam's Knowledge-base.

9.8 Paraconsistent Frame System

In Computer Science, a good solution for a given problem many times depends on a good representation. For most Artificial Intelligence applications, the choice of a knowledge representation is even more difficult, since the criteria for such choice are less clear.

Though no general consensus exists of what is knowledge representation, many schemes were proposed to represent and store knowledge. Many of such schemes have been successfully used as a foundation for the implementation of some existing systems. There are, however, several characteristics of knowledge that are not yet well understood, such as defaults and inconsistencies. Until a better comprehension of such characteristics is achieved, the representation of knowledge will remain as an active field of study.

There are several schemes to represent knowledge. The two schemes that better capture the knowledge concerning objects and their properties are semantic networks and frames.

The first of these schemes to represent knowledge, semantic networks, were originally proposed by psychology researchers, as modeling systems for the human associative memory. Later, several Computer Science researchers extended the original concept of semantic networks to facilitate the handling of more complex objects and relationships. Basically, a semantic network is a graph in which the nodes show objects, or a class, and the links show a relationship, generally binary, between objects or classes connected by the link. The nodes may be of two types: individual and generic. The first represents descriptions or affirmations concerning an individual instance of an object, while the second is related to a class or category of objects. The classes are pre-ordained in a taxonomy, and there are links representing special binary relationships such as isa—is a—and ako—a kind of. The first link type connects an individual node to a generic node and identifies an individual as belonging to a certain class. The second links two generic nodes between them and shows that a given class is a subclass of another.

The second of these knowledge representation schemes—frames—became popular in the 70s due to the appearance of the frame theory. The frame theory appeared initially as a result of an article written by Marvin Minsky. A frame system as proposed by Minsky consists in a collection of frames articulated in a semantic network. At the time, the use of frames was recommended as basic to understand visual perception, dialogues in natural language and other complex behavior. The development of languages for frame handling was partly intended for the implementation of frame-based Artificial Intelligence systems.

A frame is a representation of a complex object. It is identified by a name and consists of a set of slots. Each frame possesses at least one hierarchically superior frame, thus providing the basis of the inheritance mechanism. A special frame is the root of this inheritance hierarchy.

The inheritance hierarchy is a consequence of the classic notion of taxonomic hierarchy as a way to organize knowledge. The taxonomic hierarchy is just the beginning of inheritance reasoning. Researchers in Artificial Intelligence have added tools to represent class properties, exceptions to inherited properties, multiple superclasses and structured concepts with specific relations over the structural elements. Furthermore, the reasoning by inheritance naturally leads to simple default reasoning and nonmonotonic reasoning, and may be used to reason about prototypes and typical instances of inheritance system classes.

The two main types of existing inheritance systems are: those that do not admit exceptions to inherited properties and those that admit exceptions to inherited properties. It is easy to describe the semantics of the first type of inheritance in first order logic, in which frames may be interpreted as unary predicates and slots may be interpreted and binary predicates. The description of the semantics of the second type of inheritance system in first order logic is much more difficult, since exceptions introduce nonmonotonicity.

Since the late 70s several nonmonotonic formalisms have been proposed. Among the most widely published are, among others: Clark's predicate completion, Reiter's default logic, McDermott and Doyle's nonmonotonic logic, McCarty's circumscription, and Moore's autoepistemic logic. However, none of these formalisms deal adequately with issues like the inconsistency phenomenon.

Despite this phenomenon being increasingly common in programming environments—mainly in those possessing a certain degree of distribution—it cannot be treated, at least directly, by the Classic Logic, in which most of the current logic programming languages are based.

Thus, so as to be able to study these inconsistencies directly, one has to resort to alternative logic, it being therefore necessary to look for programming languages based on such logic.

To implement frame systems dealing with the inconsistency, the difficulty caused by the lack of a formal semantics both for paraconsistent frame systems and for inheritance reasoners dealing with inconsistencies and multiple inheritance frame systems must be taken into account.

In Ávila [8] it is presented the main features of a paraconsistent inheritance reasoner allowing to deal properly with exceptions and inconsistencies in multiple inheritance frame systems. The paraconsistent inheritance reasoner represents knowledge by means of paraconsistent frames and infers based on the inconsistency/under-determinedness degree. This reasoner, being a wide-encompassing one, also allows less complex inheritances to take place.

Furthermore, its main feature is not to eliminate contradictions, ab initio.

9.9   Paraconsistent Logics and Non-monotonic Reasoning

There are various intelligent systems including nonmonotonic reasoning in the field of Artificial Intelligence. Each system has different semantics. More than two nonmonotonic reasoning maybe required in complex intelligent systems. It is more desirable to have a common semantics for such nonmonotonic reasoning. We propose the common semantics for the nonmonotonic reasoning by annotated logics and annotated logic programs [21].

Also, annotated systems encompass deontic notions in this fashion. Roughly speaking, deontic notions can be introduced as follows in annotated logic programming. An annotation in Vector Annotated Logic Programming with Strong Negation—VALPSN [21] is a 2-dimensional vector called a *vector annotation* such that each component is a non-negative integer and the complete lattice τ of vector annotations is defined as:

$$\tau = \{(x, y) | 0 \leq x, y \leq 2\}$$

The ordering $\leq$ is defined: let $\vec{v}_1 = (x_1, y_1)$ and, $\vec{v}_2 = (x_2, y_2)$, then,

$\vec{v}_1 \preceq \vec{v}_2$ iff $x_1 \leq x_2$ and $y_1 \leq y_2$.

For a vector annotated literal, $p : (i,j)$, the first component i of the vector annotation denotes the amount of positive information to support the literal p and the second one j denotes that of negative information. For example, a vector-annotated literal $p : (2,1)$ can be informally interpreted that the literal p is known to be true of strength 2 and false of strength 1. On the other hand, an annotation in EVALPSN called an *extended vector annotation* has a form of $[(i,j), \mu]$ such that the first component (i,j) is a 2-dimensional vector as well as a vector annotation in VALPSN and the second one,

$$\mu \in \tau = \{\bot, \alpha, \beta, \gamma, *_1, *_2, *_3, \top\},$$

is an index that represents deontic notion or contradiction. The complete lattice $\bar{\tau}$ of extended vector annotations is defined as $\bar{\tau} \times \bar{\tau}$. The ordering of the lattice $\bar{\tau}$ is denoted by a symbol $\le$ and described by the Hasse's diagrams in Fig. 9.1. The intuitive meanings of the members of $\bar{\tau}$ are; $\bot$ (unknown), α (fact), β (obligation), γ (non-obligation), $*_1$ (both fact and obligation), $*_2$ (both obligation and non-obligation), $*_3$ (both fact and non-obligation) and $\top$ (inconsistent). The Hasse's diagram(cube) shows that the lattice τ_d is a tri-lattice in which the direction $\overrightarrow{\alpha\beta}$ represents *deontic truth*, the direction $\overrightarrow{\bot*_2}$ represents the amount of *deontic knowledge* and the direction $\overrightarrow{\bot\alpha}$ represents factuality. Therefore, for example, the annotation β can be intuitively interpreted to be deontically truer than the annotation γ and the annotations $\bot$ and $*_2$ are deontically neutral, i.e., neither obligation nor not-obligation. The ordering over the lattice T is denoted by a symbol $\preceq$ and defined as: let $[(i_1,j_1), \mu_1]$ and $[(i_2,j_2), \mu_2]$ be extended vector annotations,

$$[(i_1,j_1), \mu_1] \le [(i_2,j_2), \mu_2] \text{ iff } (i_1,j_1) \le (i_2,j_2) \text{ and } \mu_1 \le \mu_2.$$

There are two kinds of epistemic negations $\neg_1$ and $\neg_2$ in EVALPSN, which are defined as mappings over T_v and T_d, respectively.

Fig. 9.1 Lattice T_v and Lattice τ_d

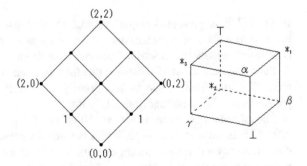

Epistemic negations $\neg_1$ and $\neg_2$ are defined as follows:

$$\neg_1([(i,j),\mu]) = [(j,i),\mu], \forall \mu \in \mathbf{T}_d$$
$$\neg_2([(i,j),\bot]) = [(i,j),\bot], \; \neg_2([(i,j),\alpha]) = [(i,j),\alpha],$$
$$\neg_2([(i,j),\beta]) = [(i,j),\gamma], \; \neg_2([(i,j),\gamma]) = [(i,j),\beta],$$
$$\neg_2([(i,j),*_1]) = [(i,j),*_3], \; \neg_2([(i,j),*_2]) = [(i,j),*_2],$$
$$\neg_2([(i,j),*_3]) = [(i,j),*_1], \; \neg_2([(i,j),\mathbf{T}]) = [(i,j),\mathbf{T}].$$

The epistemic negations $(\neg_1, \neg_2)$ followed by extended vector annotated literals can be eliminated by the above syntactic operation. The strong negation ($\sim$) in EVALPSN can be defined by the epistemic negations $\neg_1$ or $\neg_2$ as follows and interpreted as classical negation.

Deontic notions and fact can be expressed in extended vector annotations as follows:

- "fact" is expressed by an extended vector annotation, $[(m,0),\alpha]$;
- "obligation" is by an extended vector annotation, $[(m,0),\beta]$;
- "forbiddance" is by an extended vector annotation, $[(0,m),\beta]$
- "permission" is by an extended vector annotation, $[(0,m),\gamma]$,

where $m = 1$ or $m = 2$. For example, an extended vector annotated literal p : $[(2,0),\alpha]$ can be intuitively interpreted as "the literal p is a fact of strength 2", and an extended vector annotated literal q : $[(0,1),\beta]$ can be also done as "the literal q is forbidden of strength 1".

Such ideas were implemented successfully in safety control systems: railway signals, traffic intelligent signals, pipeline cleaning system, and many other applications.

9.10 Paraconsistent Electronic Circuits

In [3, 15] it is presented digital circuits (logical gates Complement, And, Or) inspired in a class of paraconsistent annotated logics Pτ. These circuits allow "inconsistent" signals in a nontrivial manner in their structure.

Such circuits consist of six states; due the existence of literal operators to each of them, the underlying logic is functionally complete; it is a many-valued and paraconsistent (at least "semantically") logic.

The simulations were made at 50 MHz, 1.2 μm, by using the software AIM-SPICE, version 1.5a. Also, it was presented a paraconsistent analyzer module combining several paraconsistent circuits, as well as a circuit that allows to detect inconsistent signals and gives a non-trivial treatment.

As far as we know, these results seem to be pioneering in using the concept of paraconsistency in the theory of electronic circuits. The applications appear to be large in horizon: it expands the scope of applications where conflicting signals are common, such as in sensor circuits in robotics, industry automation circuits, race signal control in electronic circuits, and many other fields.

9.11 Paraconsistent Artificial Neural Networks

Generally speaking, Artificial Neural Network (ANN) can be described as a computational system consisting of a set of highly interconnected processing elements, called artificial neurons, which process information as a response to external stimuli. An artificial neuron is a simplistic representation that emulates the signal integration and threshold firing behavior of biological neurons by means of mathematical structures. ANNs are well suited to tackle problems that human beings are good at solving, like prediction and pattern recognition. ANNs have been applied within several branches, among them, in the medical domain for clinical diagnosis, image analysis and interpretation signal analysis and interpretation, and drug development.

Paraconsistent Artificial Neural Network—PANN [16] is a new artificial neural network based on paraconsistent annotated logic $E\tau$ [1]. Let us present it briefly.

The atomic formulas of the paraconsistent annotated logic $E\tau$ is of the type $p_{(\mu, \lambda)}$, where $(\mu, \lambda) \in [0, 1]^2$ and $[0, 1]$ is the real unitary interval (p denotes a propositional variable). An order relation is defined on $[0, 1]^2$: $(\mu_1, \lambda_1) \leq (\mu_2, \lambda_2) \Leftrightarrow \mu_1 \leq \mu_2$ and $\lambda_1 \leq \lambda_2$, constituting a lattice that will be symbolized by τ. A detailed account of annotated logics is to be found in [1].

$p_{(\mu, \lambda)}$ can be intuitively read: "It is assumed that p's belief degree (or favorable evidence) is μ and disbelief degree (or contrary evidence) is λ." Thus, (1.0, 0.0) intuitively indicates total belief, (0.0, 1.0) indicates total disbelief, (1.0, 1.0) indicates total inconsistency, and (0.0, 0.0) indicates total paracompleteness (absence of information). The operator $\sim: |\tau| \rightarrow |\tau|$ defined in the lattice $\sim[(\mu, \lambda)] = (\lambda, \mu)$ works as the "meaning" of the logical negation of $E\tau$.

The consideration of the values of the belief degree and of disbelief degree is made, for example, by specialists who use heuristics knowledge, probability or statistics.

We can consider several important concepts (all considerations are taken with $0 \leq \mu, \lambda \leq 1$):

Segment DB—segment perfectly defined: $\mu + \lambda - 1 = 0$
Segment AC—segment perfectly undefined: $\mu - \lambda = 0$
Uncertainty Degree: $G_{un}(\mu, \lambda) = \mu + \lambda - 1$; Certainty Degree: $G_{ce}(\mu, \lambda) = \mu - \lambda$;

With the uncertainty and certainty degrees we can get the following 12 regions of output: *extreme states* that are, False, True, Inconsistent and Paracomplete, and *non-extreme states*. All the states are represented in the lattice of the next figure: such lattice τ can be represented by the usual Cartesian system.

These states can be described with the values of the certainty degree and uncertainty degree by means of suitable equations. In this work we have chosen the resolution 12 (number of the regions considered according to the Fig. 9.2), but the resolution is totally dependent on the precision of the analysis required in the output and it can be externally adapted according to the applications.

So, such limit values called Control Values are:

V_{cic} = maximum value of uncertainty control = C_3
V_{cve} = maximum value of certainty control = C_1
V_{cpa} = minimum value of uncertainty control = C_4
V_{cfa} = minimum value of certainty control = C_2

For the discussion in the present paper we have used: $C_1 = C_3 = \frac{1}{2}$ and $C_1 = C_3 = -\frac{1}{2}$ (Table 9.3).

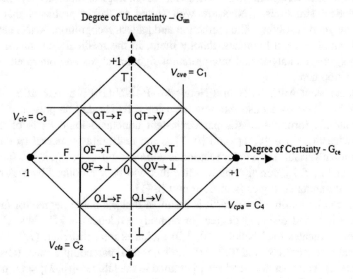

Fig. 9.2 Extreme and non-extreme states

Table 9.3 Extreme and Non-extreme states

Extreme States	Symbol	Non-extreme states	Symbol
True	V	Quasi-true tending to inconsistent	QV→T
False	F	Quasi-true tending to paracomplete	QV→⊥
Inconsistent	T	Quasi-false tending to inconsistent	QF→T
Paracomplete	⊥	Quasi-false tending to paracomplete	QF→⊥
		Quasi-inconsistent tending to true	QT→V
		Quasi-inconsistent tending to false	QT→F
		Quasi-paracomplete tending to true	Q⊥→V
		Quasi-paracomplete tending to false	Q⊥→F

In what follows, we give an idea of how is built the artificial cells of the PANN.

The main concern in any analysis is to know how to measure or to determine the certainty degree regarding a proposition, if it is False or True. Therefore, for this, we take into account only the certainty degree G_{ce}. The uncertainty degree G_{un} indicates the measure of the inconsistency or paracompleteness. If the certainty degree is low or the uncertainty degree is high, it generates an indefinition.

The resulting certainty degree G_{ce} is obtained as follows:

If: $V_{cfa} \leq G_{un} \leq V_{cve}$ or $V_{cic} \leq G_{un} \leq V_{cpa} \Rightarrow G_{ce}$ = Indefinition

For: $V_{cpa} \leq G_{un} \leq V_{cic}$

If: $G_{un} \leq V_{cfa} \Rightarrow G_{ce}$ = False with degree G_{un}

$V_{cic} \leq G_{un} \Rightarrow G_{ce}$ = True with degree G_{un}

The algorithm that expresses a basic Paraconsistent Artificial Neural Cell— PANC—is:

```
* /Definition of the adjustable values * /
    Vcve = C1 * maximum value of certainty control * /
Vcfa =C2 * / minimum value of certainty control * /
    Vcic =C3 * maximum value of uncertainty control * /
    Vcpa =C4 * minimum value of uncertainty control* /
        * Input /Variables * /
    μ, λ
        * Output /Variables *
    Digital output = S1
    Analog output = S2a
    Analog output = S2b
* /Mathematical expressions * /
begin:
0≤ μ ≤ 1 e 0≤ λ≤ 1
Gun = μ + λ - 1
Gce = μ - λ
        * / determination of the extreme states * /
if Gce ≥ C1 then S1 = V
if Gce ≥ C2 then S1 = F
if Gun ≥ C3 then S1 = T
if Gun ≤ C4 then S1 = ⊥
If not: S1 = I - Indetermination
        Gun = S2a
        Gce = S2b
```

A PANC is called *basic* PANC when given a pair (μ, λ) is used as input and resulting as output: G_{un} = resulting uncertainty degree, G_{ce} = resulting certainty degree, and X = constant of Indefinition, calculated by the equations $G_{un} = \mu + \lambda - 1$ and $G_{ce} = \mu - \lambda$ (Fig. 9.3).

A Paraconsistent Artificial Neural Cell of Learning—PANC-l is obtained from a basic PANC.

Fig. 9.3 The basic
paraconsistent artificial neural
cell

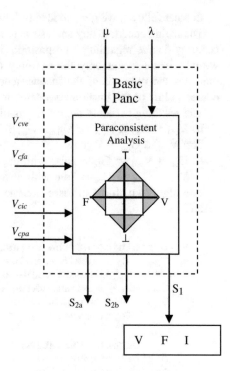

In this learning Cell, sometimes we need the action of the operator Not in the training process. Its function is to do the logical negation in the resulting output sign. For a training process, we consider initially a PANC of Analytic Connection the one not undergoing any learning process.

According to the paraconsistent analysis, a cell in these conditions has two inputs with an Indefinite value ½.

So, the basic structural equation yields the same value ½ as output, having as result an indefinition. For a detailed account see [16].

The learning cells can be used in the PANN as memory units and pattern sensors in primary layers. For instance, a PANC-l can be trained to learn a pattern by using an algorithm. For the training of a cell we can use as pattern real values between 0 and 1. The cells can also be trained to recognize values between 0 and 1.

The learning of the cells with extreme values 0 or 1 composes the primary sensorial cells. Thus, the primary sensorial cells consider as pattern a binary digit where the value 1 is equivalent to the logical state True and the value 0 is equivalent to the logical state False. The appearance of the input 0 repeated times means that the resulting belief degree is going to increase gradually in the output reaching the value 1. In these conditions we say that the cell has learned the falsehood pattern.

The same procedure is adopted when the value 1 is applied to the input repeated times. When the resulting belief degree in the output reaches the value 1 we say that the cell has learned the truth pattern. Therefore a PANC can learn two types of patterns: the truth pattern or the falsity pattern. In the learning process of a PANC, a learning factor can be introduced (LF) that is externally adjusted. Depending on the value of LF, it gives the cell a faster or slower learning. In the learning process, given an initial belief degree $\mu_r(k)$, we use the following equation to reach $\mu_r(k) = 1$, for some k.

So, for truth pattern we have

$$\mu_r(k+1) = \frac{(\mu_1 - \mu_r(k)_c)LF + 1}{2}$$

where $\mu_r(k)_c = 1 - \mu_r(k)$, and $0 \leq LF \leq 1$. For falsity pattern, we have

$$\mu_r(k+1) = \frac{(\mu_{1c} - \mu_r(k)_c)LF + 1}{2}$$

where $\mu_r(k)_c = 1 - \mu_r(k)$, $\mu_{1c} = 1 - \mu_1$, and $0 \leq LF \leq 1$

So we can say that the cell is completely learned when $\mu_r(k + 1) = 1$.

If LF = 1, we say that the cell has a natural capacity of learning. Such capacity decreases as LF approaches 0. When LF = 0, the cell loses the learning capacity and the resulting belief degree will always have the indefinition value ½.

Even after having a cell trained to recognize a certain pattern, if insistently the input receives a value totally different, the high uncertainty makes the cell unlearn the pattern gradually. The repetition of the new values implies in a decreasing of the resulting belief degree. Then, the analysis has reached an indefinition. By repeating this value, the resulting belief degree reaches 0 meaning that the cell is giving the null belief degree to the former proposition to be learned. This is equivalent to saying that the cell is giving the maximum value to the negation of the proposition, so the new pattern must be confirmed. Algorithmically, this is showed when the certainty degree G_{ce} reaches the value −1. In this condition the negation of the proposition is confirmed. This is obtained by applying the operator Not to the cell. It inverts the resulting belief degree in the output. From this moment on the PANC considers as a new pattern the new value that appeared repeatedly and unlearning the pattern learned previously. By considering two factors, LF—learning factor and UF—unlearning factor, the cell can learn or unlearn faster or slower according the application. These factors are important giving the PANN a more dynamic process.

With the certainty and uncertainty degrees and control values, it was built a logical controller called Para-analyzer and the correspondent electronic circuit Para-control. Also a suitable combination of such logical analyzer, it was built a new Artificial Neural Network dubbed Paraconsistent Artificial Neural Network [16].

The Paraconsistent Artificial Neural Networks were applied in many themes in Biomedicine as prediction of Alzheimer Disease [18, 19], Cephalometric analysis [20], speech disfluency [6], numerical characters recognition, robotics [14], among others.

9.12 Automation and Robotics

As already mentioned, logical circuits and programs can be designed based on the Para-analyzer. A hardware or a software built by using the Para-analyzer, in order to treat logical signals according the structure of the paraconsistent annotated logic Eτ, is a logical controller that we call Para-control. It was built an experimental robot based on Paraconsistent Annotated Logic which basically it has two ultra-sonic sensors (one of them capturing a favorable evidence and another, the contrary evidence) and such signals are treated according to Para-control. The first robot was dubbed Emmy (in homage to the mathematician Amalie Emmy Noether (1882–1935)) [14]. Also it was built a robot based on a software using Para-log before mentioned, which was dubbed Sofya (in homage to the mathematician Sofya Vasil'evna Kovalevskaya (=Kowalewskaja) (1850–1891)). Then several other prototypes were made with many improvements. Such robots can deal directly with conflicting and/paracomplete signals [14].

The Paracontrol is the eletric-eletronic materialization of the Para-analyzer algorithm [4] which is basically an electronic circuit, which treats logical signals in a context of logic Eτ. Such circuit compares logical values and determines domains of a state lattice corresponding to output value. Favorable evidence and contrary evidence degrees are represented by voltage. Certainty and Uncertainty degrees are determined by analogues of operational amplifiers (Fig. 9.4).

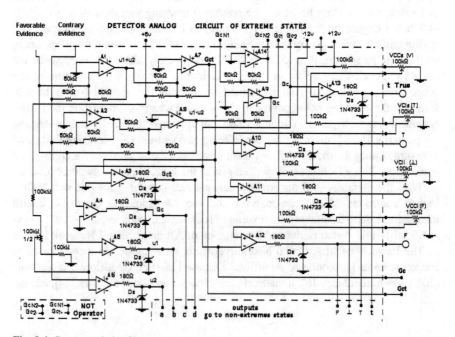

Fig. 9.4 Paracontrol circuit

The Paracontrol comprises both analog and digital systems and it can be externally adjusted by applying positive and negative voltages. The Paracontrol was tested in real-life experiments with an autonomous mobile robot Emmy, whose favorable/contrary evidences coincide with the values of ultrasonic sensors and distances are represented by continuous values of voltage.

The controller Paracontrol was applied in a series of autonomous mobile robots. In some previous works [23, 26] is presented the Emmy autonomous mobile robot. The autonomous mobile robot Emmy consists of a circular mobile platform of aluminum 30 cm in diameter and 60 cm height.

While moving in a non-structured environment the robot Emmy gets information about presence/absence of obstacles using the sonar system called Parasonic [4]. The Parasonic is able to detect obstacles in an autonomous mobile robot's path by transforming the distances to the obstacle into electric signals of the continuous voltage ranging from 0 to 5 V. The Parasonic is basically composed of two ultrasonic sensors of type POLAROID 6500 controlled by an 8051 microcontroller. The 8051 is programmed to carry out synchronization between the measurements of the two sensors and the transformation of the distance into electric voltage. Emmy has suffered improvements and the 2nd prototype is described in what follows.

The robot Emmy uses the paracontrol system to traffic in non-structured environments avoiding collisions with human beings, objects, walls, tables, etc.

The form of reception of information on the obstacles is named non-contact which is the method to obtain and to treat signals from ultra-sonic sensors or optical in order to avoid collisions.

The system of the robot's control is composed by the Para-sonic, Para-control and supporting circuits. See Fig. 9.5.

- Ultra-Sonic Sensors—the two sensors, of ultra-sonic sound waves accomplish the detection of the distance between the robot and the object through the emission of a pulse train in ultra-sonic sound waves frequency and the return reception of the signal (echo).
- Signals Treatment—The treatment of the captured signals is made through the Para-sonic. The microprocessor is programmed to transform the time elapsed between the emission of the signal and the reception of the echo in an electric signal of the 0 to 5 V for degree of belief, and from 5 to 0 V for disbelief degree. The width of each voltage is proportional at the time elapsed between the emission of a pulse train and its receivement by the sensor ones.
- Paraconsistent Analysis—The circuit logical controlling paraconsistent makes the logical analysis of the signals according to the logic $E\tau$.
- Codification—The coder circuit change the binary word of 12 digits in a code of 4 digits to be processed by the personal computer.
- Actions Processing—The microprocessor is programmed conveniently to work the relay in sequences that establish actions for the robot.
- Decodification—The circuit decoder change the binary word of 4 digits in signals to charge the relay in the programmed paths.

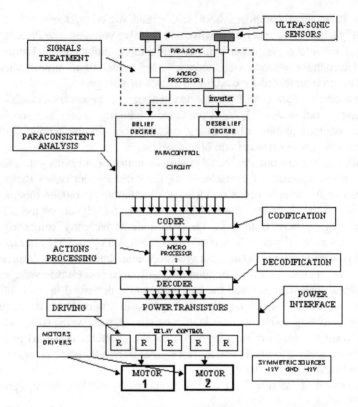

Fig. 9.5 Scheme of the system of the robot's control Emmy

- Power Interface—The power interface circuit potency interface is composed by transistors that amplify the signals making possible the relay's control by digital signals.
- Driving—The relays ON and OFF the motors M_1 and M_2 according to the decoded binary word.
- Motor Drives—The motors M_1 and M_2 move the robot according to the sequence of the relays control.
- Sources—The robot Emmy is supply by two batteries forming a symmetrical source of tension of ± 12 V DC.

As project is built in hardware besides the paracontrol it was necessary the installation of components for supporting circuits allowing the resulting signals of the paraconsistent analysis to be addressed and indirectly transformed in action.

In this first prototype of the robot Emmy it was necessary a coder and a decoder such that the referring signals to the logical states resultants of the paraconsistent analysis had its processing made by a microprocessor of 4 inputs and 4 outputs (Fig. 9.6).

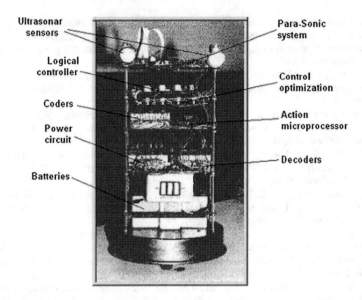

Ultrasonar sensors

Logical controller

Coders

Power circuit

Batteries

Para-Sonic system

Control optimization

Action microprocessor

Decoders

Fig. 9.6 Robot Emmy

9.13 Conclusions

As it can be seen by the previous exposition, the applications of paraconsistent systems have been very fruitful in many aspects, opening new horizons for researching. The appearance of alternative logics to the classical logic impose us some question to ponder: rationality and logicality coincide? There are in fact logics distinct from classical logic? If there are such alternative logics, there are in consequence distinct rationalities? All these issues occupy philosophers, logicians, scientists in general.

References

1. Abe, J.M.: Fundamentos da Lógica Anotada (1992) (Foundations of Annotated Logics in Portuguese). Ph.D. thesis, FFLCH/USP, São Paulo, Brazil (1992)
2. Abe, J.M.: Some aspects of paraconsistent systems and applications. Log. et Anal. **15**, 83–96 (1997)
3. Abe, J.M., Da Silva Filho, J.I.: Inconsistency and electronic circuits. In: Alpaydin, E. (ed.) Proceedings of the International ICSC Symposium on Engineering of Intelligent Systems (EIS'98), Volume 3, Artificial Intelligence, pp. 191–197. ICSC Academic Press International Computer Science Conventions Canada/Switzerland (1998)
4. Abe, J.M., da Silva Filho, J.I.: Manipulating conflicts and uncertainties in robotics. Mult. Valued Log. Soft Comput. **9**, 147–169 (2003)

5. Abe, J.M. Nakamatsu, K.: Multi-agent systems and paraconsistent knowledge. In: Nguyen, N. T., Jain, L.C. (eds.) Knowledge Processing and Decision Making in Agent-Based Systems, Book Series Studies in Computational Intelligence, vol. 167, VIII, 400 p. 92 illus., pp. 101–121. Springer (2009)

6. Abe, J.M., Prado, J.C.A., Nakamatsu, K.: Paraconsistent artificial neural network: applicability in computer analysis of speech productions. Lect. Notes Comput. Sci. **4252**, 844–850 (2006) (Springer)

7. Anderson, A.R., Belnap, Jr. N.D.: Entailment: The Logic of Relevance and Necessity, vol. I. Princeton University Press, Princeton (Anderson, A.R., Belnap, Jr., N.D., Dunn, J.M.: Entailment, vol. II (1992)) (1975)

8. Ávila, B.C.: Uma Abordagem Paraconsistente Baseada em Lógica Evidencial para Tratar Exceções em Sistemas de Frames com Múltipla Herança. Ph.D. thesis, University of São Paulo, São Paulo (1996)

9. Borkowski, L., Słupecki, J.: The logical works of J. Łukasiewicz. Stud. Logica. **8**, 7–56 (1958)

10. Da Costa, N.C.A.: Logiques classiques et non classiques: Essai sur les fondements de la logique, Dunod Masson Ho, 275 pp (1997)

11. Da Costa, N.C.A., Subrahmanian, V.S.: Paraconsistent logics as a formalism for reasoning about inconsistent knowledge bases. Artif. Intelligence Med. (Print) **1**, 167–174 (1989)

12. Da Costa, N.C.A., Abe, J.M., Subrahmanian, V.S.: Remarks on Annotated Logic. Zeitschrift f. math. Logik und Grundlagen Math. **37**, 561–570 (1991)

13. Da Costa, N.C.A., Prado, J.P.A., Abe, J.M., Ávila, B.C., Rillo, M.: Paralog: Um Prolog Paraconsistente baseado em Lógica Anotada, Coleção Documentos, Série Lógica e Teoria da Ciência, IEA-USP, **18**, 21 pp (1995)

14. Da Silva Filho, J.I.: Métodos de interpretação da Lógica Paraconsistente Anotada com anotação com dois valores LPA2v com construção de Algoritmo e implementação de Circuitos Eletrônicos, EPUSP. PhD Thesis (in Portuguese), São Paulo (1999)

15. Da Silva Filho, J.I., Abe, J.M.: Paraconsistent electronic circuits. Int. J. Comput. Anticip. Syst. **9**, 337–345 (2001)

16. Da Silva Filho, J.I., Torres, G.L., Abe, J.M.: Uncertainty Treatment Using Paraconsistent Logic—Introducing Paraconsistent Artificial Neural Networks, vol. 211, 328 pp. IOS Press, Holanda (2010)

17. Fagin, R., Halpern, J.Y., Moses, Y., Vardi, M.Y.: Reasoning About Knowledge. The MIT Press, London (1995)

18. Lopes, H.F.S., Abe, J.M., Anghinah, R.: Application of paraconsistent artificial neural networks as a method of aid in the diagnosis of alzheimer disease. J. Med. Syst. 1–9 (2009) (Springer-Netherlands)

19. Lopes, H.F.S., Abe, J.M., Kanda, P.A.M., Machado, S., Velasques, B., Ribeiro, P., Basile, L. F.H., Nitrini, R., Anghinah, R.: Improved application of paraconsistent artificial neural networks in diagnosis of alzheimer's disease. Am. J. Neurosci. **2**(1), 54–64 (2011) (Science Publications)

20. Mario, M.C., Abe, J.M., Ortega, N., Del Santo Jr., M.: Paraconsistent artificial neural network as auxiliary in cephalometric diagnosis. Artif. Organs **34**(7), 215–221 (2010) (Wiley Interscience)

21. Nakamatsu, K., Abe, J.M., Suzuki, A.: Annotated semantics for defeasible deontic reasoning, rough sets and current trends in computing. Lecture Notes in Artificial Intelligence Series, pp. 470–478, Springer (2005)

22. Nelson, D.: Constructible falsity. J. Symb. Log. **14**(1), 16–26 (1949)

23. Prado, J.P.A.(1996), Uma Arquitetura em IA Baseada em Lógica Paraconsistente, Ph.D. thesis (in Portuguese), University of São Paulo

24. Routley, R., Meyer, R.K., Plumwood, V., Brady, R.: Relevant Logics and its Rivals, vol. I. Ridgeview, Atascardero (1983)

25. Subrahmanian, V.S.: On the semantics of quantitative logic programs. In: Proceedings 4th IEEE Symposium on Logic Programming, pp. 173–182. Computer Society Press, Washington (1987)
26. Torres, C.R.: Sistema Inteligente Baseado na Lógica Paraconsistente Anotada Evidencial Eτ para Controle e Navegação de Robôs Móveis Autônomos em um Ambiente Não-estruturado. PhD. thesis (in Portuguese), Federal University of Itajuba, Brazil (2010)
27. Vasiliev, N.A.: Imaginary (non-Aristotelian) Logic. Trans. Vergauwen, R., Zaytsev, E.A. Log. et Anal. **46**(182), 127–163 (2003)

Additional Session Reading

28. Abe, J.M.: Paraconsistent Artificial Neural Networks: An Introduction. Lecture Notes In Computer Science 3214, Springer, pp. 942–948 (2004)
29. Abe, J.M.: Para-Fuzzy Logic Controller. Lecture Notes In Computer Science 3214, Springer, pp. 935–941 (2004)
30. Abe, J.M., Akama, S.: Paraconsistent Annotated Temporal Logics D*τ, pp. 217–226. Springer, Lecture Notes in Computer Science (2000)
31. Abe, J.M., Da Silva Filho, J.I.: Logic, Artificial Intelligence, and Robotics, Editors, Frontiers in Artificial Intelligence and Its Applications, vol. 71, 287 pp. IOS Press, Amsterdam, Ohmsha, Tokyo (2001)
32. Abe, J.M., Da Silva Filho, J.I.: Advances in Logic, Artificial Intelligence, and Robotics, Editors, Frontiers in Artificial Intelligence and Its Applications, vol. 85, 277 pp. IOS Press, Amsterdan, Ohmsha, Tokyo (2002)
33. Abe, J.M., Nakamatsu, K., Akama, S.: Non-Alethic Reasoning in Distributed Systems. Lecture Notes In Computer Science 3684, Springer, pp. 724–731 (2005)
34. Abe, J.M., Ortega, N., Mário, M.C., Del Santo, Jr., M.: Paraconsistent Artificial Neural Network: An Application on Cephalometric Analysis. Lecture Notes In Computer Science 3684, Springer, pp. 716–723 (2005)
35. Abe, J.M., Lopes, H.F.S., Nakamatsu, K., Akama, S.: Paraconsistent Artificial Neural Networks and EEG Analysis (2010). Lecture Notes in Computer Science, Springer, Berlin/Heidelberg, Germany, pp. LNAI 6278, 164–173 (2010)
36. Abe, J.M., Nakamatsu, K., Akama, S.: Monadic Curry System N1* (2010). Lecture Notes in Computer Science, Springer, Berlin/Heidelberg, Germany, pp. LNAI 6278, 143–153. ISSN 0302-9743 (2010)
37. Abe, J.M., Lopes, H.F.S., Nakamatsu, K.: Paraconsistent neurocomputing and brain signal analysis, Vietnam Journal of Computer Science (2014): Print, pp. 1–12. Springer, Berlin, Heidelberg (2014). ISSN 2196-8888. doi:10.1007/s40595-014-0022-9
38. Akama, S., Nakamatsu, K., Abe, J.M.: Constructive Discursive Reasoning. Lecture Notes in Computer Science, Lecture Notes in Computer Science, Springer, Berlin/Heidelberg, Germany, LNAI 6278, pp. 200–206 (2010)
39. Akama, S., Abe, J.M., Nakamatsu, K.: Contingent information: a four-valued approach. In: Advances in Intelligent Systems and Computing, vol. 326, pp. 209–217, Springer International Publishing (2015). ISSN 2194-5357. doi:10.1007/978-3-319-11680-8_17
40. Alasuutari, A., Nakamatsu, K., Abe, J.M.: Safety verification for e-business model based on paraconsistent annotated logic program bf-EVALPSN, frontiers of artificial intelligence and applications (FAIA) series. In: Proceeding of: Intelligent Decision Technologies 2014, Volume 262: Smart Digital Futures 2014, Volume: Frontiers in Artificial Intelligence and Applications, pp. 248–257, Netherlands (2014). ISSN 0922-6389 (print). doi:10.3233/978-1-61499-405-3-248
41. Ávila, B.C., Abe, J.M., Prado, J.P.A.: ParaLog-e: a paraconsistent evidential logic programming language. In: XVII International Conference of the Chilean Computer Science Society, IEEE Computer Society Press, pp. 2–8 (1997)

42. Da Costa, N.C.A., O Conhecimento Científico (2000), Discurso Editorial, São Paulo
43. Da Costa, N.C.A., Abe, J.M., Da Silva Filho, J.I., Murolo, A., Leite, C.: Lógica Paraconsistente Aplicada (in Portuguese), Editora Atlas (1999)
44. Da Silva Filho, J.I., Abe, J.M.: Paraconsistent analyzer module. Int. J. Comput. Anticip. Syst. **9**, 346–352 (2001)
45. Nakamatsu, K., Abe, J.M.: Advances in Logic-Based Intelligent Systems, Frontiers in Artificial Intelligence and Its Applications, vol. 132, 289 pp. IOS Press, Amsterdam (2005)
46. Nakamatsu, K., Abe, J.M.: The paraconsistent process order control method. Vietnam J. Comput. Sci. 1(1):29–37 (2014)
47. Nakamatsu, K., Abe, J.M.: Paraconsistent annotated logic programs and application to intelligent verification systems. Stud. Comput. Intell. (Print) **514**, 279–315 (2014). ISSN: 1860-949X. doi:10.1007/s40595-013-0002-5
48. Nakamatsu, K., Abe, J.M., Suzuki, A.: Annotated Semantics for Defeasible Deontic Reasoning. Lecture Notes in Computer Science 2005, Springer, pp. 470–478 (2000)
49. Nakamatsu, K., Suito, H., Abe, J.M., Suzuki, A.: Intelligent Real-time Traffic Signal Control Based on a Paraconsistent Logic Program EVALPSN. Lecture Notes in Computer Science, Springer, Heidelberg, vol. 2639, pp. 719–723 (2003)
50. Nakamatsu, K, Akama, S., Abe, J.M.: An Intelligent Safety Verification Based on a Paraconsistent Logic Program. Lecture Notes In Computer Science 3684, Springer, pp. 708–723 (2005)
51. Nakamatsu, K., Abe, J.M., Kountchev, R.: Introduction to Intelligent Elevator Control Based On EVALPSN. Lecture Notes in Computer Science, Springer, Berlin/Heidelberg, Germany, LNAI 6278, pp. 133–142 (2010)
52. Nakamatsu, K., Abe, J.M., Watanabe, T.: Introduction to Intelligent Network Routing Based On EVALPSN. Lecture Notes in Computer Science, Springer, Berlin/Heidelberg, Germany, LNAI 6278, pp. 123–132 (2010)
53. Nakamatsu, K., Imai, T., Abe, J.M., Watanabe, T.: Application of EVALPSN to network routing. In: Advances in Intelligent Decision Technologies, Springer, Berlin/Heidelberg, Germany, pp. 527–536 (2010)
54. Souza, S., Abe, J.M.: Nevus and Melanoma Paraconsistent Classification, Studies in Health Technology and Informatics, vol. 207, pp. 254–260, (print). IOS Press, Amsterdam, Holanda (2014). ISSN 0926-9630
55. Souza, S., Abe, J.M.: Handwritten numerical characters recognition based on paraconsistent artificial neural networks. In: Bdic et al., A. (ed.) Recent Developments in Computational Collective Intelligence, Studies in Computational Intelligence 513, pp. 93–102. © Springer International Publishing Switzerland (2014). ISSN: 1860-949X. doi:10.1007/978-3-319-01787-7_9
56. Sylvan, R., Abe, J.M.: On general annotated logics, with an introduction to full accounting logics. Bull. Symb. Log. **2**, 118–119 (1996)
57. Torres, C.R., Abe, J.M., Torres, G.L., Da Silva Filho, J.I., Martins, H.G.: A Sensing System for an Autonomous Mobile Robot Based on the Paraconsistent Artificial Neural Network. Lecture Notes in Computer Science, Springer, Berlin/Heidelberg, Germany, LNAI 6278, pp. 154–163 (2010)

Chapter 10
Annotated Logics and Intelligent Control

Seiki Akama, Jair Minoro Abe and Kazumi Nakamatsu

Abstract Annotated logics are a kind of paraconsistent (and generally paracomplete) logic, whose origin is paraconsistent logic programming. Later, these logics have been extensively studied by may researchers, and applied to many areas, in particular, Artificial Intelligence and Computer Science. Annotated logics are also suited as the foundations for intelligent control in that they can properly deal both with incomplete and inconsistent information. The chapter addresses the aspects of annotated logics as a control language for intelligent systems. After reviewing the motivation and formalization of annotated logics, we give an application to Robotics to show how they can be used for intelligent control.

Keywords Annotated logics · Intelligent control · Paracompleteness · Paraconsistency · Robotics

10.1 Introduction

It is difficult to develop an intelligent system which can simulate human intelligence, since conventional control techniques cannot be adequately applied to intelligent control. The term "intelligent control" is due to K.S. Fu in the 1970s; see Fu [35]. There seem, however, no general definitions of intelligent control. It could

S. Akama (✉)
C-Republic Corporation, 1-20-1 Higashi-Yurigaoka, Asao-Ku,
Kawasaki 215-0012, Japan
e-mail: akama@jcom.home.ne.jp

J.M. Abe
Graduate Program in Production Engineering, ICET - Paulista University,
Rua. Dr. Bacelar, 1212, Sao Paulo, SP CEP 04026-002, Brazil
e-mail: jairabe@uol.com.br

K. Nakamatsu
School of Human Science and Environment,
University of Hyogo, 1-1-12 Shinzaike-hon-cho, Himeji 670-0092, Japan
e-mail: nakamatsu@shse.u-hyogo.ac.jp

© Springer International Publishing Switzerland 2016
K. Nakamatsu and R. Kountchev (eds.), *New Approaches in Intelligent Control*,
Intelligent Systems Reference Library 107, DOI 10.1007/978-3-319-32168-4_10

be defined as follows. Intelligent control is the discipline which can emulate human intelligence by means of various techniques in Artificial Intelligence (AI).

In this sense, foundations for intelligent control are closely related to the ones for AI. The fact is not surprising because the ultimate objective of AI is to develop a system which can behave like human. Naturally, intelligent control has been benefited from neural networks, fuzzy logic, evolutionary computation, and so on. These areas may be useful to some forms of intelligent control, but may be far from satisfactory.

To formalize and implement a system based on intelligent control, we need at least two requirements. One is that it has a theoretical foundation for intelligent control. The other is the applicability to practical intelligent systems. From our point of view, none of the above areas can satisfy these requirements.

In view of the above discussion, logic-based approaches to intelligent control seem to be promising. In fact, in AI, logic-based approaches have been fully investigated But, in AI by logic most people assume *classical logic* and it is used as knowledge representation language and inference engine. Unfortunately, classical logic fails to describe intelligent control in that it cannot deal directly with incomplete and inconsistent information.

But, logic-based approaches should not be given up. This is because we can utilize *non-classical logic* instead of classical logic. In the area of formal logic, many systems of non-classical logic have been studied for many years. For example, *modal logic* extends classical logic with modal operators to express intensional concepts. Some non-classical logics like *intuitionistic logic* and *relevance logic* do not admit some classical laws.

For laying foundations for intelligent control, the so-called *paraconsistent logic*, which is a logic for inconsistent and non-trivial theories, can be considered. There are many systems of paraconsistent logics in the literature, and we believe that *annotated logics*, originally proposed by Subrahmanian [48], can serve as the basis for intelligent control.

Annotated logics were designed as theoretical foundations for paraconsistent logic programming for inconsistent knowledge; see Subrahmanian [48] and Blair and Blair and Subrahmanian [20]. Later, they have been studied as the systems of paraconsistent logic by many people; see [1, 21, 24]. The complete exposion of annotated logics can found in Abe et al. [6].

The chapter is structured as follows. In Sect. 10.2, we review paraconsistent logic. Section 10.3 formally introduces annotated logics. In Sect. 10.4, we discuss intelligent control based on annotated logics. Finally, we provide some conclusions in Sect. 10.5.

10.2 Paraconsistent Logic

Historically, the study of paraconsistent logic started in the late 1940 to formalize inconsistent but non-trivial theories, and many systems of paraconsistent logic have been proposed in the literature. Recently, we can find several interesting applications of paraconsistent logics for various areas including mathematics, philosophy and computer science.

Here, we survey some well known systems of paraconsistent logic. The survey is not complete, because there are now numerous such systems. Bu we need to know the following three.

- discursive logic
- C-systems
- relevance logic

Before reviewing these systems, we must mention some concepts. Let T be a theory whose underlying logic is L. T is called *inconsistent* when it contains theorems of the form A and $\neg A$ (the negation of A), i.e.,

$$T \vdash_L A \text{ and } T \vdash_L \neg A$$

where $\vdash_L$ denotes the provability relation in L. If T is no inconsistent, it is called *consistent*.

T is said to be *trivial*, if all formulas of the language are also theorems of T. Otherwise, T is called *non-trivial*. Then, for trivial theory T, $T \vdash_L B$ for any formula B. Note that trivial theory is not interesting since every formula is provable.

If L is classical logic (or one of several others, such as intuitionistic logic), the notions of inconsistency and triviality agree in the sense that T is inconsistent iff T is trivial. So, in trivial theories the extensions of the concepts of formula and theorem coincide.

A *paraconsistent logic* is a logic that can be used as the basis for inconsistent but non-trivial theories. In this regard, paraconsistent theories do not satisfy the *principle of non-contradiction*, i.e., $\neg(A \wedge \neg A)$.

Similarly, we can define the notion of paracomplete theory, namely T is called *paracomplete* when neither A nor $\neg A$ is a theorem. In other words,

$$T \nvdash_L A \text{ and } T \nvdash_L \neg A$$

hold in paracomplete theory. If T is not paracomplete, T is *complete*, i.e.,

$$T \vdash_L A \text{ or } T \vdash_L \neg A$$

holds. A *paracomplete logic* is a logic for paracomplete theory, in which the *principle of excluded middle*, i.e., $A \vee \neg A$ fails. In this sense, intuitionistic logic is one of the paracomplete logics.

Finally, the logic which is simultaneously paraconsistent and paracomplete is called *non-alethic logic*. Classical logic is a consistent and complete logic, and the fact may be problematic in applications.

Discursive logic, due to the Polish logician Jaśkowski [36], is a formal system J satisfying the conditions: (a) from two contradictory propositions, it should not be possible to deduce any proposition; (b) most of the classical theses compatible with (a) should be valid; (c) J should have an intuitive interpretation.

Such a calculus has, among others, the following intuitive properties remarked by Jaśkowski himself: suppose that one desires to systematize in only one deductive system all theses defended in a discussion. In general, the participants do not confer the same meaning to some of the symbols. One would have then as theses of a deductive system that formalize such a discussion, an assertion and its negation, so both are "true" since it has a variation in the sense given to the symbols. It is thus possible to regard discursive logic as one of the so-called *paraconsistent logics*.

Jaśkowski's D_2 contains propositional formulas built from logical symbols of classical logic. In addition, possibility operator $\Diamond$ in S5 is added. $\Diamond A$ reads "A is possible". Based on the possibility operator, three discursive logical symbols can be defined as follows:

discursive implication: $A \rightarrow_d B =_{def} \Diamond A \rightarrow B$
discursive conjunction: $A \wedge_d B =_{def} \Diamond A \wedge B$
discursive equivalence: $A \leftrightarrow_d B =_{def} (A \rightarrow_d B) \wedge_d (B \rightarrow_d A)$

Additionally, we can define discursive negation $\neg_d A$ as $A \rightarrow_d false$. Jaśkowski's original formulation of D_2 in [36] used the logical symbols: $\rightarrow_d, \leftrightarrow_d, \vee, \wedge, \neg$, and he later defined $\wedge_d$ in [37].

The axiomatization due to Kotas [42] has the following axioms and the rules of inference. Here, $\Box$ is the necessity operator, and is definable by $\neg\Diamond\neg$. $\Box A$ reads "A is necessary".

Axioms
(A1) $\Box(A \rightarrow (\neg A \rightarrow B))$
(A2) $\Box((A \rightarrow B) \rightarrow ((B \rightarrow C) \rightarrow (A \rightarrow C)))$
(A3) $\Box((\neg A \rightarrow A) \rightarrow A)$
(A4) $\Box(\Box A \rightarrow A)$
(A5) $\Box(\Box(A \rightarrow B) \rightarrow (\Box A \rightarrow \Box B))$
(A6) $\Box(\neg\Box A \rightarrow \Box\neg\Box A)$
Rules of Inference
(R1) substitution rule
(R2) $\Box A, \Box(A \rightarrow B)/\Box B$
(R3) $\Box A/\Box\Box A$
(R4) $\Box A/A$
(R5) $\neg\Box\neg\Box A/A$

Note that discursive implication $\rightarrow_d$ satisfies *modus ponens* in S5, but $\rightarrow$ does not. There are other axiomatizations of D_2. For example, da Costa and Dubikajtis gave an axiomatization based on the connectives $\rightarrow_d, \wedge_d, \neg$; see [27]. Semantics for discursive logic can be obtained by a Kripke semantics for modal logic S5. Jaśkowski's three conditions for J mentioned above are solved by many workers in different ways for our approach, see [9, 13]. For a comprehensive survey on discursive logic, see da Costa and Doria [28].

C-systems are paraconsistent logics due to da Costa which can be a basis for inconsistent but non-trivial theories; see da Costa [26]. The important feature of da Costa systems is to use novel interpretation, which is non-truth-functional, of negation avoiding triviality.

Here, we review C-system C_1 due to da Costa [26]. The language of C_1 is based on the logical symbols: $\wedge, \vee, \rightarrow$, and $\neg$. $\leftrightarrow$ is defined as usual. In addition, a formula A°, which is read "A is well-behaved", is shorthand for $\neg(A \wedge \neg A)$. The basic ideas of C_1 contain the following: (1) most valid formulas in the classical logic hold, (2) the law of non-contradiction $\neg(A \wedge \neg A)$ should not be valid, (3) from two contradictory formulas it should not be possible to deduce any formula.

The Hilbert system of C_1 extends the positive intuitionistic logic with the axioms for negation.

da Costa's C_1

Axioms

(DC1) $A \rightarrow (B \rightarrow A)$
(DC2) $(A \rightarrow B) \rightarrow ((A \rightarrow (B \rightarrow C)) \rightarrow (A \rightarrow C))$
(DC3) $(A \wedge B) \rightarrow A$
(DC4) $(A \wedge B) \rightarrow B$
(DC5) $A \rightarrow (B \rightarrow (A \wedge B))$
(DC6) $A \rightarrow (A \vee B)$
(DC7) $B \rightarrow (A \vee B)$
(DC8) $(A \rightarrow C) \rightarrow ((B \rightarrow C) \rightarrow ((A \vee B) \rightarrow C))$
(DC9) $B^\circ \rightarrow ((A \rightarrow B) \rightarrow ((A \rightarrow \neg B) \rightarrow \neg A))$
(DC10) $(A^\circ \wedge B^\circ) \rightarrow (A \wedge B)^\circ \wedge (A \vee B)^\circ \wedge (A \rightarrow B)^\circ$
(DC11) $A \vee \neg A$
(DC12) $\neg\neg A \rightarrow A$

Rules of Inference

(MP) $\vdash A, \vdash A \rightarrow B \Rightarrow \vdash B$

Here, (DC1)–(DC8) are axioms of the positive intuitionistic logic. (DC9) and (DC10) play a role for the formalization of paraconsistency.

A semantics for C_1 can be given by a two-valued valuation; see da Costa and Alves [22]. We denote by $\mathcal{F}$ the set of formulas of C_1. A valuation is a mapping v from $\mathcal{F}$ to $\{0, 1\}$ satisfying the following:

$v(A) = 0 \Rightarrow v(\neg A) = 1$
$v(\neg\neg A) = 1 \Rightarrow v(A) = 1$
$v(B^\circ) = v(A \rightarrow B) = v(A \rightarrow \neg B) = 1 \Rightarrow v(A) = 0$
$v(A \rightarrow B) = 1 \Leftrightarrow v(A) = 0 \operatorname{or} v(B) = 1$
$v(A \wedge B) = 1 \Leftrightarrow v(A) = v(B) = 1$
$v(A \vee B) = 1 \Leftrightarrow v(A) = 1 \operatorname{or} v(B) = 1$
$v(A^\circ) = v(B^\circ) = 1 \Rightarrow v((A \wedge B)^\circ) = v((A \vee B)^\circ) = v((A \rightarrow B)^\circ) = 1$

Note here that the interpretations of negation and double negation are not given by biconditional. A formula A is *valid*, written $\vDash A$, if $v(A) = 1$ for every valuation *v.* Completeness holds for C_1. It can be shown that C_1 is complete for the above semantics.

Da Costa system C_1 can be extended to C_n $(1 \leq n \leq \omega)$. Now, A^1 stands for A° and A^n stands for $A^{n-1} \wedge (A^{(n-1)})^\circ, 1 \leq n \leq \omega$.

Then, da Costa system C_n $(1 \leq n \leq \omega)$ can be obtained by (DC1)–(DC8), (DC12), (DC13) and the following:

(DC9n) $B^{(n)} \rightarrow ((A \rightarrow B) \rightarrow ((A \rightarrow \neg B) \rightarrow \neg A))$

(DC10n) $(A^{(n)} \wedge B^{(n)}) \rightarrow (A \wedge B)^{(n)} \wedge (A \vee B)^{(n)} \wedge (A \rightarrow B)^{(n)}$

Note that da Costa system C_ω has the axioms (DC1)–(DC8), (DC12) and (DC13). Later, da Costa investigate first-order and higher-order extensions of C-systems.

Relevance logic, also called *relevant logic* is a family of logics based on the notion of relevance in conditionals. Historically, relevance logic was developed to avoid the *paradox of implications*; see Anderson and Belnap [14, 15].

Anderson and Belnap formalized a relevant logic R to realize a major motivation, in which they do not admit $A \rightarrow (B \rightarrow A)$. Later, various relevance logics have been proposed. Note that not all relevance logics are paraconsistent but some are considered important as paraconsistent logics. For example, Priest developed the logic called *LP*; see Priest [44, 46].

Routley and Meyer proposed a basic relevant logic B, which is a minimal system having the so-called *Routley-Meyer semantics*. Thus, B is an important system and we review it below; see Routley et al. [47].

The language of B contains logical symbols: $\sim, \&, \vee$ and $\rightarrow$ (relevant implication). A Hilbert system for B is as follows:

Relevant Logic B
Axioms
(BA1) $A \rightarrow A$
(BA2) $(A\&B) \rightarrow A$
(BA3) $(A\&B) \rightarrow B$
(BA4) $((A \rightarrow B)\&(A \rightarrow C)) \rightarrow (A \rightarrow (B\&C))$
(BA5) $A \rightarrow (A \vee B)$
(BA6) $B \rightarrow (A \vee B)$
(BA7) $(A \rightarrow C)\&(B \rightarrow C)) \rightarrow ((A \vee B) \rightarrow C)$
(BA8) $(A\&(B \vee C)) \rightarrow (A\&B) \vee C)$
(BA9) $\sim \sim A \rightarrow A$
Rules of Inference
(BR1) $\vdash A, \vdash A \rightarrow B \Rightarrow \vdash B$
(BR2) $\vdash A, \vdash B \Rightarrow \vdash A\&B$
(BR3) $\vdash A \rightarrow B, \vdash C \rightarrow D \Rightarrow \vdash (B \rightarrow C) \rightarrow (A \rightarrow D)$
(BR4) $\vdash A \rightarrow \sim B \Rightarrow \vdash B \rightarrow \sim A$

A Hilbert system for Anderson and Belnap's R is as follows:

Relevance Logic R
Axioms
(RA1) $A \rightarrow A$
(RA2) $(A \rightarrow B) \rightarrow ((C \rightarrow A) \rightarrow C \rightarrow B))$
(RA3) $(A \rightarrow (A \rightarrow B) \rightarrow (A \rightarrow B)$
(RA4) $(A \rightarrow (B \rightarrow C)) \rightarrow (B \rightarrow (A \rightarrow C)$
(RA5) $(A\&B) \rightarrow A$
(RA6) $(A\&B) \rightarrow B$
(RA7) $((A \rightarrow B)\&(A \rightarrow C)) \rightarrow (A \rightarrow (B\&C))$
(RA8) $A \rightarrow (A \vee B)$
(RA9) $B \rightarrow (A \vee B)$
(RA10) $((A \rightarrow C)\&(B \vee C)) \rightarrow ((A \vee B) \rightarrow C))$
(RA11) $(A\&(B \vee C)) \rightarrow ((A\&B) \vee C)$
(RA12) $(A \rightarrow \sim A) \rightarrow \sim A$
(RA13) $(A \rightarrow \sim B)) \rightarrow (B \rightarrow \sim A)$
(RA14) $\sim \sim A \rightarrow A$
Rules of Inference
(RR1) $\vdash A, \vdash A \rightarrow B \Rightarrow \vdash B$
(RR2) $\vdash A, \vdash B \Rightarrow \vdash A\&B$

Routley et al. considered some axioms of R are too strong and formalized rules instead of axioms. Notice that B is a paraconsistent but R is not.

Next, we give a Routley-Meyer semantics for B. A model structure is a tuple $\mathcal{M} = \langle K, N, R, *, v \rangle$, where K is a non-empty set of worlds, $N \subseteq K$, $R \subseteq K^3$ is a ternary relation on K, $*$ is a unary operation on K, and v is a valuation function from a set of worlds and a set of propositional variables $\mathcal{P}$ to $\{0, 1\}$.

There are some restrictions on . v satisfies the condition that $a \leq b$ and $v(a, p)$ imply $v(b, p) = 1$ for any $a, b \in K$ and any $p \in \mathcal{P}$. $a \leq b$ is a pre-order relation defined by $\exists x (x \in N \text{ and } Rxab)$. The operation $*$ satisfies the condition $a^{**} = a$.

For any propositional variable p, the truth condition $\vDash$ is defined: $a \vDash p$ iff $v(a, p) = 1$. Here, $a \vDash p$ reads "p is true at a". $\vDash$ can be extended for any formulas in the following way:

$$a \vDash \sim A \iff a^* \nvDash A$$
$$a \vDash A\&B \iff a \vDash A \text{ and } a \vDash B$$
$$a \vDash A \vee B \iff a \vDash A \text{ or } a \vDash B$$
$$a \vDash A \rightarrow B \iff \forall bc \in K(Rabc \text{ and } b \vDash A \Rightarrow c \vDash B)$$

A formula A is *true* at a in $\mathcal{M}$ iff $a \vDash A$. A is *valid*, written $\vDash A$, iff A is true on all members of N in all model structures.

Routley et al. provides the completeness theorem for B with respect to the above semantics using canonical models; see [47].

A model structure for R needs the following conditions.

$$R0aa$$
$$Rabc \Rightarrow Rbac$$
$$R^2(ab)cd \Rightarrow R^2a(bc)d\,Raaa$$
$$a^{**} = a$$
$$Rabc \Rightarrow Rac^*b^*$$
$$Rabc \Rightarrow a' \leq a \Rightarrow Ra'bc$$

where R^2abcd is shorthand for $\exists x(Raxd \text{ and } Rxcd)$. The completeness theorem for the Routley-Meyer semantics can be proved for R; see [14, 15].

The reader is advised to consult Anderson and Belnap [14], Anderson, Belnap and Dunn [15], and Routley et al. [47] for details. A more concise survey on the subject may be found in Dunn [33].

Although the above three logics are famous approaches to paraconsistent logics, there is a rich literature on paraconsistent logics. For example, Priest developed the logic called LP; see Priest [45, 47]. Arruda [16] reviewed a survey on paraconsistent logics, and Priest, Routley and Norman [43] contains interesting papers on paraconsistent logics by the 1980s. For a recent survey, we refer Priest [45]. We can also find a Handbook surveying various subjects related to paraconsistency by Beziau et al. [19].

10.3   Annotated Logics

The important problems handled by paraconsistent logics include the paradoxes of set theory, the semantic paradoxes, and some issues in dialectics. These problems are central to philosophy and philosophical logic. However, paraconsistent logics have later found interesting applications in AI, in particular, expert systems, belief, and knowledge, among others, since the 1980s; see da Costa and Subrahmanian [23].

Annotated logics were introduced by Subrahmanian to provide a foundation for paraconsistent logic programming; see Subrahmanian [48] and Blair and Subrahmanian [20]. Paraconsistent logic programming can be seen as an extension of logic programming based on classical logic.

In 1989, Kifer and Lozinskii proposed a logic for reasoning with inconsistency, which is related to annotated logics; see Kifer and Lozinskii [38, 39]. In the same year, Kifer and Subrahmanian extended annotated logics by introducing *generalized annotated logics* in the context of logic programming; see Kifer and Subrahmanian [40]. In 1990, a resolution-style automatic theorem-proving method for annotated logics was implemented; see da Costa et al. [29].

Of course, annotated logics were developed as a foundation for paraconsistent logic programming, but they have interesting features to be examined by logicians. Formally, annotated logics are non-alethic in the sense of the above terminology.

From a viewpoint of paraconsistent logicians, annotated logics were regarded as new systems.

In 1991, da Costa and others started to study annotated logics from a foundational point of view; see da Costa et al. [21, 25]. In these works, propositional and predicate annotated logics were formally investigated by presenting axiomatization, semantics and completeness results, and some applications of annotated logics were briefly surveyed.

In 1992, Jair Minoro Abe wrote Ph.D. thesis on the foundations of annotated logics under Prof. Newton C.A. da Costa at University of Sao Paulo; see Abe [1]. Abe proposed annotated modal logics which extend annotated logics with modal operator in Abe [2]; also see Akama and Abe [10].

Some formal results including decidability annotated logics were presented in Abe and Akama [5]. Abe and Akama also investigated predicate annotated logics by the method of ultraproducts in Abe and Akama [4]. Abe [3] studied an algebraic study of annotated logics. For other works, see [11, 12].

Below, we formally introduce annotated logics. For the purpose of intelligent control, propositional annotated logics $P\tau$ is introduced. We show several formal results in annotated logics without proofs, and their proofs can be found in Abe et al. [13].

The language of the propositional annotated logics $P\tau$. We denote by L the language of $P\tau$. Annotated logics are based on some arbitrary fixed finite lattice called a *lattice of truth-values* denoted by $\tau = \langle |\tau|, \leq, \sim \rangle$, which is the complete lattice with the ordering $\leq$ and the operator $\sim : |\tau| \to |\tau|$.

Here, $\sim$ gives the "meaning" of atomic-level negation of $P\tau$. We also assume that $\top$ is the top element and $\bot$ is the bottom element, respectively. In addition, we use two lattice-theoretic operations: $\lor$ for the least upper bound and $\land$ for the greatest lower bound.[1]

Definition 10.1 (*Symbols*) The symbols of $P\tau$ are defined as follows:

1. Propositional symbols: $p, q, \ldots$ (possibly with subscript)
2. Annotated constants: $\mu, \lambda, \ldots \in |\tau|$
3. Logical connectives: $\land$ (conjunction), $\lor$ (disjunction), $\to$ (implication), and $\neg$ (negation)
4. Parentheses: (and)

Definition 10.2 (*Formulas*) Formulas are defined as follows:

1. If p is a propositional symbol and $\mu \in |\tau|$ is an annotated constant, then p_μ is a formula called an *annotated atom*.
2. If F is a formula, then $\neg F$ is a formula.
3. If F and G are formulas, then $F \land G, F \lor G, F \to G$ are formulas.

---

[1]We employ the same symbols for lattice-theoretical operations as the corresponding logical connectives.

4. If p is a propositional symbol and $\mu \in |\tau|$ is an annotated constant, then a formula of the form $\neg^k p_\mu$ $(k \geq 0)$ is called a *hyper-literal*. A formula which is not a hyper-literal is called a *complex formula*.

Here, some remarks are in order. The annotation is attached only at the atomic level. An annotated atom of the form p_μ can be read "it is believed that p's truth-value is at least μ". In this sense, annotated logics incorporate the feature of many-valued logics.

A hyper-literal is special kind of formula in annotated logics. In the hyper-literal of the form $\neg^k p_\mu$, $\neg^k$ denotes the k's repetition of $\neg$. More formally, if A is an annotated atom, then $\neg^0 A$ is A, $\neg^1 A$ is $\neg A$, and $\neg^k A$ is $\neg(\neg^{k-1} A)$. The convention is also use for $\sim$.

Next, we define some abbreviations.

Definition 10.3 Let A and B be formulas. Then, we put:

$$A \leftrightarrow B =_{def} (A \rightarrow B) \wedge (B \rightarrow A)$$
$$\neg_* A =_{def} A \rightarrow (A \rightarrow A) \wedge \neg(A \rightarrow A)$$

Here, $\leftrightarrow$ is called the *equivalence* and $\neg_*$ *strong negation*, respectively.

Observe that strong negation in annotated logics behaves classically in that it has all the properties of classical negation.

We turn to a semantics for $P\tau$. We here describe a *model-theoretic semantics* for $P\tau$. Let $\mathbf{P}$ is the set of propositional variables. An *interpretation* I is a function $I : \mathbf{P} \rightarrow \tau$. To each interpretation I, we associate a *valuation* $v_I : \mathbf{F} \rightarrow \mathbf{2}$, where $\mathbf{F}$ is a set of all formulas and $\mathbf{2} = \{0, 1\}$ is the set of truth-values. Henceforth, the subscript is suppressed when the context is clear.

Definition 10.4 (*Valuation*) A valuation v is defined as follows:
If p_λ is an annotated atom, then

$$v(p_\lambda) = 1 \text{ iff } I(p) \geq \lambda,$$
$$v(p_\lambda) = 0 \text{ otherwise},$$
$$v(\neg^k p_\lambda) = v(\neg^{k-1} p_{\sim \lambda}), \text{ where } k \geq 1.$$

If A and B are formulas, then

$$v(A \wedge B) = 1 \text{ iff } v(A) = v(B)) = 1,$$
$$v(A \vee B) = 0 \text{ iff } v(A) = v(B) = 0,$$
$$v(A \rightarrow B) = 0 \text{ iff } v(A) = 1 \text{ and } v(B) = 0.$$

If A is a complex formula, then

$$v(\neg A) = 1 - v(A).$$

Say that the valuation v *satisfies* the formula A if $v(A) = 1$ and that v *falsifies* A if $v(A) = 0$. For the valuation v, we can obtain the following lemmas.

Lemma 10.1 *Let p be a propositional variable and $\mu \in |\tau|(k \geq 0)$, then we have:*

$$v(\neg^k p_\mu) = v(p_{\sim^k \mu}).$$

Lemma 10.2 *Let p be a propositional variable, then we have:*

$$v(p_\perp) = 1$$

Lemma 10.3 *For any complex formula A and B and any formula F, the valuation v satisfies the following:*

1. $v(A \leftrightarrow B) = 1$ iff $v(A) = v(B)$
2. $v((A \rightarrow A) \wedge \neg(A \rightarrow A)) = 0$
3. $v(\neg_* A) = 1 - v(A)$
4. $v(\neg F \leftrightarrow \neg_* F) = 1$

We here define the notion of semantic consequence relation denoted by $\vDash$. Let Γ be a set of formulas and F be a formula. Then, F is a semantic consequence of Γ, written $\Gamma \vDash F$, iff for every v such that $v(A) = 1$ for each $A \in \Gamma$, it is the case that $v(F) = 1$. If $v(A) = 1$ for each $A \in \Gamma$, then v is called a model of Γ. If Γ is empty, then $\Gamma \vDash F$ is simply written as $\vDash F$ to mean that F is valid.

Lemma 10.4 *Let p be a propositional variable and $\mu, \lambda \in |\tau|$. Then, we have:*

1. $\vDash p_\perp$
2. $\vDash p_\mu \rightarrow p_\lambda, \mu \geq \lambda$
3. $\vDash \neg^k p_\mu \leftrightarrow p_{\sim^k \mu}, k \geq 0$

The consequence relation $\vDash$ satisfies the next property.

Lemma 10.5 *Let A, B be formulas. Then, if $\vDash A$ and $\vDash A \rightarrow B$ then $\vDash B$.*

Lemma 10.6 *Let F be a formula and p a propositional variable. $(\mu_i)_{i \in J}$ be an annotated constant, where J is an indexed set. Then, if $\vDash F \rightarrow p_\mu$, then $F \rightarrow p_{\mu_i}$, where $\mu = \vee \mu_i$.*

As a corollary to Lemma 10.6, we can obtain the following lemma.

Lemma 10.7 $\vDash p_{\lambda_1} \wedge p_{\lambda_2} \wedge \ldots \wedge p_{\lambda_m} \rightarrow p_\lambda$, *where $\lambda = \vee_{i=1}^{m} \lambda_i$.*

Next, we discuss some results related to paraconsistency and paracompleteness.

Definition 10.5 (*Complementary property*) A truth-value $\mu \in \tau$ has the *complementary property* if there is a λ such that $\lambda \leq \mu$ and $\sim \lambda \leq \mu$. A set $\tau' \subseteq \tau$ has the *complementary property* iff there is some $\mu \in \tau'$ such that μ has the complementary property.

Definition 10.6 (*Range*) Suppose I is an interpretation of the language L. The *range* of I, denoted $range(I)$, is defined to be $range(I) = \{\mu \mid (\exists A \in B_L)I(A) = \mu\}$, where B_L denotes the set of all ground atoms in L.

For $P\tau$, ground atoms correspond to propositional variables. If the range of the interpretation I satisfies the complementary property, then the following theorem can be established.

Theorem 10.1 *Let I be an interpretation such that $range(I)$ has the complementary property. Then, there is a propositional variable p and $\mu \in |\tau|$ such that*

$$v(p_\mu) = v(\neg p_\mu) = 1.$$

Theorem 10.1 states that there is a case in which for some propositional variable it is both true and false, i.e., inconsistent. The fact is closely tied with the notion of paraconsistency.

Definition 10.7 ($\neg$**-inconsistency**) We say that an interpretation I is $\neg$-*inconsistent* iff there is a propositional variable p and an annotated constant $\mu \in |\tau|$ such that $v(p_\mu) = v(\neg p_\mu) = 1$.

Therefore, $\neg$-inconsistency means that both A and $\neg A$ are simultaneously true for some atomic A. Below, we formally define the concepts of non-triviality, paraconsistency and paracompleteness.

Definition 10.8 (*Non-triviality*) We say that an interpretation I is *non-trivial* iff there is a propositional variable p and an annotated constant $\mu \in |\tau|$ such that $v(p_\mu) = 0$.

By Definition 10.8, we mean that not every atom is valid if an interpretation is non-trivial.

Definition 10.9 (*Paraconsistency*) We say that a interpretation I is *paraconsistent* iff it is both $\neg$-inconsistent and non-trivial. $P\tau$ is called *paraconsistent* iff there is an interpretation of I of $P\tau$ such that I is paraconsistent.

Definition 10.9 allows the case in which both A an $\neg A$ are true, but some formula B is false in some paraconsistent interpretation I.

Definition 10.10 (*Paracompleteness*) We say that an interpretation I is *paracomplete* iff there is a propositional variable p and a annotated constant $\lambda \in |\tau|$ such that $v(p_\lambda) = v(\neg p_\lambda) = 0$. $P\tau$ is called *paracomplete* iff there is an interpretation I of $P\tau$ such that I is paracomplete.

From Definition 10.10, we can see that in the paracomplete interpretation I, both A and $\neg A$ are false. We say that $P\tau$ is *non-alethic* iff it is both paraconsistent and paracomplete. Intuitively speaking, paraconsistent logic can deal with inconsistent information and paracomplete logic can handle incomplete information. This means that non-alethic logics like annotated logics can serve as logics for expressing both

Fig. 10.1 The lattice *FOUR*

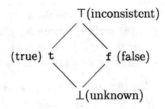

inconsistent and incomplete information. This is one of the starting points of our study of annotated logics.

As the following Theorems 10.2 and 10.3 indicate, paraconsistency and para-completeness in $P\tau$ depend on the cardinality of τ.

Theorem 10.2 *$P\tau$ is paraconsistent iff $card(\tau) \geq 2$, where $card(\tau)$ denotes the cardinality (cardinal number) of the set τ.*

Theorem 10.3 *$P\tau$ is paracomplete iff $card(\tau) \geq 2$.*

The above two theorems imply that to formalize a non-alethic logic based on annotated logics we need at least both the top and bottom elements of truth-values. The simplest lattice of truth-values is *FOUR* in Belnap [17, 18], which is shown in Fig. 10.1.

Definition 10.11 *(Theory)* Given an interpretation I, we can define the theory $Th(I)$ associated with I to be a set:

$$Th(I) = Cn(\{p_\mu \mid p \in \mathbf{P} \text{ and } I(p) \geq \mu\}).$$

Here, Cn is the semantic consequence relation, i.e.,

$$Cn(\Gamma) = \{F \mid F \in \mathbf{F} \text{ and } \Gamma \models F\}.$$

Here, Γ is a set of formulas.
$Th(I)$ can be extended for any set of formulas.

Theorem 10.4 *An interpretation I is $\neg$-inconsistent iff $Th(\Gamma)$ is $\neg$-inconsistent.*

Theorem 10.5 *An interpretation I is paraconsistent iff $Th(I)$ is paraconsistent.*

The next lemma states that the replacement of equivalent formulas within the scope of $\neg$ does not hold in $P\tau$ as in other paraconsistent logics.

Lemma 10.8 *Let A be any hyper-literal. Then, we have:*

1. $\models A \leftrightarrow ((A \to A) \to A)$
2. $\not\models \neg A \leftrightarrow \neg(((A \to A) \to A))$
3. $\models A \leftrightarrow (A \wedge A)$
4. $\not\models \neg A \leftrightarrow \neg(A \wedge A)$
5. $\models A \leftrightarrow (A \vee A)$
6. $\not\models \neg A \leftrightarrow \neg(A \vee A)$

As obvious from the above proofs, (1), (3) and (5) hold for any formula A. But, (2), (4) and (6) cannot be generalized for any A.

By the next theorem, we can find the connection of $P\tau$ and the positive fragment of classical propositional logic C.

Theorem 10.6 *If $F_1, \ldots, F_n$ are complex formulas and $K(A_1, \ldots, A_n)$ is a tautology of C, where $A_1, \ldots, A_n$ are the sole propositional variable occurring in the tautology, then $K(F_1, \ldots, F_n)$ is valid in $P\tau$. Here, $K(F_1, \ldots, F_n)$ is obtained by replacing each occurrence of A_i, $1 \leq i \leq n$, in K by F_i.*

Next, we consider the properties of strong negation $\neg_*$.

Theorem 10.7 *Let A, B be any formulas. Then,*

1. $\models (A \to B) \to ((A \to \neg_*B) \to \neg_*A)$
2. $\models A \to (\neg_*A \to B)$
3. $\models A \lor \neg_*A$

Theorem 10.7 tells us that strong negation has all the basic properties of classical negation. Namely, (1) is a principle of *reductio ad abusurdum*, (2) is the related principle of the law of non-contradiction, and (3) is the law of excluded middle. Note that $\neg$ does not satisfy these properties. It is also noticed that for any complex formula $A \models \neg A \leftrightarrow \neg_*A$ but that for any hyper-literal $Q \not\models \neg Q \leftrightarrow \neg_*Q$.

From these observations, $P\tau$ is a paraconsistent and paracomplete logic, but adding strong negation enables us to perform classical reasoning.

Next, we provide an axiomatization of $P\tau$ in the Hilbert style. There are many ways to axiomatize a logical system, one of which is the *Hilbert system*.

Hilbert system can be defined by the set of *axioms* and *rules of inference*. Here, an axiom is a formula to be postulated as valid, and rules of inference specify how to prove a formula.

We are now ready to give a Hilbert style axiomatization of $P\tau$, called $\mathcal{A}\tau$. Let A, B, C be arbitrary formulas, F, G be complex formulas, p be a propositional variable, and λ, μ, λ_i be annotated constant. Then, the postulates are as follows (cf. Abe [1]):

Postulates for $A\tau$

$(\to_1)$ $(A \to (B \to A))$
$(\to_2)$ $(A \to (B \to C)) \to ((A \to B) \to (A \to C))$
$(\to_3)$ $((A \to B) \to A) \to A$
$(\to_4)$ $A, A \to B / B$
$(\land_1)$ $(A \land B) \to A$
$(\land_2)$ $(A \land B) \to B$
$(\land_3)$ $A \to (B \to (A \land B))$
$(\lor_1)$ $A \to (A \lor B)$
$(\lor_2)$ $B \to (A \lor B)$
$(\lor_3)$ $(A \to C) \to ((B \to C) \to ((A \lor B) \to C))$
$(\neg_1)$ $(F \to G) \to ((F \to \neg G) \to \neg F)$

$(\neg_2)$ $F \to (\neg F \to A)$
$(\neg_3)$ $F \lor \neg F$
(τ_1) $p_\perp$
(τ_2) $\neg^k p_\lambda \leftrightarrow \neg^{k-1} p_{\sim\lambda}$
(τ_3) $p_\lambda \to p_\mu$, where $\lambda \geq \mu$
(τ_4) $p_{\lambda_1} \land p_{\lambda_2} \land \cdots \land p_{\lambda_m} \to p_\lambda$, where $\lambda = \bigvee_{j \in J} \lambda_j$

Here, except $(\to_4)$, these postulates are axioms. $(\to_4)$ is a rule of inferences called *modus ponens* (MP).

In da Costa et al. [24], a different axiomatization is given, but it is essentially the same as ours. There, the postulates for implication are different. Namely, although $(\to_1)$ and $(\to_3)$ are the same (although the naming differs), the remaining axiom is:

$$(A \to B) \to ((A \to (B \to C)) \to (A \to C))$$

It is well known that there are many ways to axiomatize the implicational fragment of classical logic C. In the absence of negation, we need the so-called *Pierce's law* $(\to_3)$ for C.

In $(\neg_1), (\neg_2), (\neg_3)$, F and G are complex formulas. In general, without this restriction on F and G, these are not sound rules due to the fact that they are not admitted in annotated logics.

da Costa et al. [24] fuses (τ_1) and (τ_2) as the single axiom in conjunctive form. But, we separate it in two axioms for our purposes. Also there is a difference in the final axiom. They present it for infinite lattices as

$A \to p_{\lambda_j}$ for every $j \in J$, then $A \to p_\lambda$, where $\lambda = \bigvee_{j \in J} \lambda_j$.
If τ is a finite lattice, this is equivalent to the form of (τ_2).

As usual, we can define a *syntactic consequence relation* in $P\tau$. Let Γ be a set of formulas and G be a formula. Then, G is a syntactic consequence of Γ, written $\Gamma \vdash G$, iff there is a finite sequence of formulas $F_1, F_2, \ldots, F_n$, where F_i belongs to Γ, or F_i is an axiom ($1 \leq i \leq n$), or F_j is an immediate consequence of the previous two formulas by $(\to_4)$. This definition can extend for the transfinite case in which n is an ordinal number. If $\Gamma = \emptyset$, i.e. $\vdash G$, G is a *theorem* of $P\tau$.

Let Γ, Δ be sets of formulas and A, B be formulas. Then, the consequence relation $\vdash$ satisfies the following conditions.

1. if $\Gamma \vdash A$ and $\Gamma \subset \Delta$ then $\Delta \vdash A$.
2. if $\Gamma \vdash A$ and $\Delta, A \vdash B$ then $\Gamma, \Delta \vdash B$.
3. if $\Gamma \vdash A$, then there is a finite subset $\Delta \subset \Gamma$ such that $\Delta \vdash A$.

In the Hilbert system above, the so-called *deduction theorem* holds.

Theorem 10.8 (Deduction theorem) *Let Γ be a set of formulas and A, B be formulas. Then, we have:*

$$\Gamma, A \vdash B \Rightarrow \Gamma \vdash A \to B.$$

The following theorem shows some theorems related to strong negation.

Theorem 10.9 *Let A and B be any formula. Then,*

1. $\vdash A \lor \neg_* A$
2. $\vdash A \to (\neg_* A \to B)$
3. $\vdash (A \to B) \to ((A \to \neg_* B) \to \neg_* A)$

From Theorems 10.9 and 10.10 follows.

Theorem 10.10 *For arbitrary formulas A and B, the following hold:*

1. $\vdash \neg_*(A \land \neg_* A)$
2. $\vdash A \leftrightarrow \neg_* \neg_* A$
3. $\vdash (A \land B) \leftrightarrow \neg_*(\neg_* A \lor \neg_* B)$
4. $\vdash (A \to B) \leftrightarrow (\neg_* A \lor B)$
5. $\vdash (A \lor B) \leftrightarrow \neg_*(\neg_* A \land \neg_* B)$

Theorem 10.10 implies that by using strong negation and a logical connective other logical connectives can be defined as in classical logic. If $\tau = \{t, f\}$, with its operations appropriately defined, we can obtain classical propositional logic in which $\neg_*$ is classical negation.

Now, we provide some formal results of $P\tau$ including completeness and decidability.

Lemma 10.9 *Let p be a propositional variable and $\mu, \lambda, \theta \in |\tau|$. Then, the following hold:*

1. $\vdash p_{\lambda \lor \mu} \to p_\lambda$
2. $\vdash p_{\lambda \lor \mu} \to p_\mu$
3. $\lambda \geq \mu$ and $\lambda \geq \theta \Rightarrow \vdash p_\lambda \to p_{\mu \lor \theta}$
4. $\vdash p_\mu \to p_{\mu \land \theta}.$
5. $\vdash p_\theta \to p_{\mu \land \theta}.$
6. $\lambda \leq \mu$ and $\lambda \leq \theta \Rightarrow \vdash p_{\mu \land \theta}$
7. $\vdash p_\mu \leftrightarrow p_{\mu \lor \mu}, \vdash p_\mu \leftrightarrow p_{\mu \land \mu}$
8. $\vdash p_{\mu \lor \lambda} \leftrightarrow p_{\lambda \lor \mu}, \vdash p_{\mu \land \lambda} \leftrightarrow p_{\lambda \land \mu}$
9. $\vdash p_{(\mu \lor \lambda) \lor \theta} \lor \to p_{\mu \lor (\lambda \lor \theta)}, \vdash p_{(\mu \land \lambda) \land \theta} \lor \to p_{\mu \land (\lambda \land \theta)}$
10. $p_{(\mu \lor \lambda) \land \mu} \to p_\mu, p_{(\mu \land \lambda) \lor \mu} \to p_\mu$
11. $\lambda \leq \mu \Rightarrow \vdash p_{\lambda \lor \mu} \to p_\mu$
12. $\lambda \lor \mu = \mu \Rightarrow \vdash p_\mu \to p_\lambda$
13. $\mu \geq \lambda \Rightarrow \forall \theta \in |\tau| (\vdash p_{\mu \lor \theta} \to p_{\lambda \lor \theta}$ and $\vdash p_{\mu \land \theta} \to p_{\lambda \land \theta})$
14. $\mu \geq \lambda$ and $\theta \geq \varphi \Rightarrow \vdash p_{\mu \lor \theta} \to p_{\lambda \lor \varphi}$ and $p_{\mu \land \theta} \to p_{\lambda \land \varphi}$
15. $\vdash p_{\mu \land (\lambda \lor \theta)} \to p_{(\mu \land \lambda) \lor (\mu \land \theta)}, \vdash p_{\mu \lor (\lambda \land \theta)} \to p_{(\mu \lor \lambda) \land (\mu \lor \theta)}$
16. $\vdash p_\mu \land p_\lambda \leftrightarrow p_{\mu \land \lambda}$
17. $\vdash p_{\mu \lor \lambda} \to p_\mu \lor p_\lambda$

Example 10.1 Consider the complete lattice $\tau = N \cup \omega$, where N is the set of natural numbers. The ordering on τ is the usual ordering on ordinals, restricted to the set τ. Consider the set $\Gamma = \{p_0, p_1, p_2, \ldots\}$, where $p_\omega \vdash \Gamma$. It is clear that $\Gamma \vdash p_\omega$, but an infinitary deduction is required to establish this.

Definition 10.12 $\bar{\Delta} = \{A \in \mathbf{F} \mid \Delta \vdash A\}$

Definition 10.13 Δ is said to be *trivial* iff $\bar{\Delta} = \mathbf{F}$ (i.e., every formula in our language is a syntactic consequence of Δ); otherwise, Δ is said to be *non-trivial*. Δ is said to be *inconsistent* iff there is some formula A such that $\Delta \vdash A$ and $\Delta \vdash \neg A$; otherwise, Δ is *consistent*.

From the definition of triviality, the next theorem follows:

Theorem 10.11 Δ *is trivial iff* $\Delta \vdash A \wedge \neg A$ *(or* $\Delta \vdash A$ *and* $\Delta \vdash \neg_* A$*) for some formula A.*

Theorem 10.12 *Let Γ be a set of formulas, A, B be any formulas, and F be any complex formula. Then, the following hold.*

1. $\Gamma \vdash A$ *and* $\Gamma \vdash A \rightarrow B \Rightarrow \Gamma \vdash B$
2. $A \wedge B \vdash A$
3. $A \wedge B \vdash B$
4. $A, B \vdash A \wedge B$
5. $A \vdash A \vee B$
6. $B \vdash A \vee B$
7. $\Gamma, A \vdash C$ *and* $\Gamma, B \vdash C \Rightarrow \Gamma, A \vee B \vdash C$
8. $\vdash F \leftrightarrow \neg_* F$
9. $\Gamma, A \vdash B$ *and* $\Gamma, A \vdash \neg_* B \Rightarrow \Gamma \vdash \neg_* A$
10. $\Gamma, A \vdash B$ *and* $\Gamma, \neg_* A \vdash B \Rightarrow \Gamma \vdash B$.

Note here that the counterpart of Theorem 10.12 (10) obtained by replacing the occurrence of $\neg_*$ by $\neg$ is not valid.

Now, we are in a position to prove the soundness and completeness of $P\tau$. Our proof method for completeness is based on maximal non-trivial set of formulas; see Abe [1] and Abe and Akama [5]. da Costa, Subrahmanian and Vago [24] presented another proof using Zorn's Lemma.

Theorem 10.13 (Soundness) *Let Γ be a set of formulas and A be any formula. $A\tau$ is a sound axiomatization of $P\tau$, i.e., if $\Gamma \vdash A$ then $\Gamma \vDash A$.*

For proving the completeness theorem, we need some theorems.

Theorem 10.14 *Let Γ be a non-trivial set of formulas. Suppose that τ is finite. Then, Γ can be extended to a maximal (with respect to inclusion of sets) non-trivial set with respect to $\mathbf{F}$.*

Theorem 10.15 *Let Γ be a maximal non-trivial set of formulas. Then, we have the following:*

1. *if A is an axiom of $P\tau$, then $A \in \Gamma$*
2. $A, B \in \Gamma$ *iff* $A \wedge B \in \Gamma$

3. $A \vee B \in \Gamma$ iff $A \in \Gamma$ or $B \in \Gamma$
4. if $p_\lambda, p_\mu \in \Gamma$, then $p_\theta \in \Gamma$, where $\theta = max(\lambda, \mu)$
5. $\neg^k p_\mu \in \Gamma$ iff $\neg^{k-1} p_{\sim\mu} \in \Gamma$, where $k \geq 1$
6. if $A, A \rightarrow B \in \Gamma$, then $B \in \Gamma$
7. $A \rightarrow B \in \Gamma$ iff $A \notin \Gamma$ or $B \in \Gamma$

Theorem 10.16 *Let Γ be a maximal non-trivial set of formulas. Then, the characteristic function χ of Γ, that is, $\chi_\Gamma \rightarrow 2$ is the valuation function of some interpretation $I : \mathbf{P} \rightarrow |\tau|$.*

Here is the completeness theorem for $P\tau$.

Theorem 10.17 (Completeness) *Let Γ be a set of formulas and A be any formula. If τ is finite, then $A\tau$ is a complete axiomatization for $P\tau$, i.e., if $\Gamma \models A$ then $\Gamma \vdash A$.*

The decidability theorem also holds for finite lattice.

Theorem 10.18 (Decidability) *If τ is finite, then $P\tau$ is decidable.*

The completeness does not in general hold for infinite lattice. But, it holds for special case.

Definition 10.14 (*Finite annotation property*) Suppose that Γ be a set of formulas such that the set of annotated constants occurring in Γ is included in a finite substructure of τ (Γ itself may be infinite). In this case, Γ is said to have the *finite annotation property*.

Note that if τ' is a substructure of τ then τ' is closed under the operations $\sim, \vee$ and $\wedge$. One can easily prove the following from Theorem 10.17.

Theorem 10.19 (Finitary Completeness) *Suppose that Γ has the finite annotation property. If A is any formula such that $\Gamma \vdash A$, then there is a finite proof of A from Γ.*

Theorem 10.19 tells us that even if the set of the underlying truth-values of $P\tau$ is infinite (countably or uncountably), as long as theories have the finite annotation property. the completeness result applied to them, i.e., $A\tau$ is complete with respect to such theories.

In general, when we consider theories that do not possess the finite annotation property, it may be necessary to guarantee completeness by adding a new infinitary inference rule (ω-rule), similar in spirit to the rule used by da Costa [36] in order to cope with certain models in a particular family of infinitary language. Observe that for such cases a desired axiomatization of $P\tau$ is not finitary.

From the classical result of compactness, we can state a version of the compactness theorem.

Theorem 10.20 (Weak Compactness) *Suppose that Γ has the finite annotation property. If A is any formula such that $\Gamma \vdash A$, then there is a finite subset Γ' of Γ such that $\Gamma' \vdash A$.*

Annotated logics $P\tau$ provide a general framework, and can be used to reasoning about many different logics. Below we present some examples.

The set of truth-values $FOUR = \{t, f, \bot, \top\}$, with $\neg$ defined as: $\neg t = f, \neg f = t, \neg\bot = \bot, \neg\top = \top$. Four-valued logic based on $FOUR$ was originally due to Belnap [17, 18] to model internal states in a computer. Subrahmanian [48] formalized an annotated logic with *FOUR* as a foundation for paraconsistent logic programming; also see Blair and Subrahmanian [20].

Their annotated logic may be used for reasoning about inconsistent knowledge bases. For example, we may allow logic programs to be finite collections of formulas of the form:

$$(A : \mu_0) \leftrightarrow (B_1 : \mu_1)\&\ldots\&(B_n : \mu_n)$$

where A and B_i ($1 \leq i \leq n$) are atoms and μ_j ($0 \leq j \leq n$) are truth-values in *FOUR*.

Intuitively, such programs may contain "intuitive" inconsistencies–for example, the pair

$$((p : f), (p : t))$$

is inconsistent. If we append this program to a consistent program P, then the resulting union of these two programs may be inconsistent, even though the predicate symbols p occurs nowhere in program P.

Such inconsistencies can easily occur in knowledge based systems, and should not be allowed to trivialize the meaning of a program. However, knowledge based systems based on classical logic cannot handle the situation since the program is trivial. In Blair and Subrahmanian [20], it is shown how the four-valued annotated logic may be used to describe this situation. Later, Blair and Subrahmanian's annotated logic was extended as *generalized annotated logics* by Kifer and Subrahmanian [40, 41].

There are also other examples which can be dealt with by annotated logics. The set of truth-values *FOUR* with negation defined as boolean complementation forms an annotated logic.

The unit interval $[0, 1]$ of truth-values with $\neg x = 1 - x$ is considered as the base of annotated logic for qualitative or fuzzy reasoning. In this sense, probabilistic and fuzzy logics could be generalized as annotated logics. The interval $[0, 1] \times [0, 1]$ of truth-values can be used for annotated logics for evidential reasoning. Here, the assignment of the truth-value (μ_1, μ_2) to proposition p may be thought of as saying that the degree of belief in p is μ_1, while the degree of disbelief is μ_2. Negation can be defined as $\neg(\mu_1, \mu_2) = (\mu_2, \mu_1)$.

Note that the assignment of $[\mu_1, \mu_2]$ to a proposition p by an interpretation I does not necessarily satisfy the condition $\mu_1 + \mu_2 \leq 1$. This contrasts with probabilistic reasoning. Knowledge about a particular domain may be gathered from different experts (in that domain), and these experts may different views. Some of these views may lead to a "strong" belief in a proposition; likewise, other experts may have a "strong" disbelief in the same proposition. In such a situation, it seems appropriate to report the existence of conflicting opinions, rather than use ad hoc means to resolve this conflict.

10.4 Intelligent Control Based on Annotated Logics

Annotated logics found several applications in various fields, and we here sketch intelligent control based on annotated logics in detail. As discussed in Sect. 10.2, annotated logics fit into foundations for intelligent control with the desired requirements.

For Robotics, we employ evidential annotated logics $E\tau$ for evidential reasoning, which are a variant of $P\tau$. Atomic formulas of $E\tau$ are of the type $p_{(\mu,\lambda)}$, where $(\mu, \lambda) \in [0, 1]^2$ which is a lattice with the ordering $(\mu_1, \lambda_1) \leq (\mu_2, \lambda_2) \Leftrightarrow \mu_1 \leq \mu_2$ and $\lambda_1 \leq \lambda_2$.

Based on $E\tau$, we introduce the uncertainty degree G_{un} and the certainty degree G_{ce} as follows:

$$G_{un}(\mu, \lambda) = \mu + \lambda - 1$$
$$G_{ce}(\mu, \lambda) = \mu - \lambda$$

We consider the following 12 output states in Table 10.1:

The control values are:

- V_{cic} = maximum value of uncertainty control = Ft_{ct}
- V_{cve} = maximum value of certainty control = Ft_{ce}
- V_{cpa} = minimum value of uncertainty control = $-Ft_{ct}$
- V_{cpa} = minimum value of certainty control = $-Ft_{ce}$

For the present discussion, we will use the following: $Ft_{ct} = Ft_{ce} = \frac{1}{2}$.

We also set $C_1 = C_3 = \frac{1}{2}$ and $C_2 = C_4 = -\frac{1}{2}$.

All states are represented in Fig. 10.2.

In the paraconsistent annotated structure, the main aim is to know how to measure or to determine the certainty degree concerning a proposition, if it is False or True. Therefore, we take into account only the certainty degree G_{ce}. The uncertainty degree G_{un} indicates the measure of the inconsistency or

Table 10.1 Extreme and Non-extreme states

Extreme states	Symbol	Non-extreme states	Symbol
True	V	Quasi-true tending to Inconsistent	$QV \to T$
False	F	Quasi-true tending to Paracomplete	$QV \to \bot$
Inconsistent	T	Quasi-false tending to Inconsistent	$QF \to T$
Paracomplete	$\bot$	Quasi-false tending to Paracomplete	$QF \to \bot$
		Quasi-inconsistent tending to True	$QT \to V$
		Quasi-inconsistent tending to False	$QT \to F$
		Quasi-paracomplete tending to True	$Q\bot \to V$
		Quasi-paracomplete ending to False	$Q\bot \to F$

Fig. 10.2 Representation of the certainty and uncertainty degrees

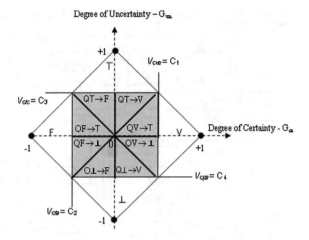

paracompleteness. If the certainty degree is low or the uncertainty degree is high, it generates an indefinition.

The *paraconsistent logical controller* (Paracontrol) is an electric-electronic materization of the para-analyzer algorithm, which is basically an electric circuit, treating logical signals in $E\tau$; see Abe and Da Silva Filho [7, 30]. Such a circuit compares logical values and determines domains of a state lattice corresponding to output value. Favorable evidence and contrary evidence degrees are represented by voltage. Certainty and uncertainty degrees are determined by analogue of operational amplifiers.

The Paracontrol comprises both analog and digital systems and it can be externally adjusted by applying positive and negative voltages. The Paracontrol was tested in real-life experiments with an autonomous mobile robot *Emmy*, whose favorable/contrary evidences coincide with the values of ultrasonic sensors and distances are represented by continuous values of voltage; see Da Silva Filho and Abe [31].[2]

Emmy consists of a circular mobile platform of aluminum 30 cm in diameter and 60 cm high. While moving in a non-structured environment, Emmy gets information about the presence/absence of obstacles using the sonar system called *Parasoninc*. Parasonic is able to detect obstacles in a robot's path by transforming the distance to the obstacle into electric signal of the continuous voltage ranging 0 to 5 V.

Parasonic is basically composed of two ultrasonic sensors of type POLAROID 6500 controlled by an 8051 micro controller. The 8051 is programmed to carry out synchronization between the measurements of the two sensors and the transformation of the distance into electric voltage.

---

[2]The name Emmy originates from the mathematician Amalie Emmy Nöwther (1882–1935).

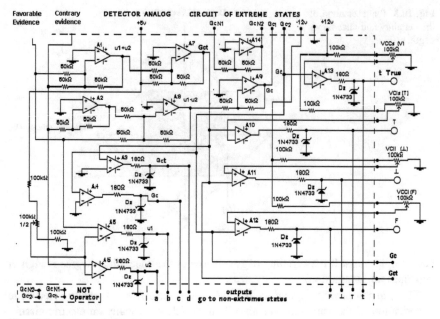

Fig. 10.3 Paracontrol circuit

Emmy uses the paracontrol system to negotiate traffic in non-constructed environments avoiding collisions with human beings, objects, walls, tables, etc. The form of reception of information on the obstacles is non-contact in nature is the method to obtain and to treat signals from ultra-sonic of optical sensors in order to avoid collisions. The circuit of Paracontrol is shown in Fig. 10.3.

The system of Emmy's control is composed by the Parasonic, Paracontrol and supporting circuits as in Fig. 10.4.

The Emmy robot control system has the following components (see Fig. 10.5):

- Ultra-Sonic Sensors: The two sensors detect the distance between the robot and the object through the emission of a pulse train of ultra-sonic sound waves and the reception of the signal (echo).
- Signal Treatment: The treatment of the captured signals is made through the Para-sonic. The microprocessor is programmed to transform the time elapsed between the emission of the signal and the reception of the echo in an electric signal o the 0 to 5 V for belief degree, and from 5 to 0 V for disbelief degree. The width of each voltage is proportional to the time that has elapsed between the emission of a pulse train and its reception by sensors.
- Paraconsistent Analysis: The paraconsistent control logic makes a logical analysis of the signals according to the logic $E\tau$.
- Codification: The coder circuit changes the binary word of 12 digits to a code of 4 digits to be processed by the personal computer.
- Action Processing: The microprocessor is programmed conveniently to work the relay in sequences to establish actions for the robot.

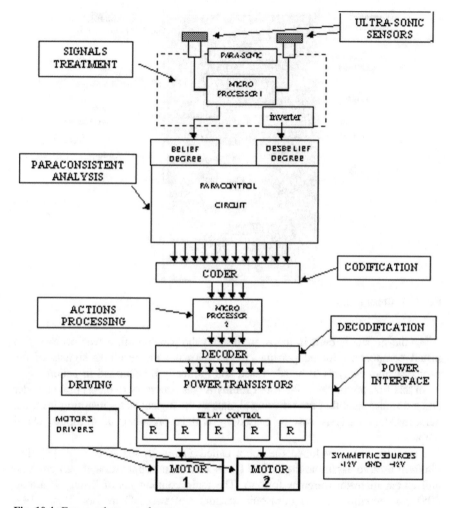

Fig. 10.4 Emmy robot control system

- Decodification: The circuit decoder changes the binary world 4 digits in signals to charge the relay in the programmed paths.
- Power Interface: The power interface circuit is composed of transistors that amplify the signals making possible the relay's control by digital signals.
- Driving: The relays ON and OFF the motors M_1 and M_2 according to the decoded binary word.
- Motor Drives: The motors M_1 and M_2 move the robot according to the sequence of the relays control.
- Sources: The robot Emmy is supplied by two batteries forming a symmetrical source of tension of activate ± 12 V DC.

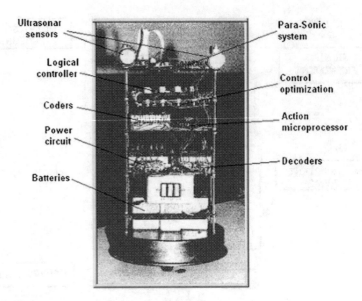

Ultrasonar sensors

Logical controller

Coders

Power circuit

Batteries

Para-Sonic system

Control optimization

Action microprocessor

Decoders

Fig. 10.5 Robot Emmy

As the project is built in hardware besides the paracontrol, it was necessary to install components for supporting circuits allowing the resulting signals of the paraconsistent analysis to be addressed and indirectly transformed in action.

In this first prototype of the robot Emmy it was necessary to incorporate a coder and a decoder such that the referring signals to the logical states resulting from the paraconsistent analysis was processed by a microprocessor of 4 inputs and 4 outputs.

Later, we also developed the robot called *Emmy II*; see Abe et al. [8]. The platform used to assemble the Emmy II robot measures approximately 23 cm high and 25 cm diameter (circular format). The main components of Emmy II are an 8051 microcontroller, two ultrasonic sensors and two DC motors. Figure 10.6 shows the basic structure of Emmy II.

In Emmy II, the signals from the 2 sensors are used to determine the favorable evidence degree μ and the contrary degree evidence degree λ regarding the proposition "There is no obstacle in front of the robot".

Fig. 10.6 Basic structure of Emmy II

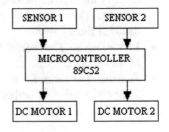

Fig. 10.7 Logical output lattice

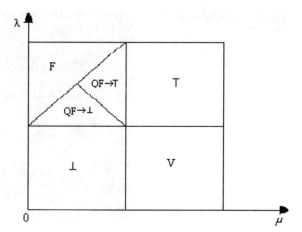

Then, the Paracontrol, recorded in the internal memory of the microcontroller is used to determine the robot movements. Also, the microcontroller is responsible for applying power to the DC motors.

The decision-making of the Emmy II regarding the movements to undertake is according to the decision state lattice shown in Fig. 10.7.

Each robot movement lasts approximately 0.4 s. The source circuitry supplies 5, 12 and 16.8 V DC to the other robot Emmy II circuits. Figures 10.8 and 10.9 are pictures of Emmy II.

The "Polaroid 6500 Series Sonar Ranging Module" was used in the Emmy II robot. This device has three inputs (Vcc, Gnd and Init) and one output (Echo). When INIT is taken high by the microcontroller, the sonar ranging module transmits 16 pulses at 49.4 kHz.

After receiving the echo pulses, which cause the ECHO output go high, the sonar ranging module sends an ECHO signal to the microcontroller. Then, with the

Fig. 10.8 Front vision of Emmy II

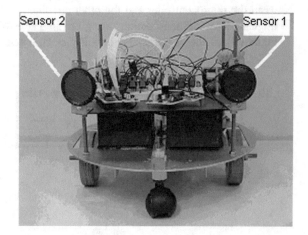

Fig. 10.9 Inferior vision of
Emmy II

Table 10.2 Logical states and action

Symbol	State	Action
V	True	Robot goes ahead
F	False	Robot goes back
⊥	Paracomplete	Robot turns right
T	Inconsistent	Robot turns left
QF → ⊥	Quasi-false tending to paracomplete	Robot turns right
QF → T	Quasi-true tending to inconsistent	Robot turns left

time interval between INIT sending and ECHO receiving, the microcontroller is
able to determine the distance between the robot and the obstacle.

The microcontroller 89C52 from 8051 family is responsible to control the
Emmy II robot. Its input/output port 1 is used to send and receive signals to and
from the Sensor Circuitry and Power Circuitry. Buffers are used in the interface
between the microcontroller and the other circuits.

Logical states and their actions are given in Table 10.2.

Figure 10.10 shows the source circuitry electric scheme of Emmy II.

Fig. 10.10 Source circuitry

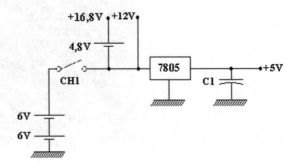

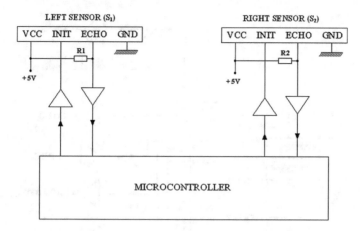

Fig. 10.11 Sensor circuit

Figure 10.11 shows the sensor circuit of Emmy II.

The microcontroller I/O Port 1 has 8 pins. The function of each pin is as follows:

- Pin 0: INIT of sonar ranging module 1 (S_1)
- Pin 1: ECHO of sonar ranging module 1 (S_1)
- Pin 2: INIT of sonar ranging module 2 (S_2)
- Pin 3: ECHO of sonar ranging module 2 (S_2)
- Pin 4: When it is taken high ($+5$ V), DC motor 1 has power supplied
- Pin 5: When it is taken high ($+5$ V), DC motor 2 has power supplied
- Pin 6: When it is taken low (0 V), DC motor 1 spins around forward while pin 4 is taken high ($+5$ V). When it is taken high ($+5$ V), DC motor 1 spins around backward while pin 4 is taken high ($+5$ V).
- Pin 7: When it is taken low (0 V), DC motor 2 spins around forward while pin 5 is taken high ($+5$ V). When it is taken high ($+5$ V), DC motor 2 spins around backward while pin 5 is taken high ($+5$ V).

Two DC motors supplied by 12 V DC are responsible for Emmy II robot movements. The Paracontrol, through the microcontroller, determines which DC motor must be supplied and in which direction it must spin.

Basically, the power interface circuitry is comprised of power field effect transistors-MOSFETs.

Figure 10.12 shows the main microcontroller connections of Emmy II.

Figure 10.13 represents a test environment.

Figure 10.14 represents power interface circuitry.

The microcontroller 89C52 internal memory can store numbers of 8 bits. If we represent these numbers in hexadecimal, it means that we can store numbers from 00 h to FFh. Internal memory location 30 h of the 89C52 microcontroller was chosen to store the favorable evidence grade value.

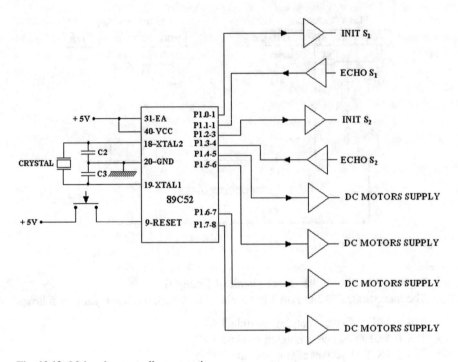

Fig. 10.12 Main microcontroller connections

Fig. 10.13 Test environment

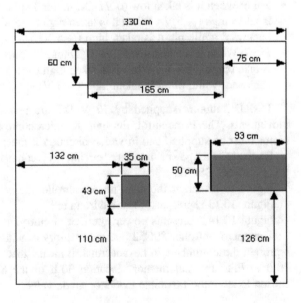

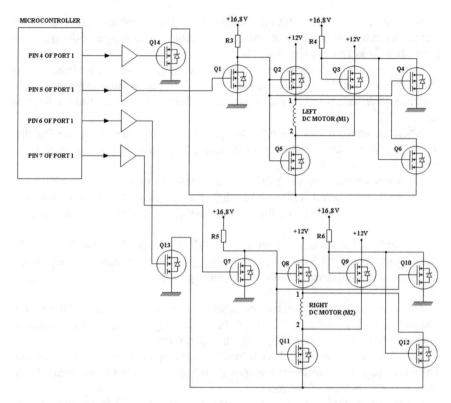

Fig. 10.14 Power interface circuitry

When the microcontroller 89C52 I/O port in pin 0 is taken high, meaning that sonar ranging module 1 transmits sonar pulses, 30 h memory position starts to increase by in 1 unit per 118 μs.

The sonar ranging module 1 ECHO is connected to microcontroller 89C52 I/O port 1 pin 1. So, when this pin is taken high, meaning that sonar echo pulses have just returned, incrementing of the 30 h internal memory position ceases.

The time for the sound to travel 4 cm is 118 μs. Hence, The value stored in the microcontroller 89C52 30 h time for the sound to travel 4 is multiplied by 2 represents the distance in centimeters between sonar ranging module 1 and the obstacle in front of it.

The maximum value that can be stored in the microcontroller 89C52 30 h internal memory position is 3Fh (to allow arithmetic calculation). SO, the sonar ranging module 1 can measure distance between 00 h and 3Fh, or 0 cm and 126 cm.

Therefore, 00 h stored in the microcontroller 89C52 30 h internal memory position means that the favorable evidence grade value (μ) on the proposition. "There is no obstacle in front o the robot" is 0.

3Fh stored in the microcontroller 89C52 30 h internal memory position means that the intermediate favorable evidence grade value (μ) on the proposition "There is no obstacle in front of the robot" is 1.

In the same way, intermediate values between 00 h and 3Fh stored in the microcontroller 89C52 30 h internal memory position means intermediate favorable evidence grade values (μ) on the proposition "There is 118 no obstacle in front of the robot".

When the microcontroller 89C52 I/O port 1 pin 2 is taken high, meaning that sonar ranging module 2 transmits sonar pulses, 31 h memory position starts to be increased by 1 unit per 118 µs.

The sonar ranging module 2 ECHO is connected to microcontroller 89C52 I/O port 1 pin 3. So, when this pin is taken high, meaning that sonar echo pulses have just returned, incrementing of the microcontroller 89C52 internal position 31 h ceases.

The time for the sound to travel 4 cm is 118 µs. Hence, the value stored in the microcontroller 89C52 31 h internal memory position multiplied by 2 represents the distance in centimeter between sonar ranging module 2 an the obstacle in front of it.

The maximum value that can be stored in the microcontroller 89C52 31 h internal memory position is 3Fh. Then, the sonar ranging module 2 can measure distances between 00 h and 3Fh, or 0 cm and 126 cm.

The microcontroller 89C52 31 h internal memory position is complemented. Thus, it represents the contrary evidence grade value (λ) on the proposition "There is no obstacle in front of the robot".

Consequently, 00 h stored in the microcontroller 89C52 31 h internal memory position means that the contrary evidence grade value (λ) on the proposition: "There is no obstacle in front of the robot" is 0.

And 4Fh stored in the microcontroller 89C52 31 h internal memory position means that the contrary evidence rate value (λ) on the proposition "There is no obstacle in front of the robot" is 1.

In the same way, intermediate values between 00 h and 3Fh stored in the microcontroller 89C52 31 h internal memory position means intermediate contrary evidence grade value (λ) on the proposition "There is no obstacle in front of the robot".

We also developed the autonomous mobile robot called *Emmy III*; see Desiderto and De Oliveira [32]. The aim of the Emmy III is to be able to move from an origin point to an end point, both predetermined, in a non-structured environment. The first prototype of Emmy III is composed of a planning system and a mechanical construction.

The planning system considers an environment to be divided in cells. The first version considers all cells to be free. Firstly, the robot must be able to move from a point to another without encountering any obstacle. This environment is divided into cells as in Elfes [34] and a planning system gives the sequence of cells the robot must travel to reach the end cell.

Fig. 10.15 Emmy III

Then, it asks for the initial point and the target point. After that a sequence of movements is given. Also a sequence of pulses is sent to the step motors that are responsible for moving the physical platform of the robot.

So, the robot moves from the initial point to the target point. The physical construction of the first prototype of the Emmy III robot is basically composed of a circular platform of approximately 286 mm diameter and two step motors.

Figure 10.15 shows the Emmy III first prototype.

The planning system is recorded in a notebook, and the communication between the notebook and the physical construction is made through the parallel port. A potency driver is responsible for getting the pulses from the note book and sending them to the step motors. Like the first prototype, the second prototype of Emmy III is basically composed of a planning system and a mechanical structure. The planning system can be recorded in any personal computer and the communication between the personal computer and the mechanical construction is done through a USB port.

The planning system considers the environment around the robot to be divided into cells. So, it is necessary to inform the planning system which the cell the robot is in and likewise for the target cell. Then, the planning system gives a sequence o cells that the robot must follow to go from the origin cell to the target cell. The planning system considers all cells to be free. The mechanical construction is basically composed of a steel structure, two DC motors and three wheels. Each motor has a wheel field in its axis and there is free wheel.

Figure 10.16 shows the mechanical structure.

Fig. 10.16 Basic structure of Emmy III

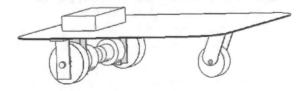

There is electronic circuitry on the steel structure. The main device of the electronic circuit is the microcontroller PIC18F4550 that is responsible for receiving the schedule from the planning system and activating the DC motors. There is also a potency driver between the microcontroller and the DC motors.

Recently, Emmy II was implemented in a new hardware structure. Basically, the robot there is similar to Emmy II, but with a modern hardware structure to receive upgrades in it functionalities; see Torres and Reis [49].

For thousands of years, humans have researched to find logical structures to 'imitate' human thinking and actions, in order to implement robots and automated processing. As the majority of automated processing and robotics are based on classical logic, in a certain sense their actions are predictable; all commands are of the type yes/no, because classical logic is two-valued.

So, robots built with classical logic are too limited in that they cannot handle conflicting and imprecise information. So, we need more flexible logical systems for robots for them to perform more complex tasks in a smooth way without becoming involved with conflicts, imprecision and paracompleteness. This means that a new class of non-classical logics should be developed and incorporated into some interesting autonomous mobile robots, which can deal with imprecise, inconsistent and paracomplete data. As we saw, annotated logics, in particular, $E\tau$, can be used to develop such robots. The fact is not surprising because annotated logics can naturally capture human reasoning.

The applications of annotated logics to robotics considered here show the power and the beauty of paraconsistent systems. They have provided other ways than usual electronic theory, opening applications of non-classical logics in order to overcome situations not covered by other systems or techniques. Therefore, we believe that annotated logics can offer theoretical foundations for intelligent control.

10.5 Conclusions

We discussed the applications annotated logics to inelligent control. After presenting formal aspecs of annotated logics, we illustrated that annotated logics can be also regarded as attractive logical systems for intelligent control. Although several systems of paraconsistent logic have been proposed so far, no systems can provide a unified framework for real applications. In this sense, annotated logics should be studied theoretically as well as practically. We expect to see further developments of the research on annotated logics for intelligent control.

Acknowledgments We are grateful to the referee for useful comments.

References

1. Abe, J.M.: On the Foundations of Annotated Logics (in Portuguese), Ph.D. Thesis, University of São Paulo, Brazil (1992)
2. Abe, J.M.: On annotated modal logics. Mathematica Japonica **40**, 553–556 (1994)
3. Abe, J.M.: Curry algebra Pτ. Logique et Analyse **161–162-163** (1998), 5–15
4. Abe, J.M., Akama, S.: Annotated logics Qτ and ultraproduct. Logique et Analyse **160**(1997), 335–343 (published in 2000)
5. Abe, J.M., Akama, S.: On some aspects of decidability of annotated systems. In: Arabnia HR (ed.) Proceedings of the International Conference on Artificial Intelligence, vol. II, pp. 789–795, CREA Press (2001)
6. Abe, J.M., Akama, S., Nakamatsu, K.: Introduction to Annotated Logics. Springer, Heidelberg (2015)
7. Abe, J.M., Da Silva Filho, J.I.: Manipulating conflicts and uncertainties in robotics. Multi-Valued Logic Soft Comput **9**, 147–169 (2003)
8. Abe, J.M., Torres, C.R., Torres, G.L., Nakamatsu, K., Kondo, M.: Intelligent paraconsistent logic controller and autonomous mobile robot Emmy II. In: Lecture Notes in Computer Science 4252, pp. 851–857. Springer, Heidelberg (2006)
9. Akama, S.: Discursive reasoning in a constructive setting. Int. J. Reasoning-Based Intell. Syst. **3**, 88–93 (2011)
10. Akama, S., Abe, J.M.: Many-valued and annotated modal logics. In: Proceedings of the 28th International Symposium on Multiple-Valued Logic, pp. 114–119, Fukuoka (1998)
11. Akama, S., Abe, J.M.: Natural deduction and general annotated logics. In: Proceedings of the 1st International Workshop on Labelled Deduction, Freiburg (1998)
12. Akama, S., Abe, J.M.: Fuzzy annotated logics. In: Proceedings of IPMU'2000, pp. 504–508, Madrid, Spain (2000)
13. Akama, S., Abe, J.M., Nakamatsu, K.: Constructive discursive logic with strong negation. Logique et Analyse **215**, 395–408 (2011)
14. Anderson, A., Belnap, N.: Entailment: the logic of relevance and necessity I. Princeton University Press, Princeton (1976)
15. Anderson, A., Belnap, N., Dunn, J.: Entailment: The Logic of Relevance and Necessity II. Princeton University Press, Princeton (1992)
16. Arruda, A.I.: A survey of paraconsistent logic. In: Mathematical Logic in Latin America. North-Holland, Amsterdam (1980)
17. Belnap, N.D.: A useful four-valued logic. In: Dunn, J.M., Epstein, G. (eds.) Modern Uses of Multi-Valued Logic, pp. 8–37. Reidel, Dordrecht (1977)
18. Belnap, N.D.: How a computer should think. In: Ryle, G. (ed.) Contemporary Aspects of Philosophy, pp. 30–55. Oriel Press (1977)
19. Beziau, J.-Y., Carnielli, W., Gabbay, D. (eds.): Handbook of Paraconsistency. College Publication, London (2007)
20. Blair, H.A., Subrahmanian, V.S.: Paraconsistent logic programming. Theoret. Comput. Sci. **68**, 135–154 (1989)
21. da Costa, N.C.A., Abe, J.M., Subrahmanian, V.S.: Remarks on annotated logic. Zeitschrift für mathematische Logik und Grundlagen der Mathematik **37**, 561–570 (1991)
22. da Costa, N.C.A., Alves, E.H.: A semantical analysis of the calculi C_n. Notre Dame J. Formal Logic **18**, 621–630 (1977)
23. da Costa, N.C.A., Subrahmanian, VS.: Paraconsistent logic as a formalism for reasoning about inconsistent knowledge. Artif. Intell. Med. **1**, 167–174 (1989)
24. da Costa, N.C.A., Subrahmanian, V.S., Vago, C.: The paraconsistent logic PT. Zeitschrift für mathematische Logik und Grundlagen der Mathematik **37**, 139–148 (1991)

25. da Costa, N.C.A.: α-models and the system T and T^*. Notre Dame J. Formal Logic, **14**, 443–454 (1974)
26. da Costa, N.C.A.: On the theory of inconsistent formal systems. Notre Dame J. Formal Logic **15**, 497–510 (1974)
27. da Costa, N.C.A., Dubikajtis, L.: On Jasśkowski's discursive logic. In: Arruda, A.I., da Costa, N.C.A., Chuaqui, R. (eds.) Non-classical logics. Model Theory and Computability, pp. 37–56. North-Holland, Amsterdam (1977)
28. da Costa, N.C.A., Doria, F.A.: On Jaśkowski's discursive logics. Stud. Logica. **54**, 33–60 (1995)
29. da Costa, N.C.A., Henschen, L.J., Lu, J.J., Subrahmanian, V.S.: Automatic theorem proving in paraconsistent logics: foundations and implementation. In: Proceedings of the 10th international conference on automated deduction, pp. 72–86. Springer, Berlin (1990)
30. Da Silva Filho, J.I.: Methodos de Aplicoes da Logica Paraconisitente Anotadade Anotacao com Dois Valores LPA2v com Construcao de Algoritmo e Implementacao de Circuitos Eletornicos (in Portugese), Ph.D. Thesis, University of São Paulo (1999)
31. Da Silva Filho, J.I., Abe, J.M.: Emmy: a paraconsistent autonomous mobile robot. In: Abe, J. M., Da Silva Filho (eds.) Frontiers in artificial intelligence and its applications, pp. 53–61. IOS Press, Amsterdam (2001)
32. Desiderato, J.M.G., De Oliveira, E.N.: Primeiro Prototipo do Dobo Móvel Autôno Emmy III (in Portuguese). University of São Paulo (2006)
33. Dunn, J.M.: Relevance logic and entailment. In: Gabbay, D., Gunthner, F. (eds.) Handbook of Philosophical Logic, vol. III, pp. 117–224. Reidel, Dordrecht (1986)
34. Elfes, A.: Using occupancy grids for mobile robot perception and navigation. Comput. Mag. **22**, 46–57 (1989)
35. Fu, K.S.: Robotics: control, sensing, vision and intelligence. McGraw-Hill Book, New York (1987)
36. Jaśkowski, S.: Propositional calculus for contradictory deductive systems (in Polish). Studia Societatis Scientiarun Torunesis, Sectio A **1**, 55–77 (1948)
37. Jaśkowski, S.: On the discursive conjunction in the propositional calculus for inconsistent deductive systems (in Polish). Studia Societatis Scientiarun Torunesis, Sectio A **8**, 171–172 (1949)
38. Kifer, M., Lozinskii, E.L.: RI: a logic for reasoning with inconsistency. In: Proceedings of LICS4, pp. 253–262 (1989)
39. Kifer, M., Lozinskii, E.L.: A logic for reasoning with inconsistency. J. Autom. Reasoning **9**, 179–215 (1992)
40. Kifer, M., Subrahmanian, V.S.: On the expressive power of annotated logic programs. In: Proceedings of the 1989 North American Conference on Logic Programming, pp. 1069–1089 (1989)
41. Kifer, M., Subrahmanian, V.S.: Theory of generalized annotated logic programming. J. Logic Programming **12**, 335–367 (1992)
42. Kotas, J.: The axiomatization of S. Jaśkowski's discursive logic, Studia Logica **33**, 195–200 (1974)
43. Priest, G., Routley, R., Norman, J. (eds.): Paraconsistent logic: essays on the inconsistent. Philosopia Verlag, München (1989)
44. Priest, G.: Logic of paradox. J. Philosophical Logic **8**, 219–241 (1979)
45. Priest, G.: Paraconsistent logic. In: Gabbay, D., Guenthner, F. (eds.) Handbook of Philosophical Logic, 2nd edn. pp. 287–393. Kluwer, Dordrecht (2002)
46. Priest, G.: In Contradiction: A Study of the Transconsistent, 2nd edn. Oxford University Press, Oxford (2006)
47. Routley, R., Plumwood, V., Meyer, R.K., Brady, R: Relevant Logics and Their Rivals, vol. 1. Ridgeview, Atascadero (1982)

48. Subrahmanian, V.: On the semantics of quantitative logic programs. In: Proceedings of the 4th IEEE Symposium on Logic Programming, pp. 173–182 (1987)
49. Torres, C.R., Reis, R.: The new hardware structure of the Emmy II robot. In: Abe, J.M. (ed.) Paraconsistent Intelligent-Based Systems. Springer, Heidelberg (2015)

Chapter 11
Paraconsistent Annotated Logic Program EVALPSN and Its Application to Intelligent Control

Kazumi Nakamatsu, Jair M. Abe and Seiki Akama

Abstract We have already proposed a paraconsistent annotated logic program called EVALPSN. In EVALPSN, an annotation called an extended vector annotation is attached to each literal. In order to deal with before-after relation between two time intervals, we also have introduced a new interpretation for extended vector annotations in EVALPSN, which is named before-after(bf)-EVALPSN. In this chapter, we introduce paraconsistent annotated logic programs EVALPSN/bf-EVALPSN and their application to intelligent control, especially logical safety verification based control with simple examples. First, the background and overview of EVALPSN are introduced, and paraconsistent annotated logics PT and the basic annotated logic program are recapitulated as the formal background of EVALPSN/bf-EVALPSN with some simple examples. Then EVALPSN is formally defined and its application to traffic signal control is introduced. EVALPSN application to pipeline valve control is also introduced with examples. Bf-EVALPSN is formally defined and its unique and useful reasoning rules are introduced with some examples. Last, this chapter is concluded with some remarks.

K. Nakamatsu (✉)
School of Human Science and Environment, University of Hyogo,
1-1-12 Shinzaike, Himeji 670-0092, Japan
e-mail: nakamatu@shse.u-hyogo.ac.jp

J.M. Abe
Graduate Program in PE, ICET, Paulista University,
R.Dr.Bacelar, 1212, Sao Paulo 04026-002, Brazil
e-mail: jairabe@uol.com.br

J.M. Abe
Institute of Advanced Studies, University of Sao Paulo,
Av. Prof. Luciano Gualberto, Travessa J, 374 Terreo,
Cidade Universitaria, Sao Paulo 05508-900, Brazil

S. Akama
C-Republic, 1-20-1, Higashi-Yurigaoka, Asao-Ku, Kawasaki 215-0012, Japan
e-mail: akama@jcom.home.ne.jp

© Springer International Publishing Switzerland 2016
K. Nakamatsu and R. Kountchev (eds.), *New Approaches in Intelligent Control*,
Intelligent Systems Reference Library 107, DOI 10.1007/978-3-319-32168-4_11

337

11.1 Introduction and Background

One of the main purposes of paraconsistent logic is to deal with inconsistency in a framework of consistent logical systems. It has been almost seven decades since the first paraconsistent logical system was proposed by Jaskowski [12]. It was four decades later that a family of paraconsistent logic called "annotated logic" was proposed by da Costa et al. [8, 49], which can deal with inconsistency by introducing many truth values called "annotations" into their syntax as attached information to atomic formulas.

The paraconsistent annotated logic by da Costa et al.was developed from the viewpoint of logic programming by Subrahmanian et al. [7, 13, 48]. Furthermore, in order to deal with inconsistency and non-monotonic reasoning in a framework of annotated logic programming, ALPSN (Annotated Logic Program with Strong Negation) and its stable model semantics was developed by Nakamatsu and Suzuki [20]. It has been shown that ALPSN can deal with some non-monotonic reasonings such as default logic [46], autoepistemic logic [15] and a non-monotonic Assumption Based Truth Maintenance System (ATMS) [9] in a framework of annotated logic programming [21–23]. Even though ALPSN can deal with non-monotonic reasoning such as default reasoning and conflicts can be represented as paraconsistent knowledge in it, it is difficult and complicated to deal with reasoning to resolve conflicts in ALPSN. On the other hands, it is known that defeasible logic can deal with conflict resolving in a logical way [5, 41, 42], although defeasible logic cannot deal with inconsistency in its syntax and its inference rules are too complicated to be implemented easily. In order to deal with conflict resolving and inconsistency in a framework of annotated logic programming, a new version of ALPSN, VALPSN(Vector Annotated Logic Program with Strong Negation) that can deal with defeasible reasoning and inconsistency was also developed by Nakamatsu et al. [16]. Moreover, it has been shown that VALPSN can be applied to conflict resolving in various systems [18, 24, 25]. It also has been shown that VALPSN provides a computational model of defeasible logic [5, 6]. Later, VALPSN was extended to EVALPSN(Extended VALPSN) by Nakamatsu et al. [26, 27] to deal with deontic notions(obligation, permission, forbiddance, etc.) and defeasible deontic reasoning [43, 44]. Recently, EVALPSN has been applied to various kinds of safety verification and intelligent control, for example, railway interlocking safety verification [30], robot action control [28, 31, 32, 39], safety verification for air traffic control [29], traffic signal control [33], discrete event control [34–36] and pipeline valve control [37, 38].

Considering the intelligent safety verification for process control, there is an occasion in which the safety verification for process order control is significant. For example, suppose a pipeline network in which two kinds of liquids, nitric acid and caustic soda are used for cleaning the pipelines. If those liquids are processed continuously and mixed in the same pipeline by accident, explosion by neutralization would be caused. In order to avoid such a dangerous accident, the safety for process order control should be strictly verified in a formal way such as EVALPSN.

However, it seems to be a little difficult to utilize EVALPSN for verifying process order control as well as the safety verification for each process in process control. We have already proposed a new version of EVALPSN called bf(before-after)-EVALPSN that can deal with before-after relations between two time intervals (processes) by using two sorts of reasoning rules. One is named Basic Before-after reasoning rule and another is Transitive Before-after reasoning rule [17, 19, 40].

This chapter mainly contributes introducing EVALPSN, bf-EVALPSN and their applications to intelligent control based on their logical reasoning. As far as we know there seems to be no other efficient computational tool that can deal with the real-time intelligent safety verification for process order control than bf-EVALPSN.

This chapter is organized as follows: firstly, in Sect. 11.1, the development background and overview of the paraconsistent annotated logic programs are introduced; in Sect. 11.2, paraconsistent annotated logics and their logic programs are reviewed as the background knowledge of EVALPSN/bf-EVALPSN, moreover EVALPSN are formally recapitulated; in Sect. 11.3, the traffic signal control system based on EVALPSN deontic defeasible reasoning and its simple simulation results by the cellular automaton method are provided as an application of EVALPSN to intelligent control in Sect. 11.4, a simple brewery pipeline network is introduced and its pipeline process control system based on EVALPSN valve safety verification is constructed as an application of EVALPSN to intelligent control systems; in Sect. 11.5, the basic concepts of bf-EVALPSN are introduced and bf-EVALPSN itself is formally defined, furthermore, an application of bf-EVALPSN to real-time intelligent safety verification for process order control is described with the pipeline examples in Sect. 11.4; in Sect. 11.6, reasoning of before-after relations in bf-EVALPSN is reviewed and a unique and useful inference method of before-after relations in bf-EVALPSN, which can be implemented as a bf-EVALPSN called "transitive bf-inference rules", is introduced with a simple example; lastly, conclusions and remarks are provided.

11.2 Paraconsistent Annotated Logic Program

This section is devoted to clarify the formal background of the paraconsistent annotated logic program EVALPSN. The more details of EVALPSN has been introduced in [17]. We assume that the reader is familiar with the basic knowledge of classical logic and logic programming [14]. In order to understand EVLPSN and its reasoning we introduce Paraconsistent Annotated Logics PT [8] in the following subsection.

11.2.1 Paraconsistent Annotated Logic PT

Here we briefly recapitulate the syntax and semantics for propositional paraconsistent annotated logics PT proposed by da Costa et al. [8].

Generally, a truth value called an *annotation* is attached to each atomic formula explicitly in paraconsistent annotated logic, and the set of annotations constitutes a complete lattice. We introduce a paraconsistent annotated logic PT with the four valued complete lattice T.

Definition 11.1 The primitive symbols of PT are:

1. propositional symbols $p, q, \ldots, p_i, q_i, \ldots$;
2. each member of T is an *annotation constant* (we may call it simply an annotation);
3. the connectives and parentheses $\wedge, \vee, \rightarrow, \neg, (,)$.

Formulas are defined recursively as follows:

1. if p is a propositional symbol and $\mu \in T$ is an annotation constant, then $p : \mu$ is an *annotated atomic formula (atom)*;
2. if F, F_1, F_2 are formulas, then $\neg F, F_1 \wedge F_2, F_1 \vee F_2, F_1 \rightarrow F_2$ are formulas.

We suppose that the four-valued lattice in Fig. 11.1 is the complete lattice T, where annotations t and f may be intuitively regarded as truth values *true* and *false*, respectively. It may be comprehensible that annotations $\perp, t, f$ and $\top$ correspond to the truth values $*, T, F$ and TF in Visser [50] and **None, T, F**, and **Both** in Belnap [4], respectively. Moreover, the complete lattice T can be viewed as a bi-lattice in which the vertical direction $\overrightarrow{\perp \top}$ indicates *knowledge amount* ordering and the horizontal direction $\overrightarrow{ft}$ does *truth* ordering [10]. We use the symbol $\leq$ to denote the ordering in terms of knowledge amount (the vertical direction $\overrightarrow{\perp \top}$) over the complete lattice T, and the symbols $\perp$ and $\top$ are used to denote the bottom and top elements, respectively. In the paraconsistent annotated logic PT, each annotated atomic formula can be interpreted epistemically, for example, $p : t$ may be interpreted epistemically as "the proposition p is known to be true".

There are two kinds of negation in the paraconsistent annotated logic PT, one of them is called *epistemic negation* and represented by the symbol $\neg$ (see Definition 2.1). The epistemic negation in PT followed by an annotated atomic formula is defined as a mapping between elements of the complete lattice T as follows:

$$\neg(\perp) = \perp, \quad \neg(t) = f, \quad \neg(f) = t, \quad \neg(\top) = \top.$$

Fig. 11.1 The 4-valued complete lattice T

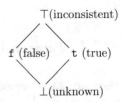

As shown in the above mapping the epistemic negation maps annotations to themselves without changing the knowledge amounts of annotations. Furthermore, the epistemic negation followed by an annotated atomic formula can be eliminated by the mapping. For example, the knowledge amount of annotation t is the same as that of annotation f as shown in the complete lattice $\mathcal{T}$, and we have the epistemic negation,[1]

$$\neg(p:\mathsf{t}) = p:\neg(\mathsf{t}) = p:\mathsf{f},$$

which shows that the knowledge amount in terms of the proposition p cannot be changed by the epistemic negation mapping. There is another negation called *ontological(strong) negation* that is defined by using the epistemic negation.

Definition 11.2 (*Strong Negation*)
   Let F be any formula,

$$\sim F =_{def} F \rightarrow ((F \rightarrow F) \wedge \neg(F \rightarrow F)).$$

The epistemic negation in Definition 11.2 is not interpreted as a mapping between annotations since it is not followed by an annotated atomic formula. Therefore, the strongly negated formula $\sim F$ can be interpreted so that if the formula F exists, the contradiction $((F \rightarrow F) \wedge \neg(F \rightarrow F))$ is implied. Usually, the strong negation is used for denying the existence of a formula following it.

The semantics for the paraconsistent annotated logics PT is defined.

Definition 11.3 Let v be the set of all propositional symbols and $\mathcal{F}$ be the set of all formulas. An interpretation I is a function,

$$I : v \rightarrow \mathcal{T}.$$

To each interpretation I, we can associate the valuation function such that

$$v_I : \mathcal{F} \rightarrow \{0, 1\},$$

which is defined as :

1. let p be a propositional symbol and μ an annotation,

$$v_I(p : \mu) = 1 \quad \textbf{iff} \quad \mu \leq I(p),$$
$$v_I(p : \mu) = 0 \quad \textbf{iff} \quad \mu \not\leq I(p);$$

---

[1]An expression $\neg p : \mu$ is conveniently used for expressing a negative annotated literal instead of $\neg(p : \mu)$ or $p : \neg(\mu)$.

2. let A and B be any formulas, and A not an annotated atom,

$$v_I(\neg A) = 1 \text{ iff } v_I(A) = 0,$$
$$v_I(\sim B) = 1 \text{ iff } v_I(B) = 0;$$

other formulas $A \to B$, $A \wedge B$, $A \vee B$ are valuated as usual.

We provide an intuitive interpretation for strongly negated annotated atoms with the complete lattice $\mathcal{T}$. For example, the strongly negated literal $\sim (p : \mathbf{t})$ implies the knowledge "p is false($\mathbf{f}$) or unknown($\perp$)" since it denies the existence of the knowledge that "p is true($\mathbf{t}$)". This intuitive interpretation is proofed by Definition 11.3 as follows: if $v_I(\sim (p : \mathbf{t})) = 1$, we have $v_I(p : \mathbf{t}) = 0$ and for any annotation $\mu \in \{\perp, \mathbf{f}, \mathbf{t}, \top\} \preceq \mathbf{t}$, we have $v_I(p : \mu) = 1$, therefore, we obtain that $\mu = \mathbf{f}$ or $\mu = \perp$.

11.2.2 EVALPSN (Extended Vector Annotated Logic Program with Strong Negation)

Generally, an annotation is explicitly attached to each literal in paraconsistent annotated logic programs as well as the paraconsistent annotated logic PT. For example, let p be a literal, μ an annotation, then $p : \mu$ is called an *annotated literal*. The set of annotations constitutes a complete lattice.

An annotation in EVALPSN has a form of $[(i,j), \mu]$ called an *extended vector annotation*. The first component (i,j) is called a *vector annotation* and the set of vector annotations, which constitutes a complete lattice,

$$\mathcal{T}_v(n) = \{(x,y) | 0 \leq x \leq n, 0 \leq y \leq n, x, y \text{ and } n \text{ are integers}\}$$

shown by the Hasse's diagram as n = 2 in Fig. 11.2. The ordering($\preceq_v$) of the complete lattice $\mathcal{T}_v(n)$ is defined as follows: let (x_1, y_1), (x_2, y_2), $\in \mathcal{T}_v(n)$,

$$(x_1, y_1) \prec_v (x_2, y_2) \text{ iff } x_1 \leq x_2 \text{ and } y_1 \leq y_2.$$

Fig. 11.2 *Lattice $\mathcal{T}_v(2)$ and Lattice $\mathcal{T}_d$*

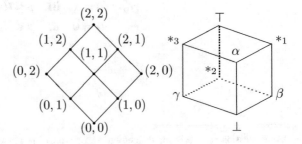

For each extended vector annotated literal $p : [(i,j), \mu]$, the integer i denotes the amount of positive information to support the literal p and the integer j denotes that of negative one. The second component μ is an index of fact and deontic notions such as obligation, and the set of the second components constitutes the following complete lattice,

$$\mathcal{T}_d = \{\bot, \alpha, \beta, \gamma, *_1, *_2, *_3, \top\}.$$

The ordering($\preceq_d$) of the complete lattice $\mathcal{T}_d$ is described by the Hasse's diagram in Fig. 11.2. The intuitive meaning of each member in $\mathcal{T}_d$ is

$\bot$(unknown),

α(fact), β(obligation), γ(non-obligation),

$*_1$ (fact and obligation),

$*_2$ (obligation and non-obligation),

$*_3$ (fact and non-obligation),   and

$\top$(inconsistency).

The complete lattice $\mathcal{T}_d$ is a quatro-lattice in which the direction $\overrightarrow{\bot\top}$ measures *knowledge amount*, the direction $\overrightarrow{\gamma\beta}$ does *deontic truth*, the direction $\overrightarrow{\bot*_2}$ does *deontic knowledge amount* and the direction $\overrightarrow{\bot\alpha}$ does *factuality*. For example, annotation β(obligation) can be intuitively interpreted to be more obligatory than annotation γ(non-obligation), and annotations $\bot$(no knowledge) and $*_2$(obligation and non-obligation) are deontically neutral, that is to say, it cannot be said whether they represent obligation or non-obligation.

The complete lattice $\mathcal{T}_e(n)$ of extended vector annotations is defined as the product,

$$\mathcal{T}_v(n) \times \mathcal{T}_d.$$

The ordering($\preceq_e$) of the complete lattice $\mathcal{T}_e(n)$ is also defined as follows: let $[(i_1,j_1), \mu_1], [(i_2,j_2), \mu_2] \in \mathcal{T}_e$,

$$[(i_1,j_1), \mu_1] \prec_e [(i_2,j_2), \mu_2] \quad \text{iff} \quad (i_1,j_1) \prec_v (i_2,j_2) \quad \text{and} \quad \mu_1 \prec_d \mu_2.$$

There are two kinds of epistemic negation ($\neg_1$ and $\neg_2$) in EVALPSN, which are defined as mappings over the complete lattices $\mathcal{T}_v(n)$ and $\mathcal{T}_d$, respectively.

Definition 11.4 (*epistemic negations $\neg_1$ and $\neg_2$ in EVALPSN*)

$$\neg_1([(i,j),\mu]) = [(j,i),\mu], \quad \forall\mu \in \mathcal{T}_d$$
$$\neg_2([(i,j),\bot]) = [(i,j),\bot], \quad \neg_2([(i,j),\alpha]) = [(i,j),\alpha],$$
$$\neg_2([(i,j),\beta]) = [(i,j),\gamma], \quad \neg_2([(i,j),\gamma]) = [(i,j),\beta],$$
$$\neg_2([(i,j),*_1]) = [(i,j),*_3], \quad \neg_2([(i,j),*_2]) = [(i,j),*_2],$$
$$\neg_2([(i,j),*_3]) = [(i,j),*_1], \quad \neg_2([(i,j),\top]) = [(i,j),\top].$$

If we regard the epistemic negations in Definition 11.4 as syntactical operations, an epistemic negation followed by a literal can be eliminated by the syntactical operation. For example, $\neg_1 p : [(2,0),\alpha] = p : [(0,2),\alpha]$ and $\neg_2 q : [(1,0),\beta] = p : [(1,0),\gamma]$. The strong negation ($\sim$) in EVALPSN is defined as well as the paraconsistent annotated logic PT.

Definition 11.5 (*well extended vector annotated literal*)

Let p be a literal. $p : [(i,0),\mu]$ and $p : [(0,j),\mu]$ are called *weva(well extended vector annotated)-literals*, where $i,j \in \{1,2,\ldots,n\}$, and $\mu \in \{\alpha, \beta, \gamma\}$.

Definition 11.6 (*EVALPSN*)

If $L_0,\ldots,L_n$ are weva-literals,

$$L_1 \wedge \cdots \wedge L_i \wedge \sim L_{i+1} \wedge \cdots \wedge \sim L_n \rightarrow L_0$$

is called an *EVALPSN clause*. An *EVALPSN* is a finite set of EVALPSN clauses.

Fact and deontic notions, "obligation", "forbiddance" and "permission" are represented by extended vector annotations,

$$[(m,0),\alpha], [(m,0),\beta], [(0,m),\beta], \text{ and } [(0,m),\gamma],$$

respectively, where m is a positive integer. For example,

$p : [(2,0),\alpha]$ is intuitively interpreted as "it is known to be true of strength 2 that p is a fact";

$p : [(1,0),\beta]$ is as "it is known to be true of strength 1 that p is obligatory";

$p : [(0,2),\beta]$ is as "it is known to be false of strength 2 that p is obligatory", that is to say, "it is known to be true of strength 2 that p is forbidden";

$p : [(0,1),\gamma]$ is as "it is known to be false of strength 1 that p is not obligatory", that is to say, "it is known to be true of strength 1 that p is permitted".

Generally, if an EVALPSN contains the strong negation $\sim$, it has stable model semantics [17] as well as other ordinary logic programs with strong negation. However, the stable model semantics may have a problem that some programs may have more than two stable models and others have no stable model. Moreover, computation of stable models takes a long time compared to usual logic programming such as PROLOG programming. Therefore, it does not seem to be so appropriate for practical application such as real time processing in general.

However, we fortunately have cases to implement EVALPSN practically, if an EVALPSN is a *stratified* program, it has a tractable model called a *perfect model* [45] and the strong negation in the EVALPSN can be treated as the *Negation as Failure* in logic programming with no strong negation. The details of stratified program and some tractable models for normal logic programs can be found in [3, 11, 45, 47], furthermore the details of the stratified EVALPSN are described in [17]. Therefore, inefficient EVALPSN stable model computation does not have to be taken into account in this chapter since all EVALPSNs that will appear in the subsequent sections are stratified.

11.3 Traffic Signal Control in EVALPSN

11.3.1 Deontic Defeasible Traffic Signal Control

Traffic jam caused by inappropriate traffic signal control is a serious issue that should be resolved. In this section, we introduce an intelligent traffic signal control system based on EVALPSN defeasible deontic reasoning, which may provide one solution for traffic jam reduction. We show how the traffic signal control is implemented in EVALPSN with taking a simple intersection example in Japan.

We suppose an intersection in which two roads are crossing described in Fig. 11.3 as an example for implementing the traffic signal control method based on

Fig. 11.3 Intersection

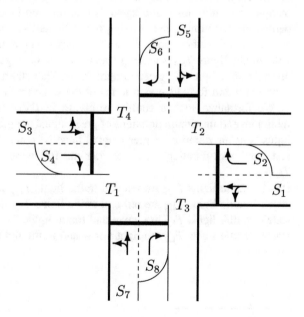

EVALPSN.[2] The intersection has four traffic lights $T_{1,2,3,4}$, which indicate four kinds of signals, green, yellow, red and right-turn arrow. Each lane connected to the intersection has a sensor to detect traffic amount. Each sensor is described by symbols $S_i (1 \le i \le 8)$ in Fig. 11.3. For example, the sensor S_6 detects the right-turn traffic amount confronting traffic light T_1. Basically, the traffic signal control is performed based on the traffic amount detected by the sensors. The chain of signaling is supposed as follows:

$$\rightarrow \text{ red } \rightarrow \text{ green } \rightarrow \text{ yellow } \rightarrow \text{ right arrow } \rightarrow \text{ red } \rightarrow .$$

For simplicity, we assume that the durations of yellow and right arrow signals are constant, and if traffic lights $T_{1,2}(T_{3,4})$ are green or right arrow, traffic lights $T_{3,4}(T_{1,2})$ are red as follows:

Signal cycle of traffic lights $T_{1,2}$
$$\rightarrow \text{ red } \rightarrow \text{ red } \rightarrow \text{ green } \rightarrow \text{ right arrow } \rightarrow \text{ red } \rightarrow,$$
Signal cycle of traffic lights $T_{3,4}$
$$\rightarrow \text{ green } \rightarrow \text{ right arrow } \rightarrow \text{ red } \rightarrow \text{ red } \rightarrow \text{ green } \rightarrow .$$

Only the turns green to right arrow and right arrow to red are controlled. The turn red to green of the front traffic signal follows the turn right arrow to red of the neighbor one. Moreover, the signaling is controlled at each unit time $t \in \{0, 1, 2, \ldots, n\}$. The traffic amount of each lane can be regarded as permission or forbiddance from turning such as green to right arrow. For example, if there are many cars waiting for traffic lights $T_{1,2}$ turning red to green, it can be regarded as permission for turning the crossing traffic lights $T_{3,4}$ green to right arrow, yellow and red. On the other hand, if there are many cars passing through the intersection with traffic lights $T_{3,4}$ signaling green, it can be regarded as forbiddance from turning traffic lights $T_{3,4}$ green to right arrow. Then, there is a conflict between those permission and forbiddance in terms of the same traffic lights $T_{3,4}$.

We formalize such a conflict resolving in EVALPSN. We assume that the minimum and maximum durations of green signal are previously given for all traffic lights, and the duration of green signal must be controlled between the minimum and maximum durations. We consider the following four states of traffic lights $T_{1,2,3,4}$,

state1   traffic lights $T_{1,2}$ are red and traffic lights $T_{3,4}$ are green,
state2   traffic lights $T_{1,2}$ are red and traffic lights $T_{3,4}$ are right arrow,
state3   traffic lights $T_{1,2}$ are green and traffic lights $T_{3,4}$ are red,
state4   traffic lights $T_{1,2}$ are right arrow and traffic lights $T_{3,4}$ are red

---

[2]The intersection is supposed to be in Japan where we need to keep left if driving a car.

Here we take the transit from the state 1 to the state 2 into account to introduce the traffic signal control properties in the state 1 and its translation into EVALPSN. The traffic signal control consists of the traffic signal control properties for the state transit the state 1 to the state 2, green light length rules, and deontic defeasible reasoning rules for traffic signal control.

We use the following EVALP literals:

- $S_i(t) : [(2,0), \alpha]$ can be informally interpreted as the traffic sensor $S_i (i = 1, 2, \ldots, 8)$ has detected traffic at time t.
- $T_{m,n}(c, t) : [(2,0), \alpha]$ can be informally interpreted as the traffic light $T_{m,n}$ indicates a signal color C at time t, where $m, n = 1, 2, 3, 4$ and c is one of signal colors green(g), red(r), or right arrow(a).
- $MIN_{m,n}(g, t) : [(2,0), \alpha]$ can be informally interpreted as the green duration of traffic lights $T_{m,n}(m, n = 1, 2, 3, 4)$ is shorter than its minimum green duration at time t.
- $MAX_{m,n}(g, t) : [(2,0), \alpha]$ can be informally interpreted as the green duration of traffic lights $T_{m,n}$ is longer than its maximum green duration at time t.
- $T_{m,n}(c, t) : [(0,k), \gamma]$ which can be informally interpreted as it is permitted for traffic lights $T_{m,n}$ to indicate signal color C at time t, where $m, n = 1, 2, 3, 4$ and c is one of the signal colors green(g), red(r), or right arrow(a); if $k = 1$, the permission is weak, and if $k = 2$, the permission is strong.
- $T_{m,n}(c, t) : [(0,k), \beta]$ can be informally interpreted as it is forbidden for traffic lights $T_{m,n}$ from indicating the signal color C at time t, where $m, n = 1, 2, 3, 4$ and c is one of the signal colors green(g), red(r), or right arrow(a); if $k = 1$, the forbiddance is weak, and if $k = 2$, the forbiddance is strong.

[Traffic Signal Control Properties in State 1]

1. If traffic sensor S_1 detects traffic amount, it has already passed the minimum green duration of traffic lights $T_{3,4}$, and neither traffic sensors S_5 nor S_7 detect traffic amount at time t, then it is weakly permitted for traffic lights $T_{3,4}$ to turn green to right arrow at time t; which is translated into the EVALPSN,

$$
\begin{aligned}
&S_1(t) : [(2,0), \alpha] \wedge \\
&T_{1,2}(r,t) : [(2,0), \alpha] \wedge T_{3,4}(g,t) : [(2,0), \alpha] \wedge \\
&\sim MIN_{3,4}(g,t) : [(2,0), \alpha] \wedge \\
&\sim S_5(t) : [(2,0), \alpha] \wedge \sim S_7(t) : [(2,0), \alpha] \\
&\rightarrow T_{3,4}(a,t) : [(0,1), \gamma].
\end{aligned} \tag{11.1}
$$

2. If traffic sensor S_3 detects traffic amount, it has already passed the minimum green duration of traffic lights $T_{3,4}$, and neither traffic sensors S_5 nor S_7 detect traffic amount at time t, then it is weakly permitted for traffic lights $T_{3,4}$ to turn green to right arrow at time t; which is translated into the EVALPSN,

$$S_3(t) : [(2,0), \alpha] \wedge$$
$$T_{1,2}(r,t) : [(2,0), \alpha] \wedge T_{3,4}(g,t) : [(2,0), \alpha] \wedge$$
$$\sim MIN_{3,4}(g,t) : [(2,0), \alpha] \wedge \tag{11.2}$$
$$\sim S_5(t) : [(2,0), \alpha] \wedge \sim S_7(t) : [(2,0), \alpha]$$
$$\rightarrow T_{3,4}(a,t) : [(0,1), \gamma].$$

3. If traffic sensor S_2 detects traffic amount, it has already passed the minimum green duration of traffic lights $T_{3,4}$, and neither traffic sensors S_5 nor S_7 detect traffic amount at time t, then it is weakly permitted for traffic lights $T_{3,4}$ to turn green to right arrow at time t, which is translated into the EVALPSN,

$$S_2(t) : [(2,0), \alpha] \wedge$$
$$T_{1,2}(r,t) : [(2,0), \alpha] \wedge T_{3,4}(g,t) : [(2,0), \alpha] \wedge$$
$$\sim MIN_{3,4}(g,t) : [(2,0), \alpha] \wedge \tag{11.3}$$
$$\sim : S_5(t) : [(2,0), \alpha] \wedge \sim S_7(t) : [(2,0), \alpha]$$
$$\rightarrow T_{3,4}(a,t) : [(0,1), \gamma],$$

4. If traffic sensor S_4 detects traffic amount, it has already passed the minimum green duration of traffic lights $T_{3,4}$, and neither traffic sensors S_5 nor S_7 detect traffic amount at time t, then it is weakly permitted for traffic lights $T_{3,4}$ to turn green to right arrow at time t; which is translated into the EVALPSN,

$$S_4(t) : [(2,0), \alpha] \wedge$$
$$T_{1,2}(r,t) : [(2,0), \alpha] \wedge T_{3,4}(g,t) : [(2,0), \alpha] \wedge$$
$$\sim MIN_{3,4}(g,t) : [(2,0), \alpha] \wedge \tag{11.4}$$
$$\sim S_5(t) : [(2,0), \alpha] \wedge \sim S_7(t) : [(2,0), \alpha]$$
$$\rightarrow T_{3,4}(a,t) : [(0,1), \gamma],$$

5. If traffic sensor S_6 detects traffic amount, it has already passed the minimum green duration of traffic lights $T_{3,4}$, and neither traffic sensors S_5 nor S_7 detect traffic amount at time t, then it is weakly permitted for traffic lights $T_{3,4}$ to turn green to right arrow at time t; which is translated into the EVALPSN,

$$S_6(t) : [(2,0), \alpha] \wedge$$
$$T_{1,2}(r,t) : [(2,0), \alpha] \wedge T_{3,4}(g,t) : [(2,0), \alpha] \wedge$$
$$\sim MIN_{3,4}(g,t) : [(2,0), \alpha] \wedge \tag{11.5}$$
$$\sim S_5(t) : [(2,0), \alpha] \wedge \sim S_7(t) : [(2,0), \alpha]$$
$$\rightarrow T_{3,4}(a,t) : [(0,1), \gamma],$$

6. If traffic sensor S_8 detects traffic amount, it has already passed the minimum green duration of traffic lights $T_{3,4}$, and neither traffic sensors S_5 nor S_7 detect traffic amount at time t, then it is weakly permitted for traffic lights $T_{3,4}$ to turn green to right arrow at time t; which is translated into the EVALPSN,

$$
\begin{aligned}
&S_6(t) : [(2,0),\alpha] \wedge \\
&T_{1,2}(r,t) : [(2,0),\alpha] \wedge T_{3,4}(g,t) : [(2,0),\alpha] \wedge \\
&\sim MIN_{3,4}(g,t) : [(2,0),\alpha] \wedge \\
&\sim S_5(t) : [(2,0),\alpha] \wedge \sim S_7(t) : [(2,0),\alpha] \\
&\rightarrow T_{3,4}(a,t) : [(0,1),\gamma],
\end{aligned}
\tag{11.6}
$$

7. If traffic sensor S_5 detects traffic amount and it has not passed the maximum green duration of traffic lights $T_{3,4}$ yet, then it is weakly forbidden for traffic lights $T_{3,4}$ to turn green to right arrow at time t; which is translated into the EVALPSN,

$$
\begin{aligned}
&S_5(t) : [(2,0),\alpha] \wedge \\
&T_{1,2}(r,t) : [(2,0),\alpha] \wedge T_{3,4}(g,t) : [(2,0),\alpha] \wedge \\
&\sim MAX_{3,4}(g,t) : [(2,0),\alpha] \\
&\rightarrow T_{3,4}(a,t) : [(0,1),\beta],
\end{aligned}
\tag{11.7}
$$

8. If traffic sensor S_7 detects traffic amount and it has not passed the maximum green duration of traffic lights $T_{3,4}$, then it is weakly forbidden for traffic lights $T_{3,4}$ to turn green to right arrow at time t; which is translated into the EVALPSN,

$$
\begin{aligned}
&S_7(t) : [(2,0),\alpha] \wedge \\
&T_{1,2}(r,t) : [(2,0),\alpha] \wedge T_{3,4}(g,t) : [(2,0),\alpha] \wedge \\
&\sim MAX_{3,4}(g,t) : [(2,0),\alpha] \\
&\rightarrow T_{3,4}(a,t) : [(0,1),\beta],
\end{aligned}
\tag{11.8}
$$

[Green light length rules for the traffic lights $T_{3,4}$]
9. If traffic lights $T_{3,4}$ are green and it has not passed the minimum duration of them yet, then it is strongly forbidden for traffic lights $T_{3,4}$ to turn green to right arrow at time t; which is translated into the EVALPSN,

$$
\begin{aligned}
&T_{3,4}(g,t) : [(2,0),\alpha] \wedge MIN_{3,4}(g,t) : [(2,0),\alpha] \\
&\rightarrow T_{3,4}(a,t) : [(0,2),\beta],
\end{aligned}
\tag{11.9}
$$

10. If traffic lights $T_{3,4}$ are green and it has already passed the maximum duration of them, then it is strongly permitted for traffic lights $T_{3,4}$ to turn green to right arrow at time t; which is translated into the EVALPSN,

$$T_{3,4}(g,t) : [(2,0), \alpha] \wedge MAX_{3,4}(g,t) : [(2,0), \alpha]$$
$$\rightarrow T_{3,4}(a,t) : [(0,2), \gamma], \tag{11.10}$$

[Deontic deasible reasoning rules]

11. If traffic lights $T_{3,4}$ are green, it is weakly permitted at least for traffic lights $T_{3,4}$ to turn green to right arrow at time t, then it is strongly obligatory for traffic lights $T_{3,4}$ to turn green to right arrow at time $t + 1$ (at the next step); which is translated into the EVALPSN,

$$T_{3,4}(g,t) : [(2,0), \alpha] \wedge T_{3,4}(a,t) : [(0,1), \gamma]$$
$$\rightarrow T_{3,4}(a,t+1) : [(2,0), \beta], \tag{11.11}$$

12. If traffic lights $T_{3,4}$ are green, it is weakly forbidden at least for traffic lights $T_{3,4}$ to turn green to right arrow at time t, then it is strongly obligatory for traffic lights $T_{3,4}$ not to turn green to right arrow at time $t + 1$ (at the next step); which is translated into the EVALPSN,

$$T_{3,4}(g,t) : [(2,0), \alpha] \wedge T_{3,4}(a,t) : [(0,1), \beta]$$
$$\rightarrow T_{3,4}(g,t+1) : [(2,0), \beta]. \tag{11.12}$$

11.3.2 Example and Simulation

Let us introduce a simple example of the EVALPSN based traffic signal control. We assume the same intersection in the previous section.

Example 11.1 Suppose that traffic lights $T_{1,2}$ are red and traffic lights $T_{3,4}$ are green. We also suppose that the minimum duration of green signal has already passed but the maximum one has not passed yet. Then, we obtain the EVALPSN,

$$T_{1,2}(r,t) : [(2,0), \alpha] \wedge T_{3,4}(g,t) : [(2,0), \alpha], \tag{11.13}$$

$$\sim MIN_{3,4}(g,t) : [(2,0), \alpha], \tag{11.14}$$

$$\sim MAX_{3,4}(g,t) : [(2,0), \alpha], \tag{11.15}$$

If traffic sensors $S_{1,3,5}$ detect traffic amount and traffic sensors $S_{2,4,6,7,8}$ do not detect traffic amount at time t, we obtain the EVALPSN,

$$S_1(t) : [(2,0), \alpha], \tag{11.16}$$

$$S_3(t) : [(2,0), \alpha], \tag{11.17}$$

$$S_5(t) : [(2,0), \alpha], \tag{11.18}$$

$$\sim S_2(t) : [(2,0), \alpha], \tag{11.19}$$

$$\sim S_4(t) : [(2,0), \alpha], \tag{11.20}$$

$$\sim S_6(t) : [(2,0), \alpha], \tag{11.21}$$

$$\sim S_7(t) : [(2,0), \alpha], \tag{11.22}$$

$$\sim S_8(t) : [(2,0), \alpha]. \tag{11.23}$$

Then, by EVALPSN clauses (11.13), (11.18), (11.15) and (11.7), the forbiddance from traffic lights $T_{3,4}$ turning to right arrow,

$$T_{3,4}(a,t) : [(0,1), \beta] \tag{11.24}$$

is derived, furthermore, by EVALPSN clauses (11.13), (11.24) and (11.12), the obligation for traffic lights $T_{3,4}$ keeping green at time $t + 1$,

$$T_{3,4}(g, t + 1) : [(2,0), \beta]$$

is obtained.

On the other hand, if traffic sensors $S_{1,3}$ detect traffic amount and traffic sensors $S_{2,4,5,6,7,8}$ do not detect traffic amount at time t, we obtain the EVALPSN,

$$S_1(t) : [(2,0), \alpha], \tag{11.25}$$

$$S_3(t) : [(2,0), \alpha], \tag{11.26}$$

$$\sim S_2(t) : [(2,0), \alpha], \tag{11.27}$$

$$\sim S_4(t) : [(2,0), \alpha], \tag{11.28}$$

$$\sim S_5(t) : [(2,0), \alpha], \tag{11.29}$$

$$\sim S_6(t) : [(2,0), \alpha], \tag{11.30}$$

$$\sim S_7(t) : [(2,0), \alpha], \tag{11.31}$$

$$\sim S_8(t) : [(2,0), \alpha]. \tag{11.32}$$

Then, by EVALPSN clauses (11.13), (11.25), (11.29), (11.31), (11.14) and (11.1), the permission for traffic lights $T_{3,4}$ turning to right arrow,

$$T_{3,4}(a,t) : [(0,1), \gamma] \tag{11.33}$$

is derived, furthermore, by EVALPSN clauses (11.13), (11.33) and (11.11), the obligation for traffic lights $T_{3,4}$ turning to right arrow at time $t + 1$,

$$T_{3,4}(a, t+1) : [(2,0), \beta]$$

is finally obtained.

Here we introduce an EVALPSN traffic control simulation system based on the cellular automaton method and its simulation results comparing to ordinary fixed-time traffic signal control. In order to evaluate the simulation results we define the concepts "step", "move times", and "stop times" as follows:

step: a time unit in the simulation system, which is a transit time that one car moves from its current cell to the next cell,

move times: shows the times that one car moves from its current cell to the next cell without stop,

stop times: shows the times that one car stops during transition from one cell to another cell,

We introduce the simulation results under the following two traffic flow conditions.

[Condition 11.1] Cars are supposed to flow into the intersection from each road with the same probabilities, right-turn 5 %, left-turn 5 % and straight 20 %. It is supposed that green signal duration is 30 steps, yellow one is 3 steps, right-arrow one is 4 steps and red one is 40 steps in the fixed-time traffic signal control. It is also supposed that green signal duration is between 14 and 30 steps in the EVALPSN traffic signal control.

[Condition 11.2] Cars are supposed to flow into the intersection with the following probabilities,

from South: right-turn 5 %, left-turn 15 % and straight 10 %;
from North: right-turn 15 %, left-turn 5 % and straight 10 %;
from West: right-turn, left-turn and straight 5 % each;
from East: right-turn and left-turn 5 % each, and straight 15 %.

Other conditions are the same as the Condition 11.1.

We measured the numbers of car stop and move times during 1000 steps, and repeated it 10 times under the same conditions. The average numbers of car stop and move times are listed in Table 11.1. The simulation results show that the number of car move times in the EVALPSN traffic signal control is larger than that in the fixed-time traffic signal control, and the number of car stop times in the EVALPSN traffic signal control is smaller than that in the fixed time one. Taking the simulation results into account, it could be concluded that the EVALPSN traffic

Table 11.1 Simulation results

	Fixed-time control		EVALPSN control	
	Stop times	Move times	Stop times	Move times
Condition 11.1	17,690	19,641	16,285	23,151
Condition 11.2	16,764	18,664	12,738	20,121

signal control is more efficient for relieving traffic congestion than the fixed-time traffic signal control.

11.4 EVALPSN Safety Verification for Pipeline Control

This section introduces EVALPSN based safety verification for pipeline valve control with a simple brewery pipeline example.

11.4.1 Pipeline Network

The pipeline network described in Fig. 11.4 is taken as an example for the brewery pipeline valve control based on EVALPSN safety verification. In Fig. 11.4, the arrows represent the directions of liquid flows, home-plate pentagons show brewery tanks, and cross figures do valves. In the pipeline network, we suppose physical entities:

Fig. 11.4 Pipeline example

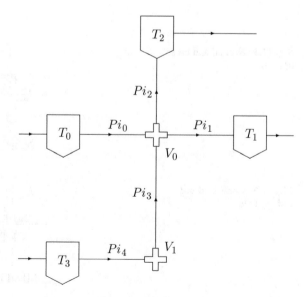

- four tanks, T_0, T_1, T_2, and T_3;
- five pipes, Pi_0, Pi_1, Pi_2, Pi_3, and Pi_4;
  (a pipe includes neither valves nor tanks)
- two valves, V_0, and V_1;

and logical entities that we suppose :

- four processes, Pr_0, Pr_1, Pr_2, and Pr_3;
  (a process is defined as a set of sub-processes and valves)
- five sub-processes, $SPr_0, SPr_1, SPr_2, SPr_3$, and SPr_4.

For example, process Pr_0 consists of sub-processes SPr_0, SPr_1 and valve V_0. Each entity is supposed to have logical or physical states. Sub-processes have two states *locked*(l) and *free*(f), then "the sub-process is locked" means that the sub-process is supposed to be locked(logically reserved) by one sort of liquid, and "free" means unlocked. Processes have two states *set*(s) and *unset*(xs), then "the process is set" means that all the sub-processes in the process are locked, and "unset" means not set.

Here we also assume that valves in the network can control two liquid flows in the normal and cross directions as shown in Fig. 11.5. Valves have two controlled states, *controlled mix*(cm) representing that the valve is controlled to mix the liquid flows in the normal and cross directions, and *controlled separate*(cs) representing that the valve is controlled to separate the liquid flow in the normal and cross directions as shown in Fig. 11.6. We suppose that there are five sorts of cleaning liquid:

cold water(cw), warm water(ww), hot water(hw),

nitric acid(na) and caustic soda(cs)

Fig. 11.5 Normal and cross directions

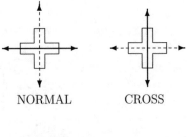

NORMAL   CROSS

Fig. 11.6 Controlled mix and separate

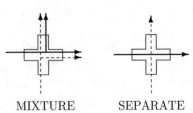

MIXTURE   SEPARATE

We also consider the following four brewery and cleaning processes in the pipeline network :

- Process Pr_0 a beer process,

$$T_0 \rightarrow V_0(\text{cs}) \rightarrow T_1$$

- Process Pr_1 a cleaning process with nitric acid,

$$
\begin{array}{c}
T_2 \\
\uparrow \\
V_0(\text{cs}) \\
\uparrow \\
T_3 \rightarrow V_1(\text{cm})
\end{array}
$$

- Process Pr_2 a cleaning process with cold water,

$$
\begin{array}{c}
T_2 \\
\uparrow \\
V_0(\text{cs}) \\
\uparrow \\
T_3 \rightarrow V_1(\text{cm})
\end{array}
$$

- Process Pr_3 a brewery mixing process,

$$
\begin{array}{c}
T_2 \\
\uparrow \\
T_0 \rightarrow V_0(\text{cm}) \rightarrow T_1 \\
\uparrow \\
T_3 \rightarrow V_1(\text{cm})
\end{array}
$$

In order to verify the safety for the above processes, the pipeline controller issues a process request consisting of if-part and then-part before starting the process. The if-part describes the current state of the pipelines that should be used in the process, and the then-part describes the permission for setting the process. For example, the process request for process Pr_1 is described as:

> if sub-process SPr_0 is free,
> sub-process SPr_1 is free,
> valve V_0 is physically controlled separate and free,
> then process Pr_0 can be set?

Fig. 11.7 Process schedule
chart

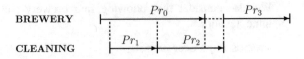

We also suppose the following process schedule for processes $Pr_{0,1,2,3}$:

- PRS-0 process Pr_0 starts before any other processes;
- PRS-1 process Pr_1 starts immediately after process Pr_0;
- PRS-2 process Pr_2 starts immediately after process Pr_1;
- PRS-3 process Pr_3 starts immediately after processes Pr_0 and Pr_2,

which are charted in Fig. 11.7.

11.4.2 Pipeline Safety Property

We introduce the safety properties, **SPr** for sub-processes, **Val** for valves, and **Pr** for processes, for assuring the pipeline valve control safety, which can avoid unexpected mix of different sorts of liquid in the valve.

- **SPr**: it is a forbidden case that the sub-process over a given pipe is simultaneously locked by different sorts of liquid.
- **Val**: it is a forbidden case that valves are controlled for unexpected mix of liquid.
- **Pr**: whenever a process is set, all its component sub-processes are locked and all its component valves are controlled consistently.

11.4.3 Predicates for Safety Verification

The EVALPSN based safety verification is carried out by verifying whether process start requests by pipeline operators contradict the safety properties or not in EVALPSN programming. Then the following three steps 1, 2 and 3 have to be executed:

1. the safety properties for the pipeline network, which should be insured when the pipeline network is locked, and some control methods for the network are translated into EVALPSN clauses, and they have to be stored as EVALPSN P_{sc};
2. the if-part of the process request that is the current state of the pipeline and the then-part of the process request that is supposed to be verified are translated into EVALP clauses as EVALPs P_i and P_t, respectively;

Fig. 11.8 EVALPSN based
safety verification

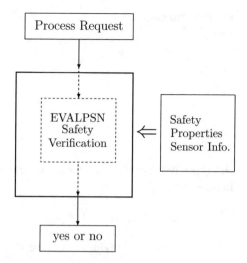

3. EVALP P_t is inquired from EVALPSN $\{P_{sc} \cup P_i\}$, then if *yes* is returned, the
   safety for the request is assured and the defeasible deontic reasoning is performed, otherwise, it is not assured; which is described in Fig. 11.8.

   In order to verify the safety for the pipeline network, the following predicates are
used in EVALPSN.

- $Pr(i, l)$ represents that the process i for the liquid l is set(s) or unset(xs), where

$$i \in \{p0, p1, p2, p3\}$$

is a process id corresponding to processes $Pr_{0,1,2,3}$, respectively,

$$l \in \{b, cw, ww, hw, na, cs\}$$

is a liquid sort, and we have the EVALP clause,

$$Pr(i, l) : [\mu_1, \mu_2],$$

where

$$\mu_1 \in \mathcal{T}_{v_1} = \{\perp_1, \mathsf{s}, \mathsf{xs}, \top_1\},$$
$$\mu_2 \in \mathcal{T}_d = \{\perp, \alpha, \beta, \gamma, *_1, *_2, *_3, \top\}.$$

The complete lattice $\mathcal{T}_{v_1}$ is a variant of the complete lattice $\mathcal{T}_v(1)$ in Fig. 11.9.
Therefore annotations $\perp_1, \mathsf{s}, \mathsf{xs}$ and $\top_1$ are for vector annotations (0,0), (1,0), (0,1)

Fig. 11.9 The complete lattice $\mathcal{T}_v(1)$

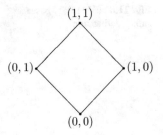

and (1,1), respectively. The epistemic negation $\neg_1$ over $\mathcal{T}_{v_1}$ is defined as the following mapping:

$$\neg_1([\bot_1, \mu_2]) = [\bot_1, \mu_2], \quad \neg_1([\mathbf{s}, \mu_2]) = [\mathbf{xs}, \mu_2],$$

$$\neg_1([\top_1, \mu_2]) = [\top_1, \mu_2], \quad \neg_1([\mathbf{xs}, \mu_2]) = [\mathbf{s}, \mu_2].$$

For example, EVALP clause $Pr(p2, b) : [\mathbf{s}, \alpha]$ can be intuitively interpreted as "it is a fact that the beer process Pr_2 is set".

• $SPr(i, j, l)$ represents that the sub-process from valve(or tank) i to valve(or tank) j occupied by liquid l is locked(1) or free(f). Moreover, if a sub-process is free, the liquid sort in the pipe is not cared, and the liquid is represented by the symbol "0" (zero). Therefore we suppose that

$$l \in \{b, cw, ww, hw, na, cs, 0\}$$

and

$$i, j \in \{v0, v1, t0, t1, t2, t3\}$$

are valve and tank ids corresponding to valves $V_{0,1}$, and tanks $T_{0,1,2,3}$. Then we have the following EVALP clause for representing sub-process states:

$$SPr(i, j, l) : [\mu_1, \mu_2],$$

where

$$\mu_1 \in \mathcal{T}_{v_2} = \{\bot_2, 1, \mathbf{f}, \top_2\},$$
$$\mu_2 \in \mathcal{T}_d = \{\bot, \alpha, \beta, \gamma, *_1, *_2, *_3, \top\}.$$

The complete lattice $\mathcal{T}_{v_2}$ is a variant of the complete lattice $\mathcal{T}_v(1)$ in Fig. 11.9. Therefore annotations $\bot_2, 1, \mathbf{f}$ and $\top_2$ are for vector annotations (0,0), (1,0), (0,1) and (1,1), respectively. The epistemic negation $\neg_1$ over T_{v_2} is defined as the following mapping:

$$\neg_1([\perp_2, \mu_2]) = [\perp_2, \mu_2], \quad \neg_1([1, \mu_2]) = [f, \mu_2],$$
$$\neg_1([\top_2, \mu_2]) = [\top_2, \mu_2], \quad \neg_1([f, \mu_2]) = [1, \mu_2].$$

For example, EVALP clause $SPr(v0, t1, b) : [f, \gamma]$ can be intuitively interpreted as "the sub-process from valve V_0 to tank T_1 is permitted to be locked by the beer process".

• $Val(i, l_n, l_c)$ represents that valve i occupied by the two sorts of liquid $l_n, l_c \in \{b, cw, ww, hw, na, cs, 0\}$ is controlled separate(cs) or mix(cm), where $i \in \{v0, v1\}$ is a valve id. We suppose that there are two directed liquid flows in the *normal* and *cross* directions in valves as shown in Fig. 11.5. Therefore, the second argument l_n represents the liquid flowing in the normal direction and the third argument l_c represents the liquid flowing in the cross direction. Generally, if a valve is released from the locked(controlled) state, the liquid flow in the valve is represented by the symbol 0 that means "free". Then we have the following EVALP clause for representing valve states,

$$Val(i, l_n, l_c) : [\mu_1, \mu_2],$$

where

$$\mu_1 \in T_{v_3} = \{\perp_3, cm, cs, \top_3\},$$
$$\mu_2 \in T_d = \{\perp, \alpha, \beta, \gamma, *_1, *_2, *_3, \top\},$$

The complete lattice T_{v_3} is a variant of the complete lattice $T_v(1)$ in Fig. 11.9. Therefore annotations $\perp_3$, cm, cs and $\top_3$ are for vector annotations (0,0), (1,0), (0,1) and (1,1), respectively. The epistemic negation $\neg_1$ over T_{v_3} is defined as the following mapping:

$$\neg_1([\perp_3, \mu_2]) = [\perp_3, \mu_2], \quad \neg_1([cs, \mu_2]) = [cm, \mu_2],$$
$$\neg_1([\top_3, \mu_2]) = [\top_3, \mu_2], \quad \neg_1([cm, \mu_2]) = [cs, \mu_2].$$

We suppose that if a process finishes, all valves included in the process are controlled separate(closed). For example, EVALP clause $Val(v0, 0, 0) : [cs, \alpha]$ can be intuitively interpreted as "valve V_0 has been released from controlled separate state"; EVALP clause $Val(v0, b, cw) : [cs, \beta]$ can be intuitively interpreted as both "it is forbidden for valve V_0 is controlled mix with beer b in the normal direction and cold water cw in the cross direction", and "it is obligatory for valve V_0 to be controlled separate with beer b in the normal direction and cold water cw in the cross direction"; and EVALP clause $Val(v0, 0, b) : [cs, \alpha]$ can be intuitively interpreted as "it is a fact that valve V_0 is controlled separate with the free flow 0 in the normal direction and beer b in the cross direction.

• $Eql(l_1, l_2)$ represents that liquids l_1 and l_2 are the same(sa) or different(di), where

$$l_1, l_2 \in \{b, cw, ww, hw, na, cs, 0\}.$$

We have the following EVALP clause for distinguishing liquid:

$$Eql(l_1, l_2) : [\mu_1, \mu_2],$$

where

$$\mu_1 \in \mathcal{T}_{v_4} = \{\perp_4, \mathtt{sa}, \mathtt{di}, \top_4\},$$
$$\mu_2 \in \mathcal{T}_d = \{\perp, \alpha, \beta, \gamma, *_1, *_2, *_3, \top\}.$$

The complete lattice $\mathcal{T}_{v_4}$ is a variant of the complete lattice $\mathcal{T}_v(1)$ in Fig. 11.9. Therefore annotations $\perp_4, \mathtt{sa}, \mathtt{di}$ and $\top_4$ are for vector annotations $(0,0)$, $(1,0)$, $(0,1)$ and $(1,1)$, respectively. The epistemic negation $\neg_1$ is defined as the following mapping:

$$\neg_1([\perp_4, \mu_2]) = [\perp_4, \mu_2], \quad \neg_1([\mathtt{di}, \mu_2]) = [\mathtt{sa}, \mu_2],$$
$$\neg_1([\top_4, \mu_2]) = [\top_4, \mu_2], \quad \neg_1([\mathtt{sa}, \mu_2]) = [\mathtt{di}, \mu_2].$$

Now we consider process release conditions when processes have finished and define some more predicates to represent the conditions. We suppose that if the terminal tank T_i of process Pr_j is filled with one sort of liquid, the finish signal $Fin(pj)$ of process Pr_j is issued.

• $Tan(ti, l)$ represents that tank T_i has been filled fully(fu) with liquid l or empty (em). Then we have the following EVALP clause for representing tank states:

$$Tan(ti, l) : [\mu_1, \mu_2],$$

where $i \in \{0, 1, 2, 3\}$ $l \in \{b, cw, ww, hw, na, cs, 0\}$

$$\mu_1 \in \mathcal{T}_{v_5} = \{\perp_5, \mathtt{fu}, \mathtt{em}, \top_5\},$$
$$\mu_2 \in \mathcal{T}_d = \{\perp, \alpha, \beta, \gamma, *_1, *_2, *_3, \top\}.$$

The complete lattice $\mathcal{T}_{v_5}$ is a variant of the complete lattice $\mathcal{T}_v(1)$ in Fig. 11.9. Therefore annotations $\perp_5, \mathtt{fu}, \mathtt{em}$ and $\top_5$ are for vector annotations $(0,0)$, $(1,0), (0,1)$ and $(1,1)$, respectively. The epistemic negation $\neg_1$ over $\mathcal{T}_{v_5}$ is defined as the following mapping:

$$\neg_1([\bot_5, \mu_2]) = [\bot_5, \mu_2], \quad \neg_1([\texttt{fu}, \mu_2]) = [\texttt{em}, \mu_2],$$
$$\neg_1([\top_5, \mu_2]) = [\top_5, \mu_2], \quad \neg_1([\texttt{em}, \mu_2]) = [\texttt{fu}, \mu_2].$$

Note that annotation $\bot_5$ can be intuitively interpreted to represent "filled with some amount of liquid but not fully", that is to say, "no information in terms of fullness". For example, EVALP clause $Tan(t2, 0) : [\texttt{em}, \alpha]$ can be interpreted as "it is a fact that tank T_2 is empty".

- $Str(pi)$ represents that the start signal for process Pr_i is issued (is) or not (ni).
- $Fin(pj)$ represents that the finish signal for process Pr_j has been issued (is) or not (ni). Then we have the following EVALP clauses for representing start/finish information:

$$Str(pi) : [\mu_1, \mu_2], \quad Fin(pi) : [\mu_1, \mu_2],$$

where $i, j \in \{0, 1, 2, 3\}$,

$$\mu_1 \in \mathcal{T}_{v_6} = \{\bot_6, \texttt{ni}, \texttt{is}, \top_6\},$$

$$\mu_2 \in \mathcal{T}_d = \{\bot, \alpha, \beta, \gamma, *_1, *_2, *_3, \top\}.$$

The complete lattice $\mathcal{T}_{v_6}$ is a variant of the complete lattice $\mathcal{T}_v(1)$ in Fig. 11.9. Therefore annotations $\bot_6, \texttt{is}, \texttt{ni}$ and $\top_6$ are for vector annotations (0,0), (1,0), (0,1) and (1,1), respectively. The epistemic negation $\neg_1$ over $\mathcal{T}_{v_6}$ is defined as the following mapping:

$$\neg_1([\bot_6, \mu_2]) = [\bot_6, \mu_2], \quad \neg_1([\texttt{is}, \mu_2]) = [\texttt{ni}, \mu_2],$$

$$\neg_1([\top_6, \mu_2]) = [\top_6, \mu_2], \quad \neg_1([\texttt{ni}, \mu_2]) = [\texttt{is}, \mu_2].$$

For example, EVALP clause $Fin(p3) : [\texttt{ni}, \alpha]$ can be interpreted as "it is a fact that the finish signal for process Pr_3 has not been issued yet".

11.4.4   Safety Property in EVALPSN

Here, we provide the formalization of all safety properties **SPr**, **Val** and **Pr** in EVALPSN.

SPr can be intuitively interpreted as derivation rules of forbiddance. If a sub-process from valve(or tank) i to valve(or tank) j is locked by one sort of liquid, it is forbidden for the sub-process to be locked by different sorts of liquid simultaneously. Thus, we have the following EVALPSN clause for representing such forbiddance for sub-processes:

$$SPr(i,j,l_1) : [l, \alpha] \wedge : Eql(l_1, l_2) : [\text{sa}, \alpha]$$
$$\rightarrow SPr(i,j,l_2) : [\text{f}, \beta], \tag{11.34}$$

where $l_1, l_2 \in \{b, cw, ww, hw, na, cs\}$. Moreover, in order to derive permission for locking sub-processes we need the following EVALPSN clause:

$$\sim SPr(i,j,l) : [\text{f}, \beta] \rightarrow SPr(i,j,l) : [\text{f}, \gamma] \tag{11.35}$$

where $l \in \{b, cw, ww, hw, na, cs\}$.

Val also can be intuitively interpreted as derivation rules of forbiddance. We have to consider two cases: one is for deriving the forbiddance from changing the control state of the valve, and another one is for deriving the forbiddance from mixing different sorts of liquid without changing the control state of the valve.

Case 1 If a valve is controlled separate, it is forbidden for the valve to be controlled mix, conversely, if a valve is controlled mixture, it is forbidden for the valve to be controlled separate. Thus, generally we have the following EVALPSN clauses:

$$Val(i, l_n, l_c) : [\text{cs}, \alpha] \wedge \sim Eql(l_n, 0) : [\text{sa}, \alpha] \wedge$$
$$\sim Eql(l_c, 0) : [\text{sa}, \alpha] \rightarrow Val(i, l_n, l_c) : [\text{cs}, \beta], \tag{11.36}$$

$$Val(i, l_n, l_c) : [\text{cm}, \alpha] \wedge \sim Eql(l_n, 0) : [\text{sa}, \alpha] \wedge$$
$$\sim Eql(l_c, 0) : [\text{sa}, \alpha] \rightarrow Val(i, l_n, l_c) : [\text{cm}, \beta], \tag{11.37}$$

where $l_n, l_c \in \{b, cw, ww, hw, na, cs, 0\}$.

Case 2 In this case, we consider another forbiddance derivation case in which different sorts of liquid are mixed even if the valve control state is not changed. We have the following EVALPSN clauses:

$$Val(i, l_{n_1}, l_{c_1}) : [\text{cs}, \alpha] \wedge \sim Eql(l_{n_1}, l_{n_2}) : [\text{sa}, \alpha] \wedge$$
$$\sim Eql(l_{n_1}, 0) : [\text{sa}, \alpha] \rightarrow Val(i, l_{n_2}, l_{c_2}) : [\text{cm}, \beta], \tag{11.38}$$

$$Val(i, l_{n_1}, l_{c_1}) : [\text{cs}, \alpha] \wedge \sim Eql(l_{c_1}, l_{c_2}) : [\text{sa}, \alpha] \wedge$$
$$\sim Eql(l_{c_1}, 0) : [\text{sa}, \alpha] \rightarrow Val(i, l_{n_2}, l_{c_2}) : [\text{cm}, \beta], \tag{11.39}$$

$$Val(i, l_{n_1}, l_{c_1}) : [\text{cm}, \alpha] \wedge \sim Eql(l_{n_1}, l_{n_2}) : [\text{sa}, \alpha]$$
$$\rightarrow Val(i, l_{n_2}, l_{c_2}) : [\text{cs}, \beta], \tag{11.40}$$

$$Val(i, l_{n_1}, l_{c_1}) : [\text{cm}, \alpha] \wedge \sim Eql(l_{c_1}, l_{c_2}) : [\text{sa}, \alpha]$$
$$\rightarrow Val(i, l_{n_2}, l_{c_2}) : [\text{cs}, \beta], \tag{11.41}$$

where $l_{n_1}, l_{c_1} \in \{b, cw, ww, hw, na, cs, 0\}$ and $l_{n_2}, l_{c_2} \in \{b, cw, ww, hw, na, cs\}$.

Here note that EVALPSN clause $\sim Eql(l_n, 0) : [\text{sa}, \alpha]$ shows that there does not exist information such that the normal direction with liquid l_n in the valve is free (not controlled). As well as the case of sub-processes, in order to derive permission for controlling valves, we need the following EVALPSN clauses:

$$\sim Val(i, l_n, l_c) : [\text{cm}, \beta] \rightarrow Val(i, l_n, l_c) : [\text{cm}, \gamma], \tag{11.42}$$

$$\sim Val(i, l_n, l_c) : [\text{cs}, \beta] \rightarrow Val(i, l_n, l_c) : [\text{cs}, \gamma], \tag{11.43}$$

where $l_n, l_c \in \{b, cw, ww, hw, na, cs, 0\}$.

Pr can be intuitively interpreted as derivation rules of permission and directly translated into EVALPSN clauses as if-then rules "if all the components of the process can be locked or controlled consistently, then the process can be set". For example, if the beer process Pr_0 consists of the sub-process from tank T_0 to valve V_0, valve V_0 with controlled separate by beer in the normal direction, and the sub-process from valve V_0 to tank T_1, then we have the following EVALP clause to derive the permission for setting process Pr_0.

ProcessPr$_0$
$$\begin{aligned} &SPr(t0, v0, b) : [\text{f}, \gamma] \wedge SPr(v0, t1, b) : [\text{f}, \gamma] \wedge \\ &Val(v0, b, l) : [\text{cm}, \gamma] \wedge Tan(t0, b) : [\text{fu}, \alpha] \wedge \\ &Tan(t1, 0) : [\text{em}, \alpha] \rightarrow Pr(p0, b) : [\text{xs}, \gamma]. \end{aligned} \tag{11.44}$$

We also have the following EVALP clauses for setting the other processes.

ProcessPr$_1$
$$\begin{aligned} &SPr(t3, v1, na) : [\text{f}, \gamma] \wedge SPr(v1, v0, na) : [\text{f}, \gamma] \wedge \\ &SPr(v0, t2, na) : [\text{f}, \gamma] \wedge Val(v0, l, na) : [\text{cm}, \gamma] \wedge \\ &Val(v1, na, 0) : [\text{cs}, \gamma] \wedge Tan(t3, na) : [\text{fu}, \alpha] \wedge \\ &Tan(t2, 0) : [\text{em}, \alpha] \rightarrow Pr(p1, na) : [\text{xs}, \gamma]. \end{aligned} \tag{11.45}$$

ProcessPr$_2$
$$\begin{aligned} &SPr(t3, v1, cw) : [\text{f}, \gamma] \wedge SPr(v1, v0, cw) : [\text{f}, \gamma] \wedge \\ &SPr(v0, t2, cw) : [\text{f}, \gamma] \wedge Val(v0, l, cw) : [\text{cm}, \gamma] \wedge \\ &Val(v1, cw, 0) : [\text{cs}, \gamma] \wedge Tan(t3, cw) : [\text{fu}, \alpha] \wedge \\ &Tan(t2, 0) : [\text{em}, \alpha] \rightarrow Pr(p2, cw) : [\text{xs}, \gamma]. \end{aligned} \tag{11.46}$$

ProcessPr$_3$

$$SPr(t0, v0, b) : [\texttt{f}, \gamma] \wedge SPr(t3, v1, b) : [\texttt{f}, \gamma] \wedge$$
$$SPr(v0, t1, b) : [\texttt{f}, \gamma] \wedge SPr(v0, t2, b) : [\texttt{f}, \gamma] \wedge$$
$$SPr(v1, v0, b) : [\texttt{f}, \gamma] \wedge Val(v0, b, b) : [\texttt{cs}, \gamma] \wedge \qquad (11.47)$$
$$Val(v1, b, 0) : [\texttt{cs}, \gamma] \wedge Tan(t0, b) : [\texttt{fu}, \alpha] \wedge$$
$$Tan(t1, 0) : [\texttt{em}, \alpha] \wedge Tan(t3, b) : [\texttt{fu}, \alpha] \wedge$$
$$Tan(t2, 0) : [\texttt{em}, \alpha] \rightarrow Pr(p3, b) : [\texttt{xs}, \gamma].$$

We suppose that $l \in \{b, cw, ww, hw, na, cs, 0\}$ in the above safety verification EVALPSN clauses for processes $Pr_{0,1,2,3}$.

11.4.5 Process Release Control in EVALPSN

Here we consider conditions for releasing process lock after the process has finished. For example, a process release condition can be expressed by "liquid l in tank T_j has been transferred into tank T_k in process Pr_i after process Pr_i has started, and the finish signal $Fin(pi)$ for process Pr_i is obtained". If the above condition is satisfied, the locked process Pr_i is allowed to be unset, and each component of the process is also allowed to be free. The release conditions for processes $Pr_{0,1,2,3}$ are formalized in the following EVALPSN clauses.

Process Pr_0

$$Str(p0) : [\texttt{is}, \alpha] \wedge Tan(t0, b) : [\texttt{em}, \alpha] \wedge$$
$$Tan(t1, b) : [\texttt{fu}, \alpha] \wedge Fin(p0) : [\texttt{is}, \alpha] \rightarrow \qquad (11.48)$$
$$Pr(p0, b) : [\texttt{s}, \gamma],$$

$$Pr(p0, b) : [\texttt{s}, \gamma] \rightarrow SPr(t0, v0, 0) : [\texttt{l}, \gamma], \qquad (11.49)$$

$$Pr(p0, b) : [\texttt{s}, \gamma] \rightarrow SPr(v0, t1, 0) : [\texttt{l}, \gamma], \qquad (11.50)$$

$$Pr(p0, b) : [\texttt{s}, \gamma] \rightarrow Val(v0, 0, l) : [\texttt{cm}, \gamma]. \qquad (11.51)$$

Process Pr_1

$$Str(p1) : [\texttt{is}, \alpha] \wedge Tan(t3, na) : [\texttt{em}, \alpha] \wedge$$
$$Tan(t2, na) : [\texttt{fu}, \alpha] \wedge Fin(p1) : [\texttt{is}, \alpha] \rightarrow \qquad (11.52)$$
$$Pr(p1, na) : [\texttt{s}, \gamma],$$

$$Pr(p1,na):[\mathtt{s},\gamma] \to SPr(v0,t2,0):[\mathtt{1},\gamma], \tag{11.53}$$

$$Pr(p1,na):[\mathtt{s},\gamma] \to SPr(v1,v0,0):[\mathtt{1},\gamma], \tag{11.54}$$

$$Pr(p1,na):[\mathtt{s},\gamma] \to SPr(t3,v1,0):[\mathtt{1},\gamma], \tag{11.55}$$

$$Pr(p1,na):[\mathtt{s},\gamma] \to Val(v0,l,0):[\mathtt{cm},\gamma], \tag{11.56}$$

$$Pr(p1,na):[\mathtt{s},\gamma] \to Val(v1,0,0):[\mathtt{cm}\gamma]. \tag{11.57}$$

Process Pr_2

$$\begin{aligned} &Str(p2):[\mathtt{is},\alpha] \wedge Tan(t3,cw):[\mathtt{em},\alpha]\wedge\\ &Tan(t2,cw):[\mathtt{fu},\alpha] \wedge Fin(p2):[\mathtt{is},\alpha] \to\\ &Pr(p2,cw):[\mathtt{s},\gamma], \end{aligned} \tag{11.58}$$

$$Pr(p2,cw):[\mathtt{s},\gamma] \to SPr(v0,t2,0):[\mathtt{1},\gamma], \tag{11.59}$$

$$Pr(p2,cw):[\mathtt{s},\gamma] \to SPr(v1,v0,0):[\mathtt{1},\gamma], \tag{11.60}$$

$$Pr(p2,cw):[\mathtt{s},\gamma] \to SPr(t3,v1,0):[\mathtt{1},\gamma], \tag{11.61}$$

$$Pr(p2,cw):[\mathtt{s},\gamma] \to Val(v0,l,0):[\mathtt{cm},\gamma], \tag{11.62}$$

$$Pr(p2,cw):[\mathtt{s},\gamma] \to Val(v1,0,0):[\mathtt{cm},\gamma]. \tag{11.63}$$

Process Pr_3

$$\begin{aligned} &Str(p3):[\mathtt{is},\alpha] \wedge Tan(t0,b):[\mathtt{em},\alpha]\wedge\\ &Tan(t3,b):[\mathtt{em},\alpha] \wedge Tan(t1,b):[\mathtt{fu},\alpha]\wedge\\ &Tan(t2,b):[\mathtt{fu},\alpha] \wedge Fin(p3):[\mathtt{is},\alpha] \to\\ &Pr(p3,b):[\mathtt{s},\gamma], \end{aligned} \tag{11.64}$$

$$Pr(p3,b):[\mathtt{s},\gamma] \to SPr(t0,v0,0):[\mathtt{1},\gamma], \tag{11.65}$$

$$Pr(p3,b):[\mathtt{s},\gamma] \to SPr(v0,t1,0):[\mathtt{1},\gamma], \tag{11.66}$$

$$Pr(p3,b):[\mathtt{s},\gamma] \to SPr(v0,t2,0):[\mathtt{1},\gamma], \tag{11.67}$$

$$Pr(p3,b):[\mathtt{s},\gamma] \to SPr(v1,v0,0):[\mathtt{1},\gamma], \tag{11.68}$$

$$Pr(p3,b):[\mathtt{s},\gamma] \to SPr(t3,v1,0):[\mathtt{1},\gamma], \tag{11.69}$$

$$Pr(p3, b) : [\mathbf{s}, \gamma] \rightarrow Val(v0, l, 0) : [\mathrm{cm}, \gamma]. \tag{11.70}$$

$$Pr(p3, b) : [\mathbf{s}, \gamma] \rightarrow Val(v1, 0, 0) : [\mathrm{cm}, \gamma]. \tag{11.71}$$

We suppose that $l \in \{b, cw, ww, hw, na, cs, 0\}$ in the above process release EVALPSN clauses for processes $Pr_{0,1,2,3}$.

11.4.6   Example

In this subsection we introduce an example of EVALPSN safety verification based pipeline control for processes $Pr_{0,1,2,3}$ in the pipeline network in Fig. 11.4. According to the process schedule in Fig. 11.7, we describe the details of EVALPSN safety verification.

Initial Stage We suppose that all the sub-processes and valves in the pipeline network are free (unlocked) and no process has already started at the initial stage. In order to verify the safety for all processes $Pr_{0,1,2,3}$, the following fact EVALP clauses (detected information) are input to the pipeline safety control EVALPSN:

$$
\begin{aligned}
&SPr(t0, v0, 0) : [\mathbf{f}, \alpha], \quad Val(v0, 0, 0) : [\mathrm{cs}, \alpha], \\
&SPr(v0, t1, 0) : [\mathbf{f}, \alpha], \quad Val(v1, 0, 0) : [\mathrm{cs}, \alpha], \\
&SPr(v0, t2, 0) : [\mathbf{f}, \alpha], \\
&SPr(v1, v0, 0) : [\mathbf{f}, \alpha], \\
&SPr(t3, v1, 0) : [\mathbf{f}, \alpha], \\
&Tan(t0, b) : [\mathbf{fu}, \alpha], \quad Tan(t1, 0) : [\mathrm{em}, \alpha],
\end{aligned}
\tag{11.72}
$$

$$Tan(t2, 0) : [\mathrm{em}, \alpha], \quad Tan(t3, na) : [\mathbf{fu}, \alpha]. \tag{11.73}$$

Then all the sub-processes and valves are permitted to be locked or controlled. However the tank conditions (11.73) and (11.74) do not permit for setting processes $Pr_{2,3}$. The beer process Pr_0 can be verified to be set as follows:

 • we have neither the forbiddance from locking sub-processes $SPr_{0,1}$, nor the forbiddance from controlling valve V_0 separate with beer in the normal direction, by EVALPSN clauses (11.34), (11.37)–(11.39) and the input fact EVALP clauses;

 • then we have the permission for locking sub-processes $SPr_{0,1}$, and controlling valve V_0 separate with beer in the normal direction and any liquid in the cross direction,

$$
\begin{aligned}
&SPr(t0, v0, b) : [\mathbf{f}, \gamma], \quad Val(v0, b, l) : [\mathrm{cm}, \gamma], \\
&SPr(v0, t1, b) : [\mathbf{f}, \gamma],
\end{aligned}
$$

where $l \in \{b, cw, ww, hw, na, cs, 0\}$, by EVALPSN clauses (11.35) and (11.42);

• moreover, we obtain the following EVALP clauses to represent tank conditions,

$$Tan(t0, b) : [\text{fu}, \alpha], \qquad Tan(t1, 0) : [\text{em}, \alpha] \; ;$$

• thus we have the permission for setting the beer process Pr_0,

$$Pr(p0, b) : [\text{xs}, \gamma],$$

by EVALPSN clause (11.44).

According to the process schedule, the beer process Pr_0 has to start first, then the nitric acid process Pr_1 has to be verified its safety and processed parallel to process Pr_0 as soon as possible. We show the safety verification for process Pr_1 at the next stage.

2nd Stage The beer process Pr_0 has already started but not finished yet, then in order to verify the safety for processes $Pr_{1,2,3}$, the following fact EVALP clauses are input to the pipeline safety control EVALPSN:

$$SPr(t0, v0, b) : [1, \alpha], \quad Val(v0, b, 0) : [\text{cs}, \alpha],$$
$$SPr(v0, t1, b) : [1, \alpha], \quad Val(v1, 0, 0) : [\text{cs}, \alpha],$$
$$SPr(v0, t2, 0) : [\text{f}, \alpha],$$
$$SPr(v1, v0, 0) : [\text{f}, \alpha],$$
$$SPr(t3, v1, 0) : [\text{f}, \alpha],$$
$$Tan(t2, 0) : [\text{em}, \alpha], \quad Tan(t3, na) : [\text{fu}, \alpha].$$

The above tank conditions permit neither the cold water process Pr_2 nor the beer process Pr_3 to be set. We show that only the nitric acid process Pr_1 can be verified to be set as follows:

• we have neither the forbiddance from locking three sub-processes $SPr_{2,3,4}$, the forbiddance from controlling valves V_0 separate with any liquid in the normal direction and nitric acid in the cross direction, nor the forbiddance from controlling valve V_1 mix(open) with nitric acid in the normal direction and no liquid in the cross direction, by EVALPSN clauses (11.34), (11.36)–(11.41) and the above fact EVALP clauses;

• therefore we have the permission for locking sub-processes $SPr_{2,3,4}$, and controlling valves V_0 and V_1 as described before,

$$SPr(v0, t2, na) : [\text{f}, \gamma], \quad Val(v0, b, na) : [\text{cm}, \gamma],$$
$$SPr(v1, v0, na) : [\text{f}, \gamma], \quad Val(v1, na, 0) : [\text{cs}, \gamma],$$
$$SPr(t3, v1, na) : [\text{f}, \gamma],$$

by EVALPSN clauses (11.35), (11.42) and (11.43);

• moreover we have the tank conditions,

$$Tan(t3, na) : [\mathtt{fu}, \alpha], \qquad Tan(t2, 0) : [\mathtt{em}, \alpha];$$

• thus we have the permission for setting the nitric acid process Pr_1,

$$Pr(p1, na) : [\mathtt{xs}, \gamma],$$

by EVALPSN clause (11.45).

Both the beer process Pr_0 and the nitric acid process Pr_1 have already started, then processes $Pr_{2,3}$ have to be verified their safety. We will show it at the next stage.

3rd Stage In order to verify the safety for the cold water process Pr_2 and the beer process Pr_3, the following fact EVALP clauses are input to the pipeline saafety control EVALPSN:

$$SPr(t0, v0, b) : [1, \alpha], \quad Val(v0, b, na) : [\mathtt{cs}, \alpha],$$
$$SPr(v0, t1, b) : [1, \alpha], \quad Val(v1, na, 0) : [\mathtt{cm}, \alpha],$$
$$SPr(v0, t2, na) : [1, \alpha],$$
$$SPr(v1, v0, na) : [1, \alpha],$$
$$SPr(t3, v1, na) : [1, \alpha].$$

Apparently, neither the cold water process Pr_2 nor the beer process Pr_3 is permitted to be set, since there is no tank condition in the input fact EVALP clauses. We show the safety verification for process Pr_2 as an example:

• we have the forbiddance from locking sub-processes $SPr_{2,3,4}$, the forbiddance from controlling valve V_0 separate with beer in the normal direction and cold water in the cross direction, and the forbiddance from controlling valve V_1 mix with cold water in the normal direction and no liquid in the cross direction,

$$SPr(v0, t2, cw) : [\mathtt{f}, \beta], \quad Val(v0, b, cw) : [\mathtt{cm}, \beta],$$
$$SPr(v1, v0, cw) : [\mathtt{f}, \beta], \quad Val(v1, cw, 0) : [\mathtt{cs}, \beta],$$
$$SPr(t3, v1, cw) : [\mathtt{f}, \beta],$$

by EVALPSN clauses (11.34), (11.39) and (11.40) and the input fact EVALP clauses;

The finish condition for the nitric acid process Pr_1 is that tank T_2 is fully filled with nitric acid and tank T_3 is empty. If the nitric acid process Pr_1 has finished and its finish conditions are satisfied, process Pr_1 is permitted to be released (unset) and has been released in fact. It is also supposed that tank T_3 is filled with cold water and tank T_2 is empty as preparation for the cold water process Pr_2 immediately after process Pr_1 has finished. Then the cold water process Pr_2 has to be verified and start according to the process schedule. We show the safety verification for the cold water process Pr_2 at the next stage.

4th Stage If the nitric acid process Pr_1 has finished and its finishing condition is satisfied, sub-processes $SPr_{2,3,4}$, valve V_1, and the cross direction of valve V_0 are permitted to be released by EVALP clauses (11.52)–(11.57). They have been released in fact. Then, since only the beer process Pr_0 is being processed, other three processes $Pr_{1,2,3}$ have to be verified. In order to do that, the following fact EVALP clauses are input to the pipeline safety control EVALPSN:

$$SPr(t0, v0, b) : [1, \alpha], \quad Val(v0, b, 0) : [\text{cs}, \alpha],$$
$$SPr(v0, t1, b) : [1, \alpha], \quad Val(v1, 0, 0) : [\text{cs}, \alpha],$$
$$SPr(v0, t2, 0) : [\text{f}, \alpha],$$
$$SPr(v1, v0, 0) : [\text{f}, \alpha],$$
$$SPr(t3, v1, 0) : [\text{f}, \alpha],$$
$$Tan(t2, 0) : [\text{em}, \alpha], \quad Tan(t3, cw) : [\text{fu}, \alpha].$$

Since the beer process Pr_0 is still being processed, neither the nitric acid process Pr_1 nor the beer process Pr_3 is permitted to be set, and only the cold water process Pr_2 is permitted to be set as well as process Pr_1 at the 2nd stage. Therefore we have the permission for setting the cold water process Pr_2,

$$Pr(p2, cw) : [\text{xs}, \gamma]$$

by EVALPSN clause (11.46).

Now, both the beer process Pr_0 and the cold water process Pr_2 are being processed. Then apparently any other processes are not permitted to be set. Moreover even if one of processes $Pr_{0,2}$ has finished, the beer process Pr_3 is not permitted to be set until both processes $Pr_{0,2}$ have finished. We show the safety verification for the beer process Pr_3 at the following stages.

5th Stage If neither the beer process Pr_0 nor the cold water process Pr_2 has finished, we have to verify the safety for the nitric acid process Pr_1 and the beer process Pr_3. The following fact EVALP clauses are input to the pipeline safety control EVALPSN:

$$SPr(t0, v0, b) : [1, \alpha], \quad Val(v0, b, cw) : [\text{cs}, \alpha],$$
$$SPr(v0, t1, b) : [1, \alpha], \quad Val(v1, cw, 0) : [\text{cm}, \alpha],$$
$$SPr(v0, t2, cw) : [1, \alpha],$$
$$SPr(v1, v0, cw) : [1, \alpha],$$
$$SPr(t3, v1, cw) : [1, \alpha].$$

Then, since all the sub-processes and valves are locked and controlled, neither processes Pr_1 nor Pr_3 is permitted to be set. It is shown that the beer process Pr_3 is not permitted to be set as follows:

• we have the forbiddance from locking sub-processes $SPr_{2,3,4}$ in process Pr_3 and controlling valves $V_{0,1}$,

$$SPr(v0, t2, b):[\mathtt{f}, \beta], \quad Val(v0, b, b):[\mathtt{cs}, \beta],$$
$$SPr(t3, v1, b):[\mathtt{f}, \beta], \quad Val(v1, b, 0):[\mathtt{cs}, \beta],$$
$$SPr(v1, v0, b):[\mathtt{f}, \beta],$$

by EVALPSN clauses (11.34), (11.36) and (11.40) and the input fact EVALP clauses;

• therefore we cannot derive the permission for setting process Pr_3.

The finish condition for the cold water process Pr_2 is that tank T_2 is fully filled with cold water and tank T_3 is empty. If the cold water process Pr_2 has finished and its finish condition is satisfied, process Pr_2 is permitted to be released(unset). It is also supposed that tank T_3 is filled with beer and tank T_2 is empty as preparation for the beer process Pr_3 immediately after process Pr_2 has finished. Then the beer process Pr_3 has to be verified and start according to the process schedule, but the beer process Pr_3 cannot be permitted to be set. We show the safety verification for process Pr_3 at the next stage.

6th Stage If the cold water process Pr_2 has finished and its finish condition is satisfied, the three sub-processes $SPr_{2,3,4}$, the valve V_1, and the cross direction of the valve V_0 are permitted to be released by EVALP clauses (11.52)–(11.57). They have been released in fact. Then, since only the beer process Pr_0 is being processed, processes $Pr_{1,2,3}$ have to be verified their safety. In order to do that, the following fact EVALP clauses are input to the pipeline safety control EVALPSN:

$$SPr(t0, v0, b):[\mathtt{1}, \alpha], \quad Val(v0, b, 0):[\mathtt{cs}, \alpha],$$
$$SPr(v0, t1, b):[\mathtt{1}, \alpha], \quad Val(v1, 0, 0):[\mathtt{cs}, \alpha],$$
$$SPr(v0, t2, 0):[\mathtt{f}, \alpha],$$
$$SPr(v1, v0, 0):[\mathtt{f}, \alpha],$$
$$SPr(t3, v1, 0):[\mathtt{f}, \alpha],$$
$$Tan(t2, 0):[\mathtt{em}, \alpha], \quad Tan(t3, b):[\mathtt{fu}, \alpha].$$

Since the beer process Pr_0 is still being processed, the beer process Pr_3 is not verified its safety due to the tank conditions and safety property **Val** for valve V_0. The safety verification is carried out as follows:

- we have the forbiddance from controlling valve V_0 mix,

$$Val(v0, b, b) : [cs, \beta],$$

by EVALPSN clause (11.36);

- therefore we cannot have the permission for setting the beer process Pr_3 then. On the other hand, even if the beer process Pr_0 has finished with the cold water process Pr_2 still being processed, the beer process Pr_3 is not permitted to be set. If both processes $Pr_{0,2}$ have finished, the beer process Pr_3 is assured its safety and set. Then process Pr_3 starts according to the process schedule. We omit the rest of the safety verification stages.

11.5 Before-After EVALPSN

In this section, we review an extended version of EVALPSN named bf (before-after)-EVALPSN formally, which can deal with before-after relations between two processes(time intervals) and introduce how to implement bf-EVALPSN aiming at the real-time safety verification for process order control [19, 40].

11.5.1 Before-After Relation in EVALPSN

First of all, we introduce a special literal $R(pi, pj, t)$ whose vector annotation represents the before-after relation between processes $Pr_i(pi)$ and $Pr_j(pj)$ at time t, where processes can be regarded as time intervals in general, and literal $R(pi, pj, t)$ is called a *bf-literal*.[3]

Definition 11.7 (*bf-EVALPSN*) An extended vector annotated literal $R(p_i, p_j, t) : [\mu_1, \mu_2]$ is called a *bf-EVALP* literal, where μ_1 is a vector annotation and $\mu_2 \in \{\alpha, \beta, \gamma\}$. If an EVALPSN clause contains bf-EVALP literals, it is called a *bf-EVALPSN clause* or just a *bf-EVALP clause* if it contains no strong negation. A bf-EVALPSN is a finite set of EVALPSN clauses and bf-EVALPSN clauses.

We provide some paraconsistent interpretations of vector annotations for representing bf-relations, which are called *bf-annotations*. Strictly speaking, bf-relations between time intervals are classified into 15 sorts according to bf-relations between start/finish times of two time intervals. We define the 15 sorts of bf-relations in bf-EVALPSN with regarding processes as time intervals.

---

[3]Hereafter, expression "**before-after**" is abbreviated as just "bf" in this chapter.

Fig. 11.10 Bf-relations, before/after

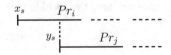

Fig. 11.11 Bf-relations, disjoint before/after

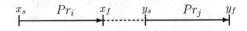

Suppose that there are two processes, Pr_i with its start/finish times x_s and x_f, and Pr_j with its start/finish times y_s and y_f.

Before (be)/After (af)

Firstly, we define the most basic bf-relations *before/after* according to the bf-relation between each start time of two processes, which are represented by bf-annotations be/af, respectively. If one process has started before/after another one starts, then the bf-relations between those processes are defined as "before(be)/after(af)", respectively. The bf-relations also are described in Fig. 11.10 with the condition that process Pr_i has started before process Pr_j starts. The bf-relation between their start/finish times is denoted by the inequality $\{x_s < y_s\}$.[4] For example, a fact at time t "process Pr_i has started before process Pr_j starts" can be represented by bf-EVALP clause

$$R(pi, pj, t) : [\text{be}, \alpha].$$

Disjoint Before (db)/After (da)

Bf-relations *disjoint before/after* between processes Pr_i and Pr_j are represented by bf-annotations db/da, respectively. The expression "disjoint before/after" implies that there is a timelag between the earlier process finish time and the later one start time. They are also described in Fig. 11.11 with the condition that process Pr_i has finished before process Pr_j starts. The bf-relation between their start/finish times is denoted by the inequality $\{x_f < y_s\}$. For example, an obligation at time t "process Pr_i must start after process Pr_j has finished" can be represented by bf-EVALP clause

$$R(pi, pj, t) : [\text{da}, \beta].$$

Immediate Before (mb)/After (ma)

Bf-relations *immediate before/ after* between processes Pr_i and Pr_j are represented by bf-annotations mb/ma, respectively. The expression "immediate before/after" implies that there is no timelag between the earlier process finish time and the later one start time. The bf-relations are also described in Fig. 11.12 with the condition

---

[4]If time t_1 is earlier than time t_2, we conveniently denote the before-after relation by the inequality $t_1 < t_2$.

Fig. 11.12 Bf-relations, immediate before/after

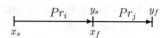

that process Pr_i has finished immediately before process Pr_j starts. The bf-relation between their start/finish times is denoted by the equality $\{x_f = y_s\}$. For example, a fact at time t "process Pr_i has finished immediately before process Pr_j starts" can be represented by bf-EVALP clause

$$R(pi, pj, t) : [\text{mb}, \alpha].$$

Joint Before (jb)**/After** (ja)
Bf-relations *joint before/after* between processes Pr_i and Pr_j are represented by bf-annotations jb/ja, respectively. The expression "joint before/after" imply that the two processes overlap and the earlier process has finished before the later one finishes. The bf-relations are also described in Fig. 11.13 with the condition that process Pr_i has started before process Pr_j starts and process Pr_i has finished before process Pr_j finishes. The bf-relation between their start/finish times is denoted by the inequalities $\{x_s < y_s < x_f < y_f\}$. For example, a fact at time t "process Pr_i has started before process Pr_j starts and finished before process Pr_j finishes" can be represented by bf-EVALP clause

$$R(pi, pj, t) : [\text{jb}, \alpha].$$

S-included Before (sb)**/After** (sa)
Bf-relations *s-included before/after* between processes Pr_i and Pr_j are represented by bf-annotations sb/sa, respectively. The expression "s-included before/after" implies that one process has started before another one starts and they finish at the same time. The bf-relations are also described in Fig. 11.14 with the condition that process Pr_i has started before process Pr_j starts and they finish at the same time. The bf-relation between their start/finish times is denoted by the equality and inequalities $\{x_s < y_s < x_f = y_f\}$. For example, a fact at time t "process Pr_i has started

Fig. 11.13 Bf-relations, Joint before/after

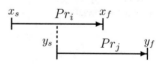

Fig. 11.14 Bf-relations, S-included before/after

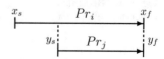

before process Pr_j starts and they finish at the same time" can be represented by
bf-EVALP clause

$$R(pi, pj, t) : [\text{sb}, \alpha].$$

Included Before (ib)/After (ia)

Bf-relations *included before/after* between processes Pr_i and Pr_j are represented by
bf-annotations ib/ia, respectively. The expression "included before/after" implies
that one process has started before another one starts and the earlier one finishes
after another one has finished. The bf-relations are also described in Fig. 11.15 with
the condition that process Pr_i has started before process Pr_j starts and finishes after
process Pr_j has finished. The bf-relation between their start/finish times is denoted
by the inequalities $\{x_s < y_s, y_f < x_f\}$. For example, an obligation at the time t
"process Pr_i must start before process Pr_j starts and process Pr_i must finish after
process Pr_j has finished" can be represented by bf-EVALP clause

$$R(pi, pj, t) : [\text{ib}, \beta].$$

F-included Before (fb)/After (fa)

The bf-relations *f-include before/ after* between processes Pr_i and Pr_j are represented by bf-annotations fb/fa, respectively. The expression "f-included
before/after" implies that two processes have started at the same time and one
process has finished before another one finishes. The bf-relations are also described
in Fig. 11.16 with the condition that processes Pr_i and Pr_j have started at the same
time and process Pr_i finishes after process Pr_j has finished. The bf-relation between
their start/finish times is denoted by the equality and inequality $\{x_s = y_s, y_f < x_f\}$.
For example, a fact at time t "processes Pr_i and Pr_j have started at the same time
and process Pr_i has finished after process Pr_j finished" can be represented by
bf-EVALP clause

$$R(pi, pj, t) : [\text{fa}, \alpha].$$

Fig. 11.15 Bf-relations,
included before/after

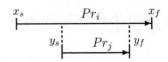

Fig. 11.16 Bf-relations,
F-included before/after

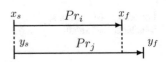

Fig. 11.17 Bf-relation, para
consistent before-after

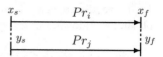

Paraconsistent Before-after (pba)

Bf-relation *paraconsistent before-after* between processes Pr_i and Pr_j is represented by bf-annotation pba. The expression "paraconsistent before-after" implies that the two processes have started at the same time and also finished at the same time. The bf-relation is also described in Fig. 11.17 with the condition that processes Pr_i and Pr_j have not only started but also finished at the same time. The bf-relation between their start/finish times is denoted by the equalities $\{x_s = y_s, y_f = x_f\}$. For example, an obligation at time t "processes Pr_i and Pr_j must not only start but also finish at the same time" can be represented by bf-EVALP clause

$$R(pi, pj, t) : [\text{pba}, \beta].$$

Here we define the epistemic negation $\neg_1$ that maps bf-annotations to themselves in bf-EVALPSN.

Definition 11.8 (*Epistemic Negation $\neg_1$ for Bf-annotations*)
The epistemic negation $\neg_1$ over the set of bf-annotations,

$$\{\text{be}, \text{af}, \text{da}, \text{db}, \text{ma}, \text{mb}, \text{ja}, \text{jb}, \text{sa}, \text{sb}, \text{ia}, \text{ib}, \text{fa}, \text{fb}, \text{pba}\}$$

is obviously defined as the following mapping:

$$\neg_1(\text{af}) = \text{be}, \quad \neg_1(\text{be}) = \text{af},$$
$$\neg_1(\text{da}) = \text{db}, \quad \neg_1(\text{db}) = \text{da},$$
$$\neg_1(\text{ma}) = \text{mb}, \quad \neg_1(\text{mb}) = \text{ma},$$
$$\neg_1(\text{ja}) = \text{jb}, \quad \neg_1(\text{jb}) = \text{ja},$$
$$\neg_1(\text{sa}) = \text{sb}, \quad \neg_1(\text{sb}) = \text{sa},$$
$$\neg_1(\text{ia}) = \text{ib}, \quad \neg_1(\text{ib}) = \text{ia},$$
$$\neg_1(\text{fa}) = \text{fb}, \quad \neg_1(\text{fb}) = \text{fa},$$
$$\neg_1(\text{pba}) = \text{pba}.$$

If we consider before-after measure over the meaningful 15 bf-annotations, obviously there exists a partial order($<_h$) based on the before-after measure, where $\mu_1 <_h \mu_2$ is intuitively interpreted that bf-annotation μ_1 denotes a more "before" degree than bf-annotation μ_2, and $\mu_1, \mu_2 \in \{\text{be}, \text{af}, \text{db}, \text{da}, \text{mb}, \text{ma}, \text{jb}, \text{ja}, \text{ib},$

$ia, sb, sa, fb, fa, pba\}$. If $\mu_1 <_h \mu_2$ and $\mu_2 <_h \mu_1$, we denote it $\mu_1 \equiv_h \mu_2$. Then we obtain the following ordering:

$$db <_h mb <_h jb <_h sb <_h ib <_h fb <_h pba <_h ia <_h ja <_h ma <_h da$$
$$\text{and}$$
$$sb \equiv_h be <_h af \equiv_h sa.$$

On the other hand, if we take before-after knowledge (information) amount of each bf-relation into account as another measure, obviously there also exists another partial order($<_v$) in terms of before-after knowledge amount, where $\mu_1 <_v \mu_2$ is intuitively interpreted that bf-annotation μ_1 has less knowledge amount in terms of bf-relation than bf-annotation μ_2. If $\mu_1 <_v \mu_2$ and $\mu_2 <_v \mu_1$, we denote it $\mu_1 \equiv_v \mu_2$. Then we obtain the following ordering:

$$be <_v \mu_1, \quad \mu_1 \in \{ db, mb, jb, sb, ib \},$$
$$af <_v \mu_2, \quad \mu_1 \in \{ da, ma, ja, sa, ia \},$$
$$db \equiv_v mb \equiv_v jb \equiv_v sb \equiv_v ib \equiv_v fb \equiv_v pba \equiv_v$$
$$fa \equiv_v ia \equiv_v sa \equiv_v ja \equiv_v ma \equiv_v da$$
$$\text{and}$$
$$be \equiv_v af.$$

If we take the before-after degree as the horizontal measure and the before-after knowledge amount as the vertical one, we obtain the complete bi-lattice $\mathcal{T}_v(12)_{bf}$ of vector annotations that includes the 15 bf-annotations.

$$
\begin{aligned}
\mathcal{T}_v(12)_{bf} = \ & \{ \perp_{12}(0,0), \cdots, be(0,8), \cdots, db(0,12), \cdots, mb(1,11), \cdots, \\
& jb(2,10), \cdots, sb(3,9), \cdots, ib(4,8), \cdots, fb(5,7), \cdots, \\
& pba(6,6), \cdots, fa(7,5), \cdots, af(8,0), \cdots, ia(8,4), \cdots, \\
& sa(9,3), \cdots, ja(10,2), \cdots, ma(11,1), \cdots, da(12,0), \cdots, \\
& \top_{12}(12,12) \},
\end{aligned}
$$

which is described as the Hasse's diagram in Fig. 11.18. We note that bf-EVALP literal

$$R(pi, pj, t) : [\mu_1(m,n), \mu_2],$$
$$\text{where} \quad \mu_2 \in \{\alpha, \beta, \gamma\} \quad \text{and}$$
$$\mu_1 \in \{be, db, mb, jb, sb, ib, fb, pba, fa, ia, sa, jb, ma, da, af\},$$

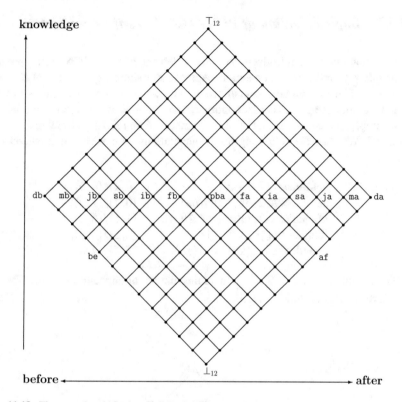

knowledge

$\top_{12}$

db mb jb sb ib fb pba fa ia sa ja ma da

be af

before $\perp_{12}$ after

Fig. 11.18 The complete bi-lattice $\mathcal{T}_v(12)_{bf}$ of Bf-annotations

is not well annotated if $m \neq 0$ and $n \neq 0$, however, the bf-EVALP literal is equivalent to the following two well annotated bf-EVALP literals:

$$R(pi,pj) : [(m,0),\mu] \quad \text{and} \quad R(pi,pj) : [(0,n),\mu].$$

Therefore such a non-well annotated bf-EVALP literal can be regarded as the conjunction of two well annotated EVALP literals. For example, suppose that there is a non-well annotated bf-EVALP clause,

$$R(pi,pj,t_1) : [(k,l),\mu_1] \rightarrow R(pi,pj,t_2) : [(m,n),\mu_2],$$

where $k \neq 0$, $l \neq 0$ $m \neq 0$ and $n \neq 0$ It can be equivalently transformed into the following two well annotated bf-EVALP clauses,

$$R(pi,pj,t_1) : [(k,0),\mu_1] \wedge R(pi,pj,t_1) : [(0,l),\mu_1] \rightarrow R(pi,pj,t_2) : [(m,0),\mu_2],$$
$$R(pi,pj,t_1) : [(k,0),\mu_1] \wedge R(pi,pj,t_2) : [(0,l),\mu_1] \rightarrow R(pi,pj,t_2) : [(0,n),\mu_2].$$

11.5.2   Implementation of Bf-EVALPSN Verification System

In this subsection we introduce how to implement bf-EVALPSN based process order safety verification with a simple example. For simplicity, we do not consider cases in which one process starts/finishes with another one starts/finishes at the same time, then the process order control system can deal with before-after relations more simply, which means that bf-annotations(relations) $\mathtt{sb}/\mathtt{sa}, \mathtt{fb}/\mathtt{fa}$ and pba are excluded. We take the following ten bf-annotations with vector annotations into account:

$$
\begin{array}{ll}
\text{before}(\mathtt{be})/\text{after}(\mathtt{af}), & (0,4)/(4,0), \\
\text{discretebefore}(\mathtt{db})/\text{after}(\mathtt{da}), & (0,7)/(7,0), \\
\text{immediatebefore}(\mathtt{mb})/\text{after}(\mathtt{ma}), & (1,6)/(6,1), \\
\text{jointbefore}(\mathtt{jb})/\text{after}(\mathtt{ja}), & (2,5)/(5,2), \\
\text{includedbefore}(\mathtt{ib})/\text{after}(\mathtt{ia}). & (3,4)/(4,3).
\end{array}
$$

The complete bi-lattice $\mathcal{T}_v(7)_{bf}$ including the ten bf-annotations is described as the Hasse's diagram in Fig. 11.19.

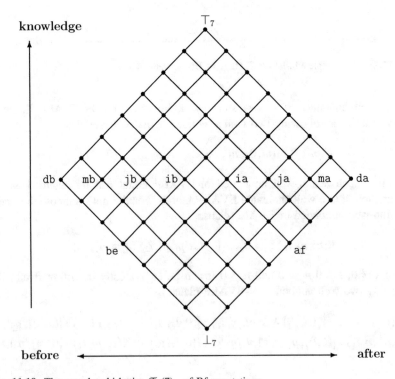

Fig. 11.19 The complete bi-lattice $\mathcal{T}_v(7)_{bf}$ of Bf-annotations

Fig. 11.20 Process timing chart

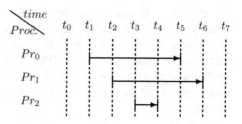

Now we show an example of implementing the real-time process order safety verification system in bf-EVALPSN.

Example 11.2 We suppose that there are four processes Pr_0(id $p0$), Pr_1(id $p1$) Pr_2(id $p2$) and the next process Pr_3(id $p3$) not appearing in Fig. 11.20. Those processes are supposed to be processed according to the processing schedule in Fig. 11.20. Then we consider three bf-relations represented by the following bf-EVALP clauses (11.74)–(11.76):

$$R(p0, p1, t_i) : [(i_1, j_1), \alpha], \tag{11.74}$$

$$R(p1, p2, t_i) : [(i_2, j_2), \alpha], \tag{11.75}$$

$$R(p2, p3, t_i) : [(i_3, j_3), \alpha], \tag{11.76}$$

which will be inferred based on each process start/finish information at time t_i ($i = 0, 1, 2, \ldots, 7$).

At time t_0, no process has started yet. Thus we have no knowledge in terms of any bf-relations. Therefore we have the bf-EVALP clauses,

$$R(p0, p1, t_0) : [(0, 0), \alpha],$$
$$R(p1, p2, t_0) : [(0, 0), \alpha],$$
$$R(p2, p3, t_0) : [(0, 0), \alpha].$$

At time t_1, only process Pr_0 has started before process Pr_1 starts, Then bf-annotations $db(0, 7), mb(1, 6), jb(2, 5)$ or $ib(3, 4)$ could be the final bf-annotation to represent the bf-relation between processes Pr_0 and Pr_1, thus the greatest lower bound be $(0, 4)$ of the set of vector annotations $\{(0, 7), (1, 6), (2, 5), (3, 4)\}$ becomes the vector annotation of bf-literal $R(p0, p1, t_1)$. Other bf-literals have the bottom vector annotation $(0, 0)$. Therefore we have the bf-EVALP clauses,

$$R(p0, p1, t_1) : [(0, 4), \alpha],$$
$$R(p1, p2, t_1) : [(0, 0), \alpha],$$
$$R(p2, p3, t_1) : [(0, 0), \alpha].$$

At time t_2, the second process Pr_1 also has started before process Pr_0 finishes. Then bf-annotations $jb(2,5)$ or $ib(3,4)$ could be the final bf-relation to represent the bf-relation between processes Pr_0 and Pr_1. Thus the greatest lower bound $(2,4)$ of the set of vector annotations $\{(2,5),(3,4)\}$ has to be the vector annotation of bf-literal $R(p0,p1,t_2)$. In addition, bf-literal $p : \mu$ has bf-annotation be $(0,4)$ as well as bf-literal $R(p0,p1,t_1)$ since process Pr_1 has also started before process Pr_2 starts. On the other hand, bf-literal $R(p2,p3,t2)$ has the bottom vector annotation $(0,0)$ since process Pr_3 has not started yet. Therefore we have the bf-EVALP clauses,

$$R(p0,p1,t_2) : [(2,4),\alpha],$$
$$R(p1,p2,t_2) : [(0,4),\alpha],$$
$$R(p2,p3,t_2) : [(0,0),\alpha].$$

At time t_3, process Pr_2 has started before both processes Pr_0 and Pr_1 finish. Then both bf-literals $R(p0,p1,t_3)$ and $R(p1,p2,t_3)$ have the same vector annotation $(2,4)$ as well as bf-literal $R(p0,p1,t_2)$. Moreover bf-literal $R(p2,p3,t_3)$ has bf-annotation be$(0,4)$ as well as bf-literal $R(p0,p1,t_1)$. Therefore we have the bf-EVALP clauses,

$$R(p0,p1,t_3) : [(2,4),\alpha],$$
$$R(p1,p2,t_3) : [(2,4),\alpha],$$
$$R(p2,p3,t_3) : [(0,4),\alpha].$$

At time t_4, process Pr_2 has finished before both processes Pr_0 and Pr_1 finish. Then bf-literal $R(p0,p1,t_4)$ still has the same vector annotation $(2,4)$ as well as the previous time t_3. In addition bf-literal $R(p1,p2,t_4)$ has its final bf-annotation $ib(3,4)$. For the final bf-relation between processes Pr_2 and Pr_3 there are still two alternatives: (1) if process Pr_3 starts immediately after process Pr_2 has finished, bf-literal $R(p2,p3,t_4)$ has its final bf-annotation mb $(1,6)$; (2) if process Pr_3 does not start immediately after process Pr_2 has finished, bf-literal $R(p2,p3,t_4)$ has its final bf-annotation $db(0,7)$. Either way, we have the knowledge that process Pr_2 has just finished at time t_4, which can be represented by vector annotation $(0,6)$ that is the greatest lower bound of the set of vector annotations $\{(1,6),(0,7)\}$. Therefore we have the bf-EVALP clauses,

$$R(p1,p2,t_4) : [(2,4),\alpha],$$
$$R(p2,p3,t_4) : [(3,4),\alpha],$$
$$R(p3,p4,t_4) : [(0,6),\alpha].$$

At time t_5, process Pr_0 has finished before processes Pr_1 finishes. Then bf-literal $R(p0, p1, t_5)$ has its final bf-annotation jb(2, 5), and bf-literal $R(p2, p3, t_5)$ also has its final bf-annotation jb(0, 7) because process Pr_3 has not started yet. Therefore we have the bf-EVALP clauses,

$$R(p1, p2, t_5) : [\text{jb}(2, 5), \alpha],$$
$$R(p2, p3, t_5) : [\text{ib}(3, 4), \alpha],$$
$$R(p3, p4, t_5) : [\text{db}(0, 7), \alpha],$$

and all the bf-relations have been determined at time t_5 before process Pr_1 finishes and process Pr_3 starts.

In Example 11.2, we have shown how the vector annotations of bf-literals are updated according to the start/finish information of processes in real-time. We will introduce the real-time safety verification for process order control based on bf-EVALPSN with small examples in the subsequent subsection.

11.5.3 Safety Verification in Bf-EVALPSN

First of all we introduce the basic idea of bf-EVALPSN safety verification for process order with a simple example.

Suppose that two processes Pr_0 and Pr_1 are being processed repeatedly, and process Pr_1 must be processed immediately before process Pr_0 starts as shown in Fig. 11.21. In bf-EVALPSN process order safety verification, the safety for process order is verified based on the safety properties to be assured in the process schedule. In order to verify the safety for the process order in Fig. 11.21, we assume two safety properties **SP-0** and **SP-1** for processes Pr_0 and Pr_1, respectively:

SP-0 process Pr_0 must start immediately after process Pr_1 has finished,
SP-1 process Pr_1 must start in a while after (disjoint after) process Pr_0 has finished.

Then safety properties **SP-0** and **SP-1** should be verified immediately before processes Pr_0 and Pr_1 start, respectively.

In order to verify the bf-relation "immediate after" with safety property **SP-0**, it shoud be verified whether process Pr_1 has finished immediately before process Pr_0 starts or not, and the safety verification should be carried out immediately after process Pr_1 has finished. Then bf-literal $R(p0, p1, t)$ must have vector annotation

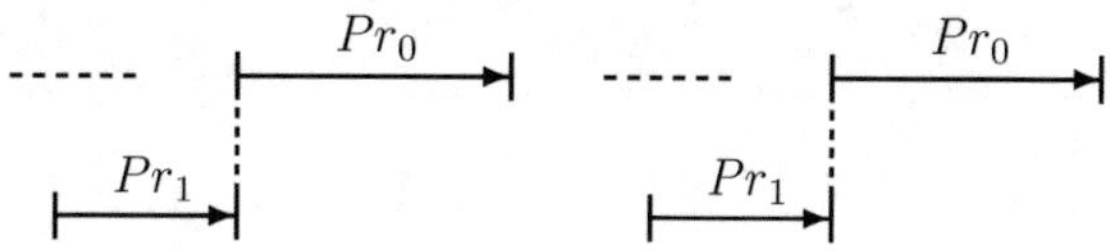

Fig. 11.21 Process schedule example

$(6,0)$, which means that process Pr_1 has finished but process Pr_0 has not started yet. Therefore safety property **SP-0** is translated to the bf-EVALPSN-clauses,

$$\mathbf{SP-0}$$

$$R(p0,p1,t):[(6,0),\alpha] \wedge \sim R(p0,p1,t):[(7,0),\alpha] \tag{11.77}$$
$$\rightarrow st(p0,t):[\mathtt{f}(0,1),\gamma],$$

$$\sim st(p0,t):[\mathtt{f}(0,1),\gamma] \rightarrow st(p0,t):[\mathtt{f}(0,1),\beta], \tag{11.78}$$

where literal $st(pi,t)$ represents "process Pr_i starts at time t" and the set of its vector annotations constitutes the complete lattice

$$\mathcal{T}_{v(1)} = \{\perp(0,0), \mathtt{t}(1,0), \mathtt{f}(1,0), \top(1,1)\}.$$

For example, EVALP-clause $st(p0,t):[\mathtt{f}(0,1),\gamma]$ can be informally interpreted as "it is permitted for process $\top$ to start at time t. On the other hand, in order to verify bf-relation "disjoint after" with safety property **SP-1**, it should be verified whether there is a timelag between process Pr_0 finish time and process Pr_1 start time or not. Then bf-literal $R(p1,p0,t)$ must have bf-annotation $\mathtt{da}(7,0)$. Therefore safety property **SP-1** is translated into the following bf-EVALPSN clauses:

$$SP-1$$
$$R(p1,p0,t):[(7,0),\alpha] \rightarrow Start(p1,t):[(0,1),\gamma], \tag{11.79}$$

$$\sim Start(p1,t):[(0,1),\gamma] \rightarrow Start(p1,t):[(0,1),\beta]. \tag{11.80}$$

We show how to verify the process order safety based on safety properties **SP-0** and **SP-1** in bf-EVALPSN. In order to verify the process order safety, the following safety verification cycle consisting of two steps is applied repeatedly.

Safety Verification Cycle
1st Step (safety verification for starting process Pr_1)
Suppose that process Pr_1 has not started yet at time t_1. If process Pr_0 has already finished at time t_1, we have the bf-EVALP clause,

$$R(p1,p0,t_1):[(7,0),\alpha]. \tag{11.81}$$

On the other hand, if process Pr_0 has just finished at time t_1, we have the bf-EVALP clause,

$$R(p1,p0,t_1):[(6,0),\alpha]. \tag{11.82}$$

If bf-EVALP clause (11.81) is input to safety property **SP-1** consisting of Bf-EVALPSN clauses (11.79) and (11.80), we obtain the EVALP clause,

$$st(p1, t_1) : [(0, 1), \gamma]$$

and the safety for starting process Pr_1 is assured. On the other hand, if bf-EVALP clause (11.82) is input to the same safety property **SP-1**, we obtain the EVALP clause

$$st(p1, t_1) : [(0, 1), \beta],$$

then the safety for starting process Pr_1 is not assured.

2nd Step (safety verification for starting process Pr_0)

Suppose that process Pr_0 has not started yet at time t_2. If process Pr_1 has just finished at time t_2, we have the bf-EVALP clause,

$$R(p0, p1, t_2) : [(6, 0), \alpha]. \tag{11.83}$$

On the other hand, if process Pr_1 has not finished yet at time t_2, we have the bf-EVALP clause,

$$R(p0, p1, t_2) : [(4, 0), \alpha]. \tag{11.84}$$

If bf-EVALP clause (11.83) is input to safety property **SP-0** (11.78), (11.79), we obtain the EVALP clause,

$$st(p0, t_2) : [(0, 1), \gamma],$$

and the safety for starting process Pr_0 is assured. On the other hand, if bf-EVALP clause (11.84) is input to the same safety property **SP-0**, we obtain the EVALP clause,

$$st(p1, t) : [(0, 1), \beta],$$

then the safety for starting process Pr_0 is not assured.

Example 11.3 In this example we suppose the same pipeline network as shown in Fig. 11.4 and the same process schedule as shown in Fig. 11.7.

Process Pr_0,   a brewery process using
      line-1, tank $T_0 \rightarrow$ valve $V_0 \rightarrow$ tank T_1;
process Pr_1,   a cleaning process by nitric acid using
      line-2, tank $T_3 \rightarrow$ valve $V_1 \rightarrow$ Valve $V_0 \rightarrow$ tank T_2;

process Pr_2,   a cleaning process by water in line-1;
process Pr_3,   a brewery process using both line-1 and line-2 with mixing at valve
          V_0

The above four processes are supposed to be processed according to the following processing schedule. We assume the following four process order safety properties for each process:

SP-2
process Pr_0 must start before any other processes start;
SP-3
process Pr_1 must start immediately after process Pr_0 has started;
SP-4
process Pr_2 must start immediately after process Pr_1 has finished;
SP-5
process Pr_3 must start immediately after both processes Pr_0 and Pr_2 have finished.

Safety property **SP-2** is translated into the bf-EVALPSN clauses,

$$SP-2$$
$$\sim R(p0,p1,t):[(4,0),\alpha] \wedge \sim R(p0,p2,t):[(4,0),\alpha] \wedge$$
$$\sim R(p0,p3,t):[(4,0),\alpha] \rightarrow Start(p0,t):[(0,1),\gamma],$$

$$\sim Start(p0,t):[(0,1),\gamma] \rightarrow Start(p0,t):[(0,1),\beta]. \qquad (11.85)$$

As well as safety property **SP-2**, other safety properties **SP-3**, **SP-4** and **SP-5** are also translated into the bf-EVALPSN clauses,

$$SP-3$$
$$R(p1,p0,t):[(4,0),\alpha] \rightarrow Start(p1,t):[(0,1),\gamma],$$

$$\sim Start(p1,t):[(0,1),\gamma] \rightarrow Start(p1,t):[(0,1),\beta], \qquad (11.86)$$

$$SP-4$$
$$R(p2,p1,t):[(6,0),\alpha] \wedge : R(p2,p1,t):[(7,0),\alpha]$$
$$\rightarrow Start(p2,t):[(0,1),\gamma],$$

$$\sim Start(p2,t):[(0,1),\gamma] \rightarrow Start(p2,t):[(0,1),\beta], \qquad (11.87)$$

SP–5

$R(p3, p0, t) : [(6, 0), \alpha] \wedge R(p3, p2, t) : [(6, 0), \alpha] \wedge$
$\sim R(p3, p2, t) : [(7, 0), \alpha] \rightarrow Start(p3, t) : [(0, 1), \gamma],$

$R(p3, p0, t) : [(6, 0), \alpha] \wedge R(p3, p2, t) : [(6, 0), \alpha] \wedge$
$\sim R(p3, p0, t) : [(7, 0), \alpha] \rightarrow Start(p3, t) : [(0, 1), \gamma],$

$\sim Start(p3, t) : [(0, 1), \gamma] \rightarrow Start(p3, t) : [(0, 1), \beta].$ \hfill (11.88)

We introduce the safety verification stages for the process order in Fig. 11.7 as follows.

Initial Stage (t_0) No process has started at time t_0, we have no information in terms of all bf-relations between all processes Pr_0, Pr_1, Pr_2 and Pr_3, thus we have the bf-EVALP clauses,

$$R(p0, p1, t_0) : [(0, 0), \alpha], \tag{11.89}$$

$$R(p0, p2, t_0) : [(0, 0), \alpha], \tag{11.90}$$

$$R(p0, p3, t_0) : [(0, 0), \alpha]. \tag{11.91}$$

In order to verify the safety for starting the first process Pr_0, the bf-EVALP clauses (11.89)–(11.91) are input to safety property **SP-2** (11.85). Then we obtain the EVALP clause,

$$Start(p0, t_0) : [(0, 1), \gamma],$$

which expresses permission for starting process Pr_0, and its safety is assured at time t_0. Otherwise, it is not assured.

2nd Stage (t_1) Suppose that only process Pr_0 has already started at time t_1. Then we have the bf-EVALP clauses,

$$R(p1, p0, t_1) : [(4, 0), \alpha]. \tag{11.92}$$

In order to verify the safety for starting the second process Pr_1, the bf-EVALP clause (11.92) is input to safety property **SP-3** (11.86). Then we obtain the EVALP clause,

$$Start(p1, t_1) : [(0, 1), \gamma],$$

and the safety for starting process Pr_1 is assured at time t_1. Otherwise, it is not assured.

3rd Stage (t_2) Suppose that processes Pr_0 and Pr_1 have already started, and neither of them has finished yet at time t_2. Then we have the bf-EVALP clauses,

$$R(p2, p0, t_2) : [(4, 0), \alpha], \tag{11.93}$$

$$R(p2, p1, t_2) : [(4, 0), \alpha]. \tag{11.94}$$

In order to verify the safety for starting the third process Pr_2, if EVALP clause (11.94) is input to safety property **SP-4** (11.87), then we obtain the EVALP clause,

$$Start(p2, t_2) : [(0, 1), \beta],$$

and the safety for starting process Pr_2 is not assured at time t_2. On the other hand, if process Pr_1 has just finished at time t_2, then, we have the bf-EVALP clause,

$$R(p2, p1, t_2) : [(6, 0), \alpha]. \tag{11.95}$$

If bf-EVALP clause (11.95) is input to safety property **SP-4** (11.87), then we obtain the EVALP clause,

$$Start(p2, t_2) : [(0, 1), \gamma],$$

and the safety for starting process Pr_2 is assured.

4th Stage (t_3) Suppose that processes Pr_0, Pr_1 and Pr_2 have already started, processes Pr_0 and Pr_1 have already finished, and only process Pr_3 has not started yet at time t_3. Then we have the bf-EVALP clauses,

$$R(p3, p0, t_3) : [(7, 0), \alpha], \tag{11.96}$$

$$R(p3, p1, t_3) : [(7, 0), \alpha], \tag{11.97}$$

$$R(p3, p2, t_3) : [(4, 0), \alpha]. \tag{11.98}$$

In order to verify the safety for starting the last process Pr_3, if bf-EVALP clauses (11.96) and (11.98) are input to safety property **SP-5** (11.88), then we obtain the EVALP clause,

$$Start(p3, t_3) : [(0, 1), \beta],$$

and the safety for starting process Pr_3 is not assured at time t_3. On the other hand, if process Pr_2 has just finished at time t_3, then we have the bf-EVALP clause,

$$R(p3, p2, t_3) : [(6, 0), \alpha]. \tag{11.99}$$

If bf-EVALP clause (11.99) is input to safety property **SP-5** (11.88), then we obtain the EVALP clause,

$$Start(p3, t_3) : [(0, 1), \gamma],$$

and the safety for starting process Pr_3 is assured.

11.6   Reasoning in Bf-EVALPSN

In this section, we introduce the process before-after relation reasoning system in bf-EVALPSN, which consists of two inference rules in bf-EVALP. One of them is the basic inference rules for bf-relations according to the before-after relations of process start/finish times, and another one is the transitive inference rules for bf-relations, which can infer the transitive bf-relation from two continuous bf-relations.

11.6.1   Basic Reasoning for Bf-Relation

We introduce the basic inference rules of bf-relations with referring to Example 11.2 in Sect. 11.5, which are called *basic bf-inference rules*. Hereafter we call the inference rules as *ba-inf rules* shortly. First of all, in order to represent the basic bf-inference rules in bf-EVALPSN, we introduce the following literals for expressing process start/finish information again:

$fi(p_i, t)$, which is intuitively interpreted that process Pr_i finishes at time t.

Those literals are used for expressing process finish information and may have one of vector annotations $\bot(0, 0), t(1, 0), f(0, 1)$ and $\top(1, 1)$, where annotations $t(1, 0)$ and $f(0, 1)$ can be intuitively interpreted as "true" and "false", respectively. We show a group of ba-inf rules to be applied at the initial stage (time t_0) for bf-relation reasoning, which are named $(0, 0)$-*rules*.

(0,0)-rules
Suppose that no process has started yet and the vector annotation of bf-literal $R(p_i, p_j, t)$ is $(0, 0)$, which shows that there is no knowledge in terms of the bf-relation between processes Pr_i and Pr_j, then the following two ba-inf rules can be applied at the initial stage.

(0,0)-rule-1

If process Pr_i has started before process Pr_j tarts, then the vector annotation $(0,0)$ of bf-literal $R(p_i, p_j, t)$ should turn to bf-annotation $\mathtt{be}(0,8)$, which is the greatest lower bound of the set of bf-annotations

$$\{\mathtt{db}(0,12),\ \mathtt{mb}(1,11),\ \mathtt{jb}(2,10),\ \mathtt{sb}(3,9),\ \mathtt{ib}(4,8)\}.$$

(0,0)-rule-2

If both processes Pr_i and Pr_j have started at the same time, then it is reasonably anticipated that the bf-relation between processes Pr_i and Pr_j will be in the set of bf-annotations,

$$\{\mathtt{fb}(5,7),\ \mathtt{pba}(6,6),\ \mathtt{fa}(7,5)\}$$

whose greatest lower bound is $(5,5)$ (refer to Fig. 11.18). Therefore the vector annotation $(0,0)$ of bf-literal $R(p_i, p_j, t)$ should turn to vector annotation $(5,5)$.

Ba-inf rules $(0,0)$-rule-1 and $(0,0)$-rule-2 may be translated into the bf-EVALPSN clauses,

$$
\begin{aligned}
&R(p_i,p_j,t):[(0,0),\alpha] \wedge st(p_i,t):[\mathtt{t},\alpha] \wedge \sim st(p_j,t):[\mathtt{t},\alpha] \\
&\quad \rightarrow R(p_i,p_j,t):[(0,8),\alpha],
\end{aligned}
\tag{11.100}
$$

$$
\begin{aligned}
&R(p_i,p_j,t):[(0,0),\alpha] \wedge st(p_i,t):[\mathtt{t},\alpha] \wedge st(p_j,t):[\mathtt{t},\alpha] \\
&\quad \rightarrow R(p_i,p_j,t):[(5,5),\alpha].
\end{aligned}
\tag{11.101}
$$

Suppose that one of ba-inf rules $(0,0)$-rule-1 and 2 has been applied, then the vector annotation of bf-literal $R(p_i, p_j, t)$ should be one of $(0,8)$ or $(5,5)$. Therefore we need to consider two groups of ba-inf rules to be applied after ba-inf rules $(0,0)$-rule-1 and $(0,0)$-rule-2, which are named $(0,8)$-*rules* and $(5,5)$-*rules*, respectively.

(0,8)-rules

Suppose that process Pr_i has started before process Pr_j starts, then the vector annotation of bf-literal $R(p_i, p_j, t)$ should be $(0,8)$. We have the following inference rules to be applied after ba-inf rule $(0,0)$-rule-1.

(0,8)-rule-1

If process Pr_i has finished before process Pr_j starts, and process Pr_j has started immediately after process Pr_i finishes, then the vector annotation $(0,8)$ of bf-literal $R(p_i, p_j, t)$ should turn to bf-annotation $\mathtt{mb}(1,11)$.

(0,8)-rule-2

If process Pr_i has finished before process Pr_j starts, and process Pr_j has not started immediately after process Pr_i has finished, then the vector annotation $(0,8)$ of bf-literal $R(p_i, p_j, t)$ should turn to bf-annotation $\mathtt{db}(0,12)$.

<u>(0,8)-rule-3</u>
If process Pr_j starts before process Pr_i finishes, then the vector annotation $(0,8)$ of bf-literal $R(p_i, p_j, t)$ should turn to vector annotation $(2, 8)$ that is the greatest lower bound of the set of bf-annotations,

$$\{\mathtt{jb}(2, 10), \ \mathtt{sb}(3, 9), \ \mathtt{ib}(4, 8)\}.$$

Ba-inf rules $(0, 8)$-rule-1, $(0, 8)$-rule-2 and $(0, 8)$-rule-3 may be translated into the bf-EVALPSN clauses,

$$R(p_i, p_j, t) : [(0, 8), \alpha] \wedge fi(p_i, t) : [\mathbf{t}, \alpha] \wedge st(p_j, t) : [\mathbf{t}, \alpha]$$
$$\rightarrow R(p_i, p_j, t) : [(1, 11), \alpha], \tag{11.102}$$

$$R(p_i, p_j, t) : [(0, 8), \alpha] \wedge fi(p_i, t) : [\mathbf{t}, \alpha] \wedge {\sim} st(p_j, t) : [\mathbf{t}, \alpha]$$
$$\rightarrow R(p_i, p_j, t) : [(0, 12), \alpha], \tag{11.103}$$

$$R(p_i, p_j, t) : [(0, 8), \alpha] \wedge {\sim} fi(p_i, t) : [\mathbf{t}, \alpha] \wedge st(p_j, t) : [\mathbf{t}, \alpha]$$
$$\rightarrow R(p_i, p_j, t) : [(2, 8), \alpha]. \tag{11.104}$$

(5,5)-rules
Suppose that both processes Pr_i and Pr_j have already started at the same time, then the vector annotation of bf-literal $R(p_i, p_j, t)$ should be $(5, 5)$. We have the following inference rules to be applied after ba-inf rule $(0, 0)$-rule-2.

<u>(5,5)-rule-1</u>
If process Pr_i has finished before process Pr_j finishes, then the vector annotation $(5, 5)$ of bf-literal $R(p_i, p_j, t)$ should turn to bf-annotation $\mathtt{sb}(5, 7)$.
<u>(5,5)-rule-2</u>
If both processes Pr_i and Pr_j have finished at the same time, then the vector annotation $(5, 5)$ of bf-literal $R(p_i, p_j, t)$ should turn to bf-annotation $\mathtt{pba}(6, 6)$.
<u>(5,5)-rule-3</u>
If process Pr_j has finished before process Pr_i finishes, then the vector annotation $(5, 5)$ of bf-literal $R(p_i, p_j, t)$ should turn to bf-annotation $\mathtt{sa}(7, 5)$.

Ba-inf rules $(5, 5)$-rule-1, $(5, 5)$-rule-2 and $(5, 5)$-rule-3 may be translated into the bf-EVALPSN clauses,

$$R(p_i, p_j, t) : [(5, 5), \alpha] \wedge fi(p_i, t) : [\mathbf{t}, \alpha] \wedge {\sim} fi(p_j, t) : [\mathbf{t}, \alpha]$$
$$\rightarrow R(p_i, p_j, t) : [(5, 7), \alpha], \tag{11.105}$$

$$R(p_i, p_j, t) : [(5, 5), \alpha] \wedge fi(p_i, t) : [\mathbf{t}, \alpha] \wedge fi(p_j, t) : [\mathbf{t}, \alpha]$$
$$\rightarrow R(p_i, p_j, t) : [(6, 6), \alpha], \tag{11.106}$$

$$R(p_i, p_j, t) : [(5,5), \alpha] \wedge \sim fi(p_i, t) : [t, \alpha] \wedge fi(p_j, t) : [t, \alpha]$$
$$\rightarrow R(p_i, p_j, t) : [(7,5), \alpha]. \tag{11.107}$$

If ba-inf rules, $(0,8)$-rule-1, $(0,8)$-rule-2, $(5,5)$-rule-1, $(5,5)$-rule-2 and $(5,5)$-rule-3, and have been applied, bf-relations represented by bf-annotations such as $jb(2,10)/ja(10,2)$ between two processes should be derived. On the other hand, even if ba-inf rule $(0,8)$-rule-3 has been applied, no bf-annotation could be derived. Therefore a group of ba-inf rules called $(2,8)$-rules should be considered after ba-inf rule $(0,8)$-rule-3.

(2,8)-rules
Suppose that process Pr_i has started before process Pr_j starts and process Pr_j has started before process Pr_i finishes, then the vector annotation of bf-literal $R(p_i, p_j, t)$ should be $(2,8)$ and the following three rules should be considered.

$(2,8)$-rule-1 If process Pr_i finished before process Pr_j finishes, then the vector annotation $(2,8)$ of bf-literal $R(p_i, p_j, t)$ should turn to bf-annotation $jb(2,10)$.
$(2,8)$-rule-2 If both processes Pr_i and Pr_j have finished at the same time, then the vector annotation $(2,8)$ of bf-literal $R(p_i, p_j, t)$ should turn to bf-annotation $fb(3,9)$.
$(2,8)$-rule-3 If process Pr_j has finished before Pr_i finishes, then the vector annotation $(2,8)$ of bf-literal $R(p_i, p_j, t)$ should turn to bf-annotation $ib(4,8)$.

Ba-inf rules $(2,8)$-rule-1, $(2,8)$-rule-2 and $(2,8)$-rule-3 may be translated into the bf-EVALPSN clauses,

$$R(p_i, p_j, t) : [(2,8), \alpha] \wedge fi(p_i, t) : [t, \alpha] \wedge \sim fi(p_j, t) : [t, \alpha]$$
$$\rightarrow R(p_i, p_j, t) : [(2,10), \alpha], \tag{11.108}$$

$$R(p_i, p_j, t) : [(2,8), \alpha] \wedge fi(p_i, t) : [t, \alpha] \wedge fi(p_j, t) : [t, \alpha]$$
$$\rightarrow R(p_i, p_j, t) : [(3,9), \alpha] \tag{11.109}$$

$$R(p_i, p_j, t) : [(2,8), \alpha] \wedge \sim fi(p_i, t) : [t, \alpha] \wedge fi(p_j, t) : [t, \alpha]$$
$$\rightarrow R(p_i, p_j, t) : [(4,8), \alpha]. \tag{11.110}$$

The application orders (from the left to the right) of all ba-inf rules are summarized in Table 11.2.

11.6.2 Transitive Reasoning for Bf-Relations

Suppose that a bf-EVALPSN process order control system has to deal with ten processes. Then if it deals with all the bf-relations between ten processes, forty five

Table 11.2 Application orders of basic Bf-inference rules

Vector annotation	Rule	Vector annotation	Rule	Vector annotation	Rule	Vector annotation
$(0,0)$	rule-1	$(0,8)$	rule-1	$(0,12)$		
			rule-2	$(1,11)$		
			rule-3	$(2,8)$	rule-1	$(2,10)$
					rule-2	$(3,9)$
					rule-3	$(4,8)$
	rule-2	$(5,5)$	rule-1	$(5,7)$		
			rule-2	$(6,6)$		
			rule-3	$(7,5)$		

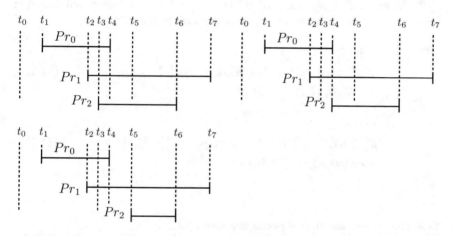

Fig. 11.22 Process time chart 1(*top left*), 2(*top right*), 3(*bottom left*)

bf-relations have to be considered. It may take much computing cost. In order to reduce such cost, we consider inference rules to derive the bf-relation between processes Pr_i and Pr_k from two bf-relations between processes Pr_i and Pr_j and between processes Pr_j and Pr_k in bf-EVALPSN, which are named *transitive bf-inference rules*. Hereafter we call transitive bf-inference rules as *tr-inf rules* for short. We introduce how to derive some of tr-inf rules and how to apply them to real-time process order control.

Suppose that three processes Pr_0, Pr_1 and Pr_2 are being processed according to the process schedule in Fig. 11.22 in which only the start time of process Pr_2 varies time t_3 to time t_5 and there is no variation of bf-relations between all the processes. The vector annotations of bf-literals $R(p0, p1, t)$, $R(p1, p2, t)$ and $R(p0, p2, t)$ at each time $t_i(i = 1, \ldots, 7)$ are shown by the three tables in Table 11.3. For each table, if we focus on the vector annotations at time t_1 and time t_2, the following tr-inf rule in bf-EVALP clause can be derived:

$$rule-1$$
$$R(p0, p1, t) : [(0, 8), \alpha] \wedge R(p1, p2, t) : [(0, 0), \alpha]$$
$$\rightarrow R(p0, p2, t) : [(0, 8), \alpha]$$

which may be reduced to the bf-EVALP clause,

$$R(p0, p1, t) : [(0, 8), \alpha] \rightarrow R(p0, p2, t) : [(0, 8), \alpha]. \tag{11.111}$$

Furthermore, if we also focus on the vector annotations at time t_3 and time t_4 in Table 11.3, the following two tr-inf rules also can be derived:

$$rule-2$$
$$R(p0, p1, t) : [(2, 8), \alpha] \wedge R(p1, p2, t) : [(2, 8), \alpha] \tag{11.112}$$
$$\rightarrow R(p0, p2, t) : [(2, 8), \alpha],$$

$$rule-3$$
$$R(p0, p1, t) : [(2, 10), \alpha] \wedge R(p1, p2, t) : [(2, 8), \alpha] \tag{11.113}$$
$$\rightarrow R(p0, p2, t) : [(2, 10), \alpha].$$

Table 11.3 Vector annotations of process time chart 1,2,3

Process time chart 1	t_0	t_1	t_2	t_3	t_4	t_5	t_6	t_7
$R(p0, p1, t)$	$(0, 0)$	$(0, 8)$	$(2, 8)$	$(2, 8)$	$(2, 10)$	$(2, 10)$	$(2, 10)$	$(2, 10)$
$R(p1, p2, t)$	$(0, 0)$	$(0, 0)$	$(0, 8)$	$(2, 8)$	$(2, 8)$	$(2, 8)$	$(4, 8)$	$(4, 8)$
$R(p0, p2, t)$	$(0, 0)$	$(0, 8)$	$(0, 8)$	$(2, 8)$	$(2, 10)$	$(2, 10)$	$(2, 10)$	$(2, 10)$
Process time chart 2	t_0	t_1	t_2	t_3	t_4	t_5	t_6	t_7
$R(p0, p1, t)$	$(0, 0)$	$(0, 8)$	$(2, 8)$	$(2, 8)$	$(2, 10)$	$(2, 10)$	$(2, 10)$	$(2, 10)$
$R(p1, p2, t)$	$(0, 0)$	$(0, 0)$	$(0, 8)$	$(0, 8)$	$(2, 8)$	$(2, 8)$	$(4, 8)$	$(4, 8)$
$R(p0, p2, t)$	$(0, 0)$	$(0, 8)$	$(0, 8)$	$(0, 8)$	$(1, 11)$	$(1, 11)$	$(1, 11)$	$(1, 11)$
Process time chart 3	t_0	t_1	t_2	t_3	t_4	t_5	t_6	t_7
$R(p0, p1, t)$	$(0, 0)$	$(0, 8)$	$(2, 8)$	$(2, 8)$	$(2, 10)$	$(2, 10)$	$(2, 10)$	$(2, 10)$
$R(p1, p2, t)$	$(0, 0)$	$(0, 0)$	$(0, 8)$	$(0, 8)$	$(0, 8)$	$(2, 8)$	$(4, 8)$	$(4, 8)$
$R(p0, p2, t)$	$(0, 0)$	$(0, 8)$	$(0, 8)$	$(0, 8)$	$(0, 12)$	$(0, 12)$	$(0, 12)$	$(0, 12)$

As well as tr-inf rules **rule-2** and **rule-3**, the following two tr-inf rules also can be derived with focusing on the variation of the vector annotations at time t_4.

$$rule-4$$
$$R(p0,p1,t) : [(2,10),\alpha] \wedge R(p1,p2,t) : [(2,8),\alpha] \qquad (11.114)$$
$$\rightarrow R(p0,p2,t) : [(1,11),\alpha],$$

$$rule-5$$
$$R(p0,p1,t) : [(2,10),\alpha] \wedge R(p1,p2,t) : [(0,8),\alpha] \qquad (11.115)$$
$$\rightarrow R(p0,p2,t) : [(0,12),\alpha].$$

Among all the tr-inf rules only tr-inf rules **rule-3** and **rule 4** have the same precedent(body),

$$R(p0,p1,t) : [(2,10),\alpha] \wedge R(p1,p2,t) : [(2,8),\alpha],$$

and different consequents(heads),

$$R(p0,p2,t) : [(2,10),\alpha] \quad \text{and} \quad R(p0,p2,t) : [(1,11),\alpha].$$

Having the same precedent may cause duplicate application of those tr-inf rules. If we take tr-inf rules **rule-3** and **rule-4** into account, obviously they cannot be uniquely applied. In order to avoid duplicate application of tr-inf rules **rule-3** and **rule-4**, we consider all correct applicable orders **order-1**, **order-2** and **order-3** for all tr-inf rules **rule-1**, …, **rule-5**.

$$order-1 : \qquad rule-1 \rightarrow rule-2 \rightarrow rule-3 \qquad (11.116)$$

$$order-2 : \qquad rule-1 \rightarrow rule-4 \qquad (11.117)$$

$$order-3 : \qquad rule-1 \rightarrow rule-5 \qquad (11.118)$$

As indicated in the above orders, tr-inf rule **rule 3** should be applied immediately after tr-inf rule **rule 2**, on the other hand, tr-inf rule **rule 4** should be done immediately after tr-inf rule **rule 1**. Thus if we take the applicable orders (11.116)–(11.118) into account, such confusion may be avoidable. Actually, tr-inf rules are not complete, that is to say there exist some cases in which bf-relations cannot be uniquely determined by only tr-inf rules.

We show an application of tr-inf rules by taking process time chart 3 in Fig. 11.22 as an example.

At time t_1, tr-inf rule **rule-1** is applied and we have the bf-EVALPSN clause,

$$R(p0, p2, t_1) : [(0, 8), \alpha].$$

At time t_2 and time t_3, no tr-inf rule can be applied and we still have the same vector annotation $(0, 8)$ of bf-literal $R(p0, p2, t_3)$.

At time t_4, only tr-inf rule **rule-5** can be applied and we obtain the bf-EVALP clause,

$$R(p0, p2, t_4) : [(0, 12), \alpha]$$

and the bf-relation between processes Pr_0 and Pr_2 has been inferred according to process order **order-3** (11.118).

We could not introduce all tr-inf rules in this subsection though, it is sure that we have many cases that can reduce bf-relation computing cost in bf-EVALPSN process order control by using tr-inf rules. In real-time process control systems, such reduction of computing cost is required and significant in practice.

As another topic, we briefly introduce anticipation of bf-relations in bf-EVALPSN. For example, suppose that three processes Pr_0, Pr_1 and Pr_2 have started sequentially, and only process Pr_1 has finished at time t as shown in Fig. 11.23. Then two bf-relations between processes Pr_0 and Pr_1 and between processes Pr_1 and Pr_2 have been already determined, and we have the following two bf-EVALP clauses with the final bf-annotations of the bf-literals,

$$R(p0, p1, t) : [\mathtt{ib}(4, 8), \alpha] \quad \text{and} \quad R(p1, p2, t) : [\mathtt{mb}(1, 11), \alpha]. \tag{119}$$

On the other hand, the bf-relation between processes Pr_0 and Pr_2 cannot be determined yet. However, if we use the tr-inf rule,

$$rule - 6$$
$$R(p0, p1, t) : [(4, 8), \alpha] \wedge R(p1, p2, t) : [(2, 10), \alpha] \tag{11.120}$$
$$\rightarrow R(p0, p2, t) : [(2, 8), \alpha],$$

we obtain vector annotation $(2, 8)$ as the bf-annotation of bf-literal $R(p0, p2, t)$. Moreover, it is logically anticipated that the bf-relation between processes Pr_0 and Pr_2 will be finally represented by one of three bf-annotations (vector annotations), $\mathtt{jb}(2, 10)$, $\mathtt{sb}(3, 9)$ and $\mathtt{ib}(4, 8)$, since the vector annotation $(2, 8)$ is the greatest lower bound of the set of vector annotations, $\{(2, 10), (3, 9), (4, 8)\}$. As mentioned

Fig. 11.23 Anticipation of bf-relation

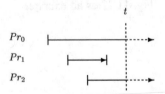

above, we can systematically anticipate the final bf-annotations from incomplete bf-annotations in bf-EVALPSN. This remarkable anticipatory feature of bf-EVALPSN reasoning could be applied to safety verification and intelligent control that may require such logical anticipation of bf-relations.

11.6.3 Transitive bf-Inference Rules

In this subsection we list up all transitive bf-inference rules (tr-inf rules) with taking their application orders into account. For simplicity, we represent the tr-inf rule,

$$R(p_i, p_j, t) : [(n_1, n_2), \alpha] \wedge R(p_j, p_k, t) : [(n_3, n_4), \alpha] \rightarrow R(p_i, p_k, t) : [(n_5, n_6), \alpha]$$

by only vector annotations and logical connectives $\wedge$ and $\rightarrow$, as follows:

$$(n_1, n_2) \wedge (n_3, n_4) \rightarrow (n_5, n_6)$$

in the list of tr-inf rules.

Transitive bf-Inference Rules

$$
\begin{aligned}
&\textbf{TR0} \quad (0,0) \wedge (0,0) \rightarrow (0,0) \\
&\textbf{TR1} \quad (0,8) \wedge (0,0) \rightarrow (0,8) \\
&\quad \textbf{TR1} - \textbf{1} \quad (0,12) \wedge (0,0) \rightarrow (0,12) \\
&\quad \textbf{TR1} - \textbf{2} \quad (1,11) \wedge (0,8) \rightarrow (0,12) \\
&\quad \textbf{TR1} - \textbf{3} \quad (1,11) \wedge (5,5) \rightarrow (1,11) \\
&\quad \textbf{TR1} - \textbf{4} \quad (2,8) \wedge (0,8) \rightarrow (0,8) \\
&\qquad \textbf{TR1} - \textbf{4} - \textbf{1} \quad (2,10) \wedge (0,8) \rightarrow (0,12) \\
&\qquad \textbf{TR1} - \textbf{4} - \textbf{2} \quad (4,8) \wedge (0,12) \rightarrow (0,8) \\
&\qquad \textbf{TR1} - \textbf{4} - \textbf{3} \quad (2,8) \wedge (2,8) \rightarrow (2,8)
\end{aligned}
$$

$$(11.121)$$

$$
\begin{aligned}
&\textbf{TR1} - \textbf{4} - \textbf{3} - \textbf{1} \ (2,10) \wedge (2,8) \rightarrow (2,10) \\
&\textbf{TR1} - \textbf{4} - \textbf{3} - \textbf{2} \ (4,8) \wedge (2,10) \rightarrow (2,8) \\
&\textbf{TR1} - \textbf{4} - \textbf{3} - \textbf{3} \ (2,8) \wedge (4,8) \rightarrow (4,8) \\
&\textbf{TR1} - \textbf{4} - \textbf{3} - \textbf{4} \ (3,9) \wedge (2,10) \rightarrow (2,10) \\
&\textbf{TR1} - \textbf{4} - \textbf{3} - \textbf{5} \ (2,10) \wedge (4,8) \rightarrow (3,9) \\
&\textbf{TR1} - \textbf{4} - \textbf{3} - \textbf{6} \ (4,8) \wedge (3,9) \rightarrow (4,8) \\
&\textbf{TR1} - \textbf{4} - \textbf{3} - \textbf{7} \ (3,9) \wedge (3,9) \rightarrow (3,9)
\end{aligned}
$$

$$(11.122)$$

$$\mathbf{TR1-4-4} \quad (3,9) \wedge (0,12) \rightarrow (0,12)$$
$$\mathbf{TR1-4-5} \quad (2,10) \wedge (2,8) \rightarrow (1,11)$$
$$\mathbf{TR1-4-6} \quad (4,8) \wedge (1,11) \rightarrow (2,8)$$
$$\mathbf{TR1-4-7} \quad (3,9) \wedge (1,11) \rightarrow (1,11)$$

(11.123)

$$\mathbf{TR1-5} \quad (2,8) \wedge (5,5) \rightarrow (2,8)$$
$$\mathbf{TR1-5-1} \quad (4,8) \wedge (5,7) \rightarrow (2,8)$$
$$\mathbf{TR1-5-2} \quad (2,8) \wedge (7,5) \rightarrow (4,8)$$
$$\mathbf{TR1-5-3} \quad (3,9) \wedge (5,7) \rightarrow (2,10)$$
$$\mathbf{TR1-5-4} \quad (2,10) \wedge (7,5) \rightarrow (3,9)$$

(11.124)

$$\mathbf{TR2} \quad (5,5) \wedge (0,8) \rightarrow (0,8)$$
$$\mathbf{TR2-1} \quad (5,7) \wedge (0,8) \rightarrow (0,12)$$
$$\mathbf{TR2-2} \quad (7,5) \wedge (0,12) \rightarrow (0,8)$$
$$\mathbf{TR2-3} \quad (5,5) \wedge (2,8) \rightarrow (2,8)$$

(11.125)

$$\mathbf{TR2-3-1} \quad (5,7) \wedge (2,8) \rightarrow (2,10)$$
$$\mathbf{TR2-3-2} \quad (7,5) \wedge (2,10) \rightarrow (2,8)$$
$$\mathbf{TR2-3-3} \quad (5,5) \wedge (4,8) \rightarrow (4,8)$$
$$\mathbf{TR2-3-4} \quad (7,5) \wedge (3,9) \rightarrow (4,8)$$

(11.126)

$$\mathbf{TR2-4} \quad (5,7) \wedge (2,8) \rightarrow (1,11)$$
$$\mathbf{TR2-5} \quad (7,5) \wedge (1,11) \rightarrow (2,8)$$

(11.127)

$$\mathbf{TR3} \quad (5,5) \wedge (5,5) \rightarrow (5,5)$$
$$\mathbf{TR3-1} \quad (7,5) \wedge (5,7) \rightarrow (5,5)$$
$$\mathbf{TR3-2} \quad (5,7) \wedge (7,5) \rightarrow (6,6)$$

(11.128)

Note that the bottom vector annotation $(0,0)$ in tr-inf rules implies that for any non-negative integers m and n bf-EVALP clause $R(p_j, p_k, t) : [(n,m), \alpha]$ satisfies $R(p_j, p_k, t) : [(0,0), \alpha]$ it.

Here we emphasize two important points (I) and (II) in terms of application of tr-inf rules.

(I) Names of tr-inf rules such as TR1-4-3 show their application orders. For example, if tr-inf rule TR1 has been applied, one of tr-inf rules TR1-1,TR1-2,... or TR1-5 should be applied at the following stage; if tr-inf rule TR1-4 has been applied after tr-inf rule TR1, one of tr-inf rules TR1-4-1,TR1-4-2, ... or TR1-4-7 should be applied at the following stage; on the other hand, if one of tr-inf rules TR1-1, TR1-2 or TR1-3 has been applied after tr-inf rule TR1, there is no tr-inf rule to be applied at the following stage because bf-annotations $db(0, 12)$ or $mb(1, 11)$ between processes Pr_i and Pr_k have been already derived.

(II) the following eight tr-inf rules,

$$
\begin{array}{llll}
\text{TR1-4-2} & (122), & \text{TR2-2} & (126), \\
\text{TR1-4-3-2} & (123), & \text{TR2-3-2} & (127), \\
\text{TR1-4-6} & (124), & \text{TR2-5} & (128), \\
\text{TR1-5-1} & (125) & \text{TR3-1} & (129)
\end{array}
$$

have no following rule to be applied at the following stage, even though they cannot derive the final bf-relations between processes represented by bf-annotations such as $jb(2, 10)/ja\,(10, 2)$. For example, suppose that tr-inf rule TR1-4-3-2 has been applied, then the vector annotation $(2, 8)$ of the bf-literal (p_i, p_k, t) just implies that the final bf-relation between processes Pr_i and Pr_k is one of three bf-annotations, $jb(2, 10)$, $sb(3, 9)$ and $ib(4, 8)$. Therefore, if one of the above eight tr-inf rules has been applied, one of ba-inf rules $(0, 8)$-rule, $(2, 8)$-rule or $(5, 5)$-rule should be applied for deriving the final bf-annotation at the following stage. For instance, if tr-inf rule TR1-4-3-2 has been applied, ba-inf rule $(2, 8)$-rule should be applied at the following stage.

Now we show a simple example of bf-relation reasoning by tr-inf rules taking the process time chart 3(bottom left) in Fig. 11.22.

Example 11.4 At time t_1, tr-inf rule TR1 is applied and we have the bf-EVALP clause,

$$R(p_i, p_k, t_1) : [(0, 8), \alpha].$$

At time t_2, tr-inf rule TR1-2 is applied, however bf-literal $R(p_i, p_k, t_2)$ has the same vector annotation $(0, 8)$ as the previous time t_1. Therefore we have the bf-EVALP clause,

$$R(p_i, p_k, t_2) : [(0, 8), \alpha].$$

At time t_3, no transitive bf-inference rule can be applied, since the vector annotations of bf-literals $R(p_i, p_j, t_3)$ and $R(p_j, p_k, t_3)$ are the same as the previous time t_2. Therefore we still have the bf-EVALP clause having the same vector annotation,

$$R(p_i, p_k, t_3) : [(0, 8), \alpha].$$

At time t_4, tr-inf rule TR1-2-1 is applied and we obtain the bf-EVALP clause having bf-annotation $db(0, 12)$,

$$R(p_i, p_k, t_4) : [(0, 12), \alpha].$$

11.7   Conclusions and Remarks

In this chapter, we have introduced paraconsistent annotated logic programs EVALPSN and bf-EVALPSN that was proposed most recently, which can deal with process before-after relations, and we also have introduced the safety verification method and intelligent process control based on EVALPSN/bf-EVALPSN as applications. The bf-EVALPSN safety verification based process order control method can be applied to various process order control systems requiring real-time processing.

An interval temporal logic has been proposed by Allen et al. for knowledge representation of properties, actions and events [1, 2]. In the interval temporal logic, predicates such as *Meets(m,n)* are used for representing primitive before-after relations between time intervals m and n, and other before-after relations are represented by six predicates such as *Before*, *Overlaps*, etc. It is well known that the interval temporal logic is a logically sophisticated tool to develop practical planning or natural language understanding systems [1, 2]. However, it does not seem to be so suitable for practical real-time processing because before-after relations between two processes cannot be determined until both of them finish. On the other hand, in bf-EVALPSN bf-relations are represented more minutely in paraconsistent vector annotations and can be determined according to start/finish information of two processes in real time. Moreover EVALPSN can be implemented on microchips as electronic circuits, although it has not introduced in this chapter. We have already shown that some EVALPSN based control systems can be implemented on a microchips in [32, 39]. Therefore bf-EVALPSN is a more practical tool for dealing with real-time process order control and its safety verification.

In addition to the suitable characteristics for real-time processing, bf-EVALPSN can deal with incomplete and paracomplete knowledge in terms of before-after relation in vector annotations, although the treatment of paracomplete knowledge has not been discussed in this chapter. Furthermore bf-EVALPSN has inference rules for transitive reasoning of before-after relations as shortly described. Therefore if we apply EVALPSN and bf-EVALPSN appropriately, various systems should intellectualize more.

References

1. Allen, J.F.: Towards a general theory of action and time. Artif. Intell. **23**, 123–154 (1984)
2. Allen, J.F., Ferguson, G.: Actions and events in interval temporal logic. J. Logic Comput. **4**, 531–579 (1994)
3. Apt, K.R., Blair, H.A., Walker, A.: Towards a theory of declarative knowledge. In: Minker, J. (ed.) Foundation of Deductive Database and Logic Programs, pp. 89–148. Morgan Kaufmann, CA (1989)
4. Belnap, N.D.: A useful four valued logic. In: Dunn, M., Epstein, G. (eds.) Modern uses of multiple-valued logic, pp. 8–37. D.Reidel Publishing, Netherlands (1977)
5. Billington, D.: Defeasible logic is stable. J. Logic Comput. **3**, 379–400 (1993)

6. Billington, D.: Conflicting literals and defeasible logic. In: Nayak, A., Pagnucco, M. (eds.) Proceedings of 2nd Australian Workshop Commonsense Reasoning, 1 December, Perth, Australia, Australian Artificial Intelligence Institute, Australia, pp. 1–15 (1997)
7. Blair, H.A., Subrahmanian, V.S.: Paraconsistent logic programming. Theoret. Comput. Sci. **68**, 135–154 (1989)
8. da Costa, N.C.A., Subrahmanian, V.S., Vago, C.: The paraconsistent logics PT. Zeitschrift für Mathematische Logic und Grundlangen der Mathematik **37**, 139–148 (1989)
9. Dressler, O.: An extended basic ATMS. In: Reinfrank, M. et al., (eds.) In: Proceedings of 2nd International Workshop on Non-monotonic Reasoning, 13–15 June, Grassau, Germany, (Lecture Notes in Computer Science LNCS 346), pp. 143–163. Springer, Heidelberg (1988)
10. Fitting, M.: Bilattice and the semantics of logic programming. J. Logic Program. **11**, 91–116 (1991)
11. Gelder, A.V., Ross, K.A., Schlipf, J.S.: The well-founded semantics for general logic programs. J. ACM **38**, 620–650 (1991)
12. Jaskowski, S.: Propositional calculus for contradictory deductive system (English translation of the original Polish paper). Stud. Logica. **24**, 143–157 (1948)
13. Kifer, M., Subrahmanian, V.S.: Theory of generalized annotated logic programming and its applications. J. Logic Program. **12**, 335–368 (1992)
14. Lloyd, J.W.: Foundations of Logic Programming, 2nd edn. Springer, Berlin (1987)
15. Moore, R.: Semantical considerations on non-monotonic logic. Artif. Intell. **25**, 75–94 (1985)
16. Nakamatsu, K.: On the relation between vector annotated logic programs and defeasible theories. Logic Logical Philos. **8**, 181–205 (2000)
17. Nakamatsu, K.: Paraconsistent annotated logic program EVALPSN and its application. In: Fulcher, J., Jain, C.L. (eds.) Computational Intelligence: A Compendium (Studies in Computational Intelligence 115), pp. 233–306. Springer, Germany (2008)
18. Nakamatsu, K., Abe, J.M.: Reasonings based on vector annotated logic programs. In: Mohammadian, M. (ed.) Computational Intelligence for Modelling, Control and Automation (CIMCA99), (Concurrent Systems Engineering Series 55), pp. 396–403. IOS Press, Netherlands (1999)
19. Nakamatsu, K., Abe, J.M.: The development of paraconsistent annotated logic program. Int. J. Reasoning-based Intell. Syst. **1**, 92–112 (2009)
20. Nakamatsu, K., Suzuki, A.: Annotated semantics for default reasoning. In: Dai, R. (ed.) Proceedings 3rd Pacific Rim International Conference on Artificial Intelligence (PRICAI94), 15–18 August, Beijing, China, pp. 180–186. International Academic Publishers, China (1994)
21. Nakamatsu, K., Suzuki, A.: A nonmonotonic ATMS based on annotated logic programs. In: Wobcke, W., et al. (eds.) Agents and Multi-agents Systems (Lecture Notes in Artificial Intelligence LNAI 1441), pp. 79–93. Springer, Berlin (1998)
22. Nakamatsu, K., Suzuki, A.: Autoepistemic theory and paraconsistent logic program. In: Nakamatsu, K., Abe, J.M. (eds.) Advances in Logic Based Intelligent Systems (Frontiers in Artificial Intelligence and Applications 132), pp. 177–184. IOS Press, Netherlands (2005)
23. Nakamatsu, K., Suzuki, A.: Annotated semantics for non-monotonic reasonings in artificial intelligence—I, II, III, IV. In: Nakamatsu, K., Abe, J.M.: (eds.) Advances in Logic Based Intelligent Systems (Frontiers in Artificial Intelligence and Applications 132), pp. 185–215. IOS Press, Netherlands (2005)
24. Nakamatsu, K., Abe, J.M., Suzuki, A.: Defeasible reasoning between conflicting agents based on VALPSN. In: Tessier, C., Chaudron, L. (eds.) In: Proceedings of AAAI Workshop Agents' Conflicts, 18 July, Orland, FL, pp. 20–27. AAAI Press, Menlo Park, CA (1999)
25. Nakamatsu, K., Abe, J.M., Suzuki, A.: Defeasible reasoning based on VALPSN and its application. In: Nayak, A., Pagnucco, M. (eds.) Proceedings of the Third Australian Commonsense Reasoning Workshop, 7 December, Sydney, Australia, pp. 114–130. University of Newcastle, Sydney, Australia (1999)

26. Nakamatsu, K., Abe, J.M., Suzuki, A.: A defeasible deontic reasoning system based on annotated logic programming. In: Dubois, D.M. (ed.) Proceedings of 4th International Conference on Computing Anticipatory Systems(CASYS2000), 7–12 August, 2000, Liege, Belgium, (AIP Conference Proceedings 573), pp. 609–620. American Institute of Physics, New York, NY (2001)

27. Nakamatsu, K., Abe, J.M., Suzuki, A.: Annotated semantics for defeasible deontic reasoning. In: Ziarko, W., Yao, Y. (eds.) Proceedings of 2nd International Conference on Rough Sets and Current Trends in Computing(RSCTC2000), 16–19 October, 2000, Banff, Canada, (Lecture Notes in Artificial Intelligence LNAI 2005), pp. 432–440. Springer, Berlin (2001)

28. Nakamatsu, K., Abe, J.M., Suzuki, A.: Extended vector annotated logic program and its application to robot action control and safety verification. In: Abraham, A., et al. (eds.) Hybrid Information Systems (Advances in Soft Computing Series), pp. 665–680. Physica-Verlag, Heidelberg (2002)

29. Nakamatsu, K., Suito, H., Abe, J.M., Suzuki, A.: Paraconsistent logic program based safety verification for air traffic control. In: El Kamel, A. et al., (eds.) Proceedings of IEEE International Conference System, Man and Cybernetics 02(SMC02), 6–9 October, Hammamet, Tunisia, IEEE SMC (CD-ROM) (2002)

30. Nakamatsu, K., Abe, J.M., Suzuki, A.: A railway interlocking safety verification system based on abductive paraconsistent logic programming. In: Abraham, A., et al. (eds.) Soft Computing Systems (HIS02) (Frontiers in Artificial Intelligence and Applications 87), pp. 775–784. IOS Press, The Netherlands (2002)

31. Nakamatsu, K., Abe, J.M., Suzuki, A.: Defeasible deontic robot control based on extended vector annotated logic programming. In: Dubois, D.M. (ed.) Proceedings of 5th International Conference on Computing Anticipatory Systems(CASYS2001) 13–18 August, 2001, Liege, Belgium, (AIP Conference Proceedings 627), pp. 490–500. American Institute of Physics, New York, NY (2002)

32. Nakamatsu, K., Mita, Y., Shibata, T.: Defeasible deontic action control based on paraconsistent logic program and its hardware application. In: Mohammadian, M. (ed.) In: Proceedings of International Conference on Computational Intelligence for Modelling Control and Automation 2003(CIMCA2003), 12–14 February, Vienna, Austria, IOS Press, Netherlands (CD-ROM) (2003)

33. Nakamatsu, K., Seno, T., Abe, J.M., Suzuki, A.: Intelligent real-time traffic signal control based on a paraconsistent logic program EVALPSN. In: Wang, G., et al. (eds.) Rough Sets, Fuzzy Sets, Data Mining and Granular Computing(RSFDGrC2003), 26–29 May, Chongqing, China, (Lecture Notes in Artificial Intelligence LNAI 2639), pp. 719–723. Springer, Berlin (2003)

34. Nakamatsu, K., Komaba, H., Suzuki, A.: Defeasible deontic control for discrete events based on EVALPSN. In: Tsumoto, S. et al., (eds.) Proceedings of 4th International Conference in Rough Sets and Current Trends in Computing(RSCTC2004), 1–5 June, Uppsala, Sweeden, (Lecture Notes in Artificial Intelligence LNAI 3066), pp. 310–315. Springer, Berlin (2004)

35. Nakamatsu, K., Ishikawa, R., Suzuki, A.: A paraconsistent based control for a discrete event cat and mouse. In: Negoita, M.G.H. et al., (eds.) Proceedings 8th International Conference on Knowledge-Based Intelligent Information and Engineering Systems(KES2004), 20–25 September, Wellington, New Zealand, (Lecture Notes in Artificial Intelligence LNAI 3214), pp. 954–960. Springer, Berlin (2004)

36. Nakamatsu, K., Chung, S.-L., Komaba, H., Suzuki, A.: A discrete event control based on EVALPSN stable model. In: Slezak, D., et al. (eds.) Rough Sets, Fuzzy Sets, Data Mining and Granular Computing(RSFDGrC2005), 31 August–3 September, Regina, Canada, (Lecture Notes in Artificial Intelligence LNAI 3641), pp. 671–681. Springer, Berlin (2005)

37. Nakamatsu, K., Abe, J.M., Akama, S.: An intelligent safety verification based on a paraconsistent logic program. In: Khosla, R. et al., (eds.) Proceedings of 9th International Conference on Knowledge-Based Intelligent Information and Engineering Systems (KES2005), 14–16 September, Melbourne, Australia, (Lecture Notes in Artificial Intelligence LNAI 3682), pp. 708–715. Springer, Berlin (2005)

38. Nakamatsu, K., Kawasumi, K., Suzuki, A. (2005) Intelligent verification for pipeline based on EVALPSN. In: Nakamatsu, K., Abe, J.M. (eds.) Advances in Logic Based Intelligent Systems (Frontiers in Artificial Intelligence and Applications 132), pp. 63–70. IOS Press, Netherlands
39. Nakamatsu, K., Mita, Y., Shibata, T.: An intelligent action control system based on extended vector annotated logic program and its hardware implementation. J. Intell. Autom. Soft. Comput. **13**, 289–304 (2007)
40. Nakamatsu, K., Abe, J.M., Akama, S.: A logical reasoning system of process before-after relation based on a paraconsistent annotated logic program bf-EVALPSN. KES J. **15**(3), 146–163 (2011)
41. Nute, D.: Defeasible reasoning. In: Stohr, E.A. et al., (eds.) Proceedings 20th Hawaii International Conference System Science(HICSS87) 1, 6–9 January, Kailua-Kona, Hawaii, pp. 470–477. University of Hawaii, Hawaii (1987)
42. Nute, D.: Basic defeasible logics. In: del Cerro, L.F., Penttonen, M. (eds.) Intensional Logics for Programming, pp. 125–154. Oxford University Press, UK (1992)
43. Nute, D.: Defeasible logic. In: Gabbay, D.M., et al. (eds.) Handbook of Logic in Artificial Intelligence and Logic Programming 3, pp. 353–396. Oxford University Press, UK (1994)
44. Nute, D.: Apparent obligatory. In: Nute, D. (ed.) Defeasible Deontic Logic (Synthese Library 263), pp. 287–316. Kluwer Academic Publisher, Netherlands (1997)
45. Przymusinski, T.C.: On the declarative semantics of deductive databases and logic programs. In: Minker, J. (ed.) Foundation of Deductive Database and Logic Programs, pp. 193–216. Morgan Kaufmann, New York, NY (1988)
46. Reiter, R.: A logic for default reasoning. Artif. Intell. **13**, 81–123 (1980)
47. Shepherdson, J.C.: Negation as failure, completion and stratification. In: Gabbay, D.M., et al. (eds.) Handbook of Logic in Artificial Intelligence and Logic Programming 5, pp. 356–419. Oxford University Press, UK (1998)
48. Subrahmanian, V.S.: Amalgamating knowledge bases. ACM Trans. Database Syst. **19**, 291–331 (1994)
49. Subrahmanian, V.S.: On the semantics of qualitative logic programs. In: Proceedings of the 1987 Symposium on Logic Programming(SLP87), August 31–September 4, IEEE Computer Society Press, San Francisco, CA, pp 173–182 (1987)
50. Visser, A.: Four valued semantics and the liar. J. Philos. Logic **13**, 99–112 (1987)

Printed in the United States
By Bookmasters